# *The History of the Society of Dairy Technology*

## *60 years of service to the Milk Industries of the UK and Ireland*

# FOREWORD

## Lord Wade of Chorlton, Patron of the Society

As the Society's Patron, I congratulate it on attaining its 60th anniversary during 2003. I am also very pleased that it has recognised this significant milestone by undertaking the publication of a book on the history of both the Society and parallel developments in the dairy industry during the past sixty eventful years.

I congratulate all those who have contributed to this book which is the result of much diligent research and great endeavour by its Editor, Dr. David Armstrong, and his many SDT member collaborators.

I commend this publication to our members and to both the specialist and general reader as a record not only of the activities of the Society in the UK and Ireland, but also the chequered history of the milk industry in these countries over the past six decades.

# CONTENTS

# INTRODUCTION

This book deals with the history of the SDT (much of it is extracts from the SDT Journals published continuously since 1947) from its beginning in the dark days of war, its expansion into all regions of the UK and Ireland and the problems which it subsequently encountered.

The SDT like the British milk industry in which many of its members worked has experienced great difficulty over the past two decades. Perhaps it is purely coincidental that this has happened since the UK joined the EEC in 1973.

Inevitably, because of closures in the industry (including State services and research centres), there has been a substantial fall in the number of Society members but, though the outlook for the industry remains uncertain, our membership base has now stabilised and there is optimism about the recruitment of new members.

One of the purposes of this book is to preserve for posterity the archives of the Society. For that reason I have included a final Chapter on papers of special interest and also included in the Appendices copies of significant SDT documents over the years.

The Society has a documents store near Bicester, Oxfordshire. The documents have now been indexed and boxed in an orderly fashion so that reference can be made to them. Many of the archives of the dairy industry in the UK have been lost because of the winding up of the Milk Marketing Boards. As a business the SDT scarcely registers in terms of size, but its contribution to the industry has been immense and it has managed to adapt and survive.

The system of classification of members into Groups has been abandoned to provide more flexibility on recruitment, and the number of Council members has been significantly reduced which cuts costs and facilitates decision making. The premises at Huntington are still an income-earning asset.

Our new Patron, Lord Wade of Chorlton, has already made a significant contribution to the well being of the Society.

As a consequence of incorporation, Custodial Trustees (and the Editor is one) are no longer required. Over the years they have been assiduous in looking after the interests of the Society.

As a footnote, but none the less significant for that, the Danish Society of Dairy Technology was founded in 1942 when that country was under German occupation.


Dr David L Armstrong
Editor
October 2003

# ACKNOWLEDGEMENTS

Many people, and mostly SDT members, have contributed to the production of this book, not all of whom I have mentioned by name but all their contributions and messages of support are much appreciated.

The cost of preparation and publication of the book has been met by the SDT supplemented by substantial contributions from the Richard and Betty Lawes Foundation (through Mr J Neville, Council member, and the Lawes/Neville family), JohnsonDiversey (through Mr. M. Walton, current President of the SDT), and Tetra Pak UK Ltd (through Mr. A. Stack, Honorary Treasurer of the SDT). These contributions are gratefully acknowledged.

I am much indebted to Dr. George Chambers (a Past President) for a well prepared contribution on the growth of the Society in Ireland, North and South, and his pertinent observations on the state of dairy research in the UK; to Dr. Phil Kelly for the detailed history of the Southern Ireland Section; to Mr Francis Wall for access to his well researched thesis on An Bord Bainne; to the Irish Co-operative Organisation Society (ICOS) for its magnificent *"Fruits of a Century"* (Editor Maurice Henry), an invaluable source of information on the growth of co-operation in the milk industry in Ireland; to the Past Presidents, Officers and members of the Society who have contributed articles on the history of the Sections, Regions and Branches, many of whom are mentioned in the text; to Drs. Mann and O'Sullivan for their individual contributions on the Society's journal and Symposia & Conferences, respectively; to Mrs Ros Gale, retired Executive Secretary for her 'recollections' contribution; to the authors of the articles in *"The.UK Milk Marketing Boards: a Concise History"* (published by the SDT); to Mr. A. Wilbey Senior Vice-President of the Society for assistance on accessing journals in the University of Reading's library; to Dr. Ian McDougall for assistance with the Northern Ireland Section, to Dr R Crawford past President, for his contributions to the history of the Scottish Section and Mr T. Brigstocke for updating the situation in England and Wales.

Dr. A.C. O'Sullivan, Executive Director of the SDT, has given freely of his time and expert knowledge and, without his contribution, it is doubtful if this book could have been published.

My thanks to Miss Joanne Quilter, Finishing Touch, Chichester who word-processed efficiently a miscellany of Journal extracts, typescripts and MSS.

I record my gratitude to my wife Beatrice who reduced the MSS to typescript, a tedious task, and tolerated the disorderly array of papers and files in our home.

Finally, the Society and I thank the printer, Halstan & Co. Ltd. of Amersham, who managed to produce the book on time in spite of a very tight schedule and many emendations and amendments.

**Patron 1993**

**His Royal Highness The Prince Phillip, Duke of Edinburgh KG KT**

The fifty years since the founding of the Society of Dairy Technology have seen a major revolution in every aspect of the dairy industry. Everything from the management of a dairy herd and the processing of dairy products to the development of new products and their marketing has changed out of all recognition.

Much of this is certainly due to the work of the Society and it deserves many congratulations on its fiftieth anniversary.

The anniversary is a time to look back and to pay tribute to all those who have made the Society what it is today, but it is also a time to take stock and to look ahead. I am sure there will be plenty for the Society to do, and I hope that it will look at future developments with an eye to their ethical and environmental acceptability as well as their economic value.

# CHAPTER ONE

## THE HISTORY OF THE SOCIETY OF DAIRY TECHNOLOGY:
## 60 YEARS OF SERVICE TO THE MILK INDUSTRIES OF THE UK & IRELAND

This year (2003) the Society of Dairy Technology will celebrate 60 years of continuous service to the milk industries of the United Kingdom and the Republic of Ireland. This is the story of those momentous years.

In 1993 the Society published a commemorative booklet on the occasion of its Golden Jubilee. The Society was honoured to have as its Patron that year HRH the Prince Phillip, Duke of Edinburgh KG KT. His remarks reproduced on the facing page showed prescience in stressing ethical and environmental acceptability as important considerations in future developments.

The introduction to the commemorative booklet is a good starting point for a history of the SDT. It records that the first informal meeting was held on 23rd November 1942 when the UK was engaged in a monumental battle for survival against the might of Hitler's Germany and when the very existence of Western Civilisation was at stake. Much of Europe was already ravaged by war and in a global conflict involving Japan and the USA destruction and death most brutal was the order of the day.

Looking back now it is incredible that a group of academics and leaders of the milk industry had the confidence to sit down and consider the future needs of the industry. "The people who attended this meeting had one thing in common, they were convinced that an organisation should be formed to facilitate contact and discussion between representatives of all sections of the dairy industry."

The founding members set up the Society as a charity with objectives and rules which were amended from time to time.

It was not until 1964 that the SDT was entered in the Central Register of Charities under the Charities Act 1960 and given the number 213901. This certificate of registration, which was sent to M. Sonn at 17 Devonshire Street, is attached as an appendix - there is no explanation of the delay in registration but it may have been prompted in the end by the desire to give the Society more permanence.

Recommendations made by the ad hoc committee were considered and approved at a general meeting on 2nd March 1943. Professor H. D. Kay was appointed President and the Society started life with 240 founder members. Sadly, the last founder member Dr. Ashton died on the 17th December 2002 (see Professor Kay, Early History of the Society.)

The objects of the Society set out then have stood the test of time and, though amendments have been made over the years, the essence of the SDT is still the application of education and experiment to improve practice in all branches of the dairy industry.

*PROFESSOR H. D. KAY*
*PRESIDENT 1943-44*

Originally there were ordinary and associate members and ordinary members were elected to one of three groups. There was also a fourth group reserved for "Associate and other members" but in 1958 this became the Students Group. Later to facilitate recruitment and simplify the procedure the grouping of members was abandoned.

The original annual subscription was one guinea, which can be contrasted with the current subscription of £58 for full members. When adjusted for the dramatic fall in the value of money over the years the increase to the current subscription is not as great as it at first glance appears. In 1943 the purchasing power of one guinea was considerable.

MANAGEMENT AND OFFICES

From the beginning of the Society much of the work of management was carried out by members on a voluntary basis. As the Society grew it became apparent that to cope successfully with the organisation of meetings, the enrolment of members, finance and the Journal, executive assistance was required.

The first number of the Journal (Vol No 1, 1947) did not have a first page giving the names of officers and council members. Vol 1 No 2 Jan 1948 did give this information. The President was Mr J Matthews and the Secretary who was paid was Mr S H Dingley. At the AGM on 27 October 1947 it was noted that "the expansion of the Society's work had necessitated a new arrangement with Mr Dingley and some increased cost in salaries, travelling, branch meetings and sundry expenses..."

MR. J. MATTHEWS
PRESIDENT 1947-48

The first Secretary (Honorary) of the Society was Mr W G Sutton who was at that time Director of the Commonwealth Bureau of Dairy Science. Professor Kay comments in his early history of the Society that Sutton was a "tower of strength in all the early secretarial and administrative work of the Society....". Later Mr A L Barton joined Sutton as Joint Honorary Secretary. It may be noted in the context of the setting up of the Society that Mr W A Nell of the Express Dairy Company gave £100 to cover initial expenses (Kay, Early History.) The first general meeting of the Society was held on 24 June 1943. Speakers were Messrs Proctor, Crossley, Matthews, Wright and Capstick. The number of members then was 1750. In 1947/48 the Executive Secretary was Mr S H Dingley, referred to in the previous paragraph.

Mr Michael Sonn was appointed Executive Secretary in 1948. He also worked for the Royal Association of British Dairy Farmers (RABDF). His main task for the RABDF was the organisation of the Dairy Show in Olympia which was, for many years, the most important event in the dairying calendar. During the time he was engaged in the Dairy Show, the Honorary Secretary looked after the secretarial work of the SDT.

MR. F. PROCTOR
PRESIDENT 1944-45

MR. E. CAPSTICK
PRESIDENT 1948-49

At the time of Sonn's appointment the SDT shared an office with the RABDF in 17 Devonshire Street when Frank Francis was the Secretary of that organisation.

In 1971/72, the Society moved its offices from Devonshire Street and rented an office at 172A Ealing Road, Wembley above a curry shop which meant that the smell of curry was ever present! The rent was £800 p.a.

The Council then decided that the time had come, as the Society was flourishing, to find a permanent headquarters. After much searching Mr. & Mrs. Barron (Mrs. Barron before her marriage was Miss Wendy McCullough a well known and respected member) found a building in Huntingdon, Cambridgeshire (72 Ermine Street). It was decided to purchase the building on 3rd September 1982 at a price of £42,500.

Crossley House

ACQUISITION OF PREMISES AT HUNTINGDON
Mr Toby Weldon (President) made this statement to the AGM setting out the background to the purchase of the premises at Huntingdon:

Crossley House, SDT office in
Huntingdon

COUNCIL'S REPORT

The President: "We come now to the President's report. It's up to me now to make a suitable statement. As you know I was taken ill in November of last year with what everyone thought was something simple and I was told to ride a bike and goodness knows what, all the wrong things it turned out until the real trouble was discovered and I've been a cripple ever since. I want to thank everybody for their cards and good wishes not forgetting the magnificent azalea that was sent to my wife and myself. I can assure you that your good wishes played a great part in keeping my spirits up".

"The other point that I was keen about was to examine why it was that subscriptions tended to run away with us and it was pretty obvious that one of the main reasons way that we were paying out £2,100 in rent for what I can only describe now in retrospect as offices which were not really the standard that the Society should own, and faced with the renewal of our lease in December 1984 it looked very much as if the rent would go up about 3 times. We therefore formed an office sub-committee to explore the possibility of new premises and indeed we were eventually successful. Now, here I would like to pay a great tribute to the sub-committee, especially to Wilf Hipkins who took the chair in my absence, to Ron Dicker the treasurer, who had problems looking after the finance, to Wendy Barron and her husband, Arthur, as Honorary Chartered Surveyor, who has been a tower of strength as an adviser to us and of course, to Ridley Rowling, who you will see paid a major role in this. We hunted all over the place for offices. We applied to all the major dairy companies, to people associated with the trade. We had offers of offices in York, Bristol, and the Agricultural grounds in Stoneleigh and the West Country, and there was a great deal of time spent exploring these…"

"But eventually we came up with new premises in Huntingdon, purpose-built for offices, which were owned by the Association of Conference Executives who were selling and moving to larger offices. It's fully equipped and carpeted, 1100 square feet of offices with its own kitchen, two separate toilets, a small garden and car parking space for several cars. We've succeeded in getting these premises, which are probably twice as much as we really need. What we do with the surplus we haven't yet decided - we might even let it to pay our overheads. But we have acquired those premises for £42,500, and I think in these days of rising property values we have done pretty well. Now, I signed the contract, we didn't have the money, but we drew up a list of major companies in the Industry we thought would be able to lend us money on a short term basis. I'm pleased to say we only got half way through the list before we obtained all the money we wanted…"

"At least he was available, Ridley, a Past President as you know, has given his time freely, he visited all these people personally and came away with a cheque for £4,000 loans from each of the offices he visited. I'm sure you'll agree we owe Ridley a special vote of thanks".

"It means we now have offices which will be our own for the outlay of the equivalent of about six years, or might be even five years, of our expected rent and its an office of which, I think, the Society can be proud, and it reaches a standard that this Society deserves to have. Now, of course we have to give this money back to the industry and consequently we are launching an appeal. The first appeal obviously must be to members. We can't appeal to Industry without first demonstrating that we can help ourselves. So we have done two things. We have taken three quarters of our reserve capital and put it into the Building fund. With incidental expenses this leaves us about £35,000 to find. I want again to emphasise that the membership has already started to send money in, and I want to emphasise equally strongly that your donations be covenanted. It doesn't matter how much you covenant because (a) you get tax relief on it and (b) we, as a Society get £46 from every £100 we get from you. That's a very substantial sum and on reflection we should really have made it clear on the appeal and not given you the option, saying covenanted if you feel like it, we'd like you particularly to covenant… Here we should say we haven't gone into this blindly, we did make considerable enquiries before we set out on this project to see what sort of response we would get. I wrote to every Past President and to each Trustee and several people in the Trade and the indications were that the support would be forthcoming. I think they realise the value of the Society to the Trade and indeed if the society didn't exist then it would have to be made to exist because without the Society there is no platform for people in research and development…"

"Of course, the main result of buying our own property is that we shall not be adding to the membership subscription in 1985 simply to pay for the expected increase in rent and in the next two years, of course, there will be a savings of some £2,500 in rent which we would have to pay if we had stayed in our present offices. So with a bit of luck, and the ingenuity and good housekeeping of our Treasurer I'm hoping that the financial basis of the Society will be on a good footing..."

"I would like to say that in the whole of this operation there have been no casualties but the way it is I'm afraid it isn't quite like that. It means that our very capable assistant secretary, Mrs Erica Cybulla, would have had to sacrifice three quarters of her salary simply to get to the new offices and this of course was an unreasonable burden to both her and the Society. We had to come to some arrangement and we did think that Erica might be able to move with us but it is not possible and it is with great regret therefore that I have to say that the one black note is we shall be losing Erica and we have made suitable arrangements with her..."

BUILDING FUND

Mr R A Dicker giving his Treasurer's report made reference to the Building Fund. (JSTD vol 36, No.1 January 1983). "These net assets are mainly the £20,000 worth of investments and in connection with the purchase of the new offices in Huntingdon, we have redeemed approximately three quarters of these, giving us £16,372 net and leaving retained investments at a guaranteed value of £5,530. The reduced value of investments is more than compensated for by the security of owning our own bricks and mortar - once it's paid for that is - and it is this which is the current concern regarding society finances."

"Trustees and Council have overwhelmingly backed the Building scheme and I am in no doubt that in addition to the greater security, we shall see a further reduction in our annual expenditure: there are higher costs at Huntingdon in terms of heating, insurance and maintenance but these, together with the loss of interest on redeemed investments, are more than offset by savings in rent

MR. F. G. WELDON
PRESIDENT 1981-82

and even rates. Furthermore, the premises have almost twice as much office space as we need and there is the very real possibility of sub-letting and gaining income. This is something which Council will be considering next year."

The background to the purchase of the premises in Huntingdon is set out in the President's (Mr Toby Weldon) letter to members of the SDT dated 30 Sept. 1982 as follows:

"The Society has occupied rented office accommodation in Wembley since 1972. The current lease expires in 1985 and it is anticipated that the rent could then jump from its current level of £2,100 p.a. to two or three times this amount. The Council has therefore been concerned for some time to find alternative accommodation on attractive terms".

"At the beginning of my term of office last November I proposed to Council that we should seriously consider purchasing freehold accommodation, or even building on a suitably located site, and that we should appeal to the dairy industry in order to raise funds for this project. Initial response from industry indicated that there would be enough support to an Appeal for us to proceed with confidence. All Past Presidents and Trustees gave their support to the scheme and your Trustees were also in agreement with £16,000 (approximately 75%) of the Society's reserves being used to partially fund the purchase of premises".

"Earlier this year extremely suitable premises in Huntingdon became available, slightly larger and at a lower price than had been envisaged. These premises were deemed entirely suitable for the Society's needs by the Office Sub-Committee, which includes Mr Arthur Barron as Honorary Chartered Surveyor, and which was authorised by Council to act on its behalf. Council agreed to the purchase of these premises in the sum of £42,500 and completion took place on 3 September 1982. It is hoped to move to the new premises at 72, Ermine Street, Huntingdon during the first week of December".

"We have been extremely fortunate in obtaining bridging loan facilities on very favourable terms from a number of major organisations in or connected with the dairy industry and the Society is extremely grateful for these facilities".

"It is now necessary to launch our Building Fund Appeal formally and individual approaches will be made, with the help of Sections, to all organisations in or connected with our dairy industry who may wish to provide donations to the Fund. However, I am writing to all Members initially since I am sure many of you may also wish to contribute personally to the purchase of our new offices".

"With legal and removal costs plus other expenses associated with the move to Huntingdon, it is anticipated that we need to raise about £35,000. We understand that donations from companies in or connected with the dairy industry would be an allowable expense against Corporation tax".

"For personal donations and where UK income tax is paid, the Society, as a Registered Charity, can obtain additional benefit by way of a tax refund from the Inland Revenue if the donations are covenanted over a minimum four-year period."

"Your Honorary Treasurer has been informed by the Inland Revenue that, to enable individual donations to be made in one payment and avoid extended borrowing by the Society, there is no objection to individual members providing one payment only, in the form of a first instalment of the covenanted donation and the balance by way of an interest free loan to the Society (repayable on demand). This loan would be used by the Society as the remaining instalments of the covenanted donation from the Member. The Society would then claim the tax refunds from the Revenue as each instalment falls due".

"I am extremely hopeful that we shall be successful with the Appeal in achieving our target. As you will realise, the purchase of our own property puts us in a much more satisfactory position financially and gives the Society much more security in the future".

This letter incorporated an appeal for donations to fund the purchase of the building. Members were asked to covenant their donations so that the Society as a Charity could reclaim the tax. The interest free "bridging" loans referred to in the letter were provided by:

| | |
|---|---|
| APV International | £3,000.00 |
| Cliffords Dairies | £2,000.00 |
| Express Dairies Ltd | £3,000.00 |
| Milk Marketing Board England and Wales | £3,000.00 |
| Milk Marketing Board for N Ireland | £3,000.00 |
| Scottish Milk Marketing Board | £1,000.00 |
| Tetra Pak Ltd | £3,000.00 |

The main loans noted as £3000 were in fact for £4000 but £1000 was repaid within a year. The balance of the loans was later converted to a gift to the Society, a generous action which was of great assistance to the finances of the Society.

Mr Wilf Hipkins, the subsequent President, sent out a further letter dated January 1983 urging members to make donations to the building fund and to covenant the donations. He admitted that donations were coming in slower than expected.

A report was made on the Building Fund and the bridging loans in the accounts for the year ending 30 June 1987, as follows:

1. "The majority of loans made to the Society at the time of the purchase of the Huntingdon premises have been either converted into a straight donation or recalled by the companies concerned. Three loans of £3,000 each remains unsecured - from Express, Tetra Pak and MMB E&W. Therefore these will be shown, as previously, in the 1986/7 balance sheet as unsecured loans which will be taken into account when determining the Society's net assets".

   "None of the companies concerned is pressing for the return of its loan, which may have been forgotten. The Society should still look for an opportune time with each company to have the loan converted into a donation; the auditors have accepted the present position and are not insisting that the Society should pursue the state of the loans".

2. "After discussions with the auditors, the Office Building Fund will be renamed as the Office Maintenance Fund in order to provide a source for the costs incurred in keeping the building to the required standard e.g. internal/external decoration, repairs etc".

   "Expenditure on the building during the 1986/87 year will be charged to the Fund. In order to provide a regular income into the Fund, thereby avoiding any major drain on the Income and

Expenditure Account in any one year, it is planned to allocate £1,000 from the Profit/Loss Account to the Maintenance Fund. This will become a regular feature in future years although the Society can alter the amount involved as better information is accumulated on the "true" cost of keeping the building in good repair. The rent from sub-letting will be included in the income under the Profit and Loss Account and not allocated to the Building Fund as this is not directly related to costs but reflects the current market state for property letting in the Huntingdon area."

The building was named Crossley House in honour of Professor Eric Crossley. The opening was recorded in the supplement of the July 1983 issue of the Journal:

"Mrs Janet Crossley, widow of the late Professor E. Crossley who died in June last year, formally 'cut the ribbon' at the Society's Huntingdon offices when the building was named Crossley House, on 24 May. She was accompanied by Professor Crossley's sister and 26 members also attended the opening which was held on one of the few sunny days experienced during May."

There were many problems with the building and with the rights of adjoining owners. The buildings, which opened out directly onto a busy main road, were not in a good structural state.

There was more space than was required for the SDT and it was decided to let out part of the property. Savory and Moore, taken over by Lloyds Chemists, rented the ground floor and converted it to a pharmacy.

For two decades the lack of Charity registration caused no problem and indeed there was no reason why it should. Problems arose out of the decision to acquire office premises. Peter Hoare who became Chairman of the Trustees wrote in a letter dated 16 June 1989:

"Until 25th August 1964 the SDT was not registered as a Charity. Up to that time the Society could lay down its own rules for its Trustees, if indeed it had any. However, having been granted Charity Status under the Charities Act 1960 SDT became legally bound to appoint Trustees who have to ensure that the Society carries on its business within the terms of the Act".

"The position since 1964 has been, therefore, that officers of the Society are elected by the Members to carry out the business of the Society (see Rule 11)".

"In contrast to this the Trustees are appointed under the terms of the Charities Act 1960 and are legally accountable to the Charity Commission for the way in which the Society is run by its officers and members."

With references to his aside "if indeed it had any", the fledgling Society gave careful attention to its constitution which is evident from the minutes of meetings.

A revision of the Society's Constitution and rules was adopted at an Extraordinary General Meeting on 30 April 1958 (JSDT vol 11 No 3 1958). The main changes concerned:

(A) CLASSES OF MEMBERSHIP

"Rule 5. The terms hitherto employed are not considered entirely logical. It is felt that the terms used should broadly correspond with the terms used in other technical bodies hence the recommendation that the main body of members should in future be known as 'Members' instead of 'Ordinary Members,' the definition of qualifications being unchanged: also that the term 'Associate Member' be replaced by 'Student Member' and apply only to a person undergoing training. The revised Constitution and Rules provides for 'Honorary Membership' a class created by resolution of the Council in 1950.

(B) COUNCIL

Rule 12 (e). Admission of the Chairman of the London Branch as a member of Council.
Rule 17 (e) and (f). New rules defining action in the event of complications arising during elections to the Council.

(C) TERMS OF OFFICE

Rule 18 (c) and (d). Limitation of the maximum unbroken term of office of the Honorary Treasurer and the Honorary Secretary to four years instead of six as at present.
Rule 43 (h). Provision of limitation of term of office for Section Chairmen not previously covered.

(D) DISPOSAL OF ASSETS

Rule 26. Introduced to cover the disposal of assets in the event of dissolution of the Society.

(E) BRANCHES

Rule 44. Introduced to control the development of Branches, no provision having been made heretofore."

It was unfortunate that the Chairman and fellow Trustees misinterpreted the requirements of the Charity Commission and presumed that the Trustees had a superior position in decision-making. Letters in the file show that the Trustees and the Council were writing to each other and taking up position as if they belonged to different organisations. Each party believed quite genuinely that they were uniquely responsible for the well being of the Society.

GUIDELINES

In 1989/90 much energy was expended in drawing up "Guidelines for communication between Trustees, Officers and Council of the Society." The opening paragraph of a draft document dated 5 Sept, 1989 set out the parameters.

<u>Guidelines for communication between Trustees, Officers and Council of the Society</u>

"These notes are meant to guide relationships and communications between the Trustees and the Officers and Council of the Society to avoid misunderstanding. They are not exhaustive, intended as a substitute for common sense, or intended to affect the different responsibilities of those involved. They should be reviewed regularly and amended if necessary".

"Trustees and Council members should always be aware of each other's responsibilities. Council members represent members either through direct election or through the Sections. Officers are elected by Council. The responsibilities of Council and the Officers for the running of the Society are set out in the Constitution and Rules. Trustees are elected by the members in accordance with Rules 23a - d. They have legal responsibilities which arise from the Society's status as a Charity: these are essentially to account to the Charities Commission for the proper financial operation of the Society. Trustees have traditionally been Past Presidents of the Society. If they attend Council, they do so as Past Presidents: they can take part in discussion but do not have a vote".

"To ensure that these different responsibilities are properly understood, Trustees and Council members should each have a copy of the Society's current Constitution and Rules, and a copy of Charities Commission Leaflet CC3. A copy of the Charities Act 1960 should be available for reference in the Society's files".

It is not clear whether these guidelines were ever formally approved by the Society.

TRUSTEES

The extent to which the Trustees, at the time, misunderstood their function is demonstrated by a paragraph in the Minutes of the Joint Meeting of Trustees and Officers on the 26th January 1990. The persons present were:

Mr R A Hoare (Chairman), Mrs S. W. Barron, Dr. H Burton, Dr. G Chambers, Dr. J Gordon, Mrs S M Moore.

"Under the heading Review of Trustees: "It was agreed that the appointment of a new Trustee with <u>accountancy skills</u> would be of considerable benefit to the Society. Several alternatives were discussed and it was finally agreed that the President would informally approach Mr W Freeman of DTF and invite him to offer himself for election. Mr Hoare agreed to be involved in the preliminary discussions if required".

"It was agreed that when the Rules and Constitution were amended a clause would be inserted whereby Trustees were obliged to resign when they became 70 years of age".

It is interesting that at that time it was proposed that Trustees should be obliged to resign at 70 years of age!

The management of the building and the Trustees was linked and led to the only acrimonious episode in the history of the Society. The Trustees had, in all good faith, usurped the responsibility of the Council especially in regard to the management of the building. The Council had to re-establish its authority and after much debate the problem was resolved amicably by establishing two classes of Trustees: the Managing Trustees (which was the President, Officers and Council) and Custodial Trustees who had a residual responsibility for the investments of the Society including the building. At the AGM on 24 October 1994 at Bristol, Mr. W. Hipkins, Chairman of the Trustees (a post he has held with distinction for many years) made a gracious speech accepting the new arrangement (JSDT Vol 48 No1 Feb 1995).

*Mr. W. J. Hipkins*
*President 1982-83*

Mr Peter Lee was appointed Executive Secretary in 1976 in succession to Mr Stan Stilton who was Secretary from 1973 until 1976. His successor in June 1989 was Mrs Ros Gale who retired in December 1998, having given many years of outstanding service recognised by electing her as an Honorary Member. Other members of the office staff over the years were Mrs Cybulla (Wembley office), Mrs Elizabeth Russell, Mrs Joan Lawford and Miss Lorraine Attwood.

With the retirement of Mrs Gale and the significant decline in number of members and subscription income Council concluded that it was no longer economic to continue with Crossley House and that the administration of the Society should be contracted out. Serious consideration was given to the sale of the building, but it was concluded that as letting the building was giving a useful income (better than could have been obtained at the time by investing the cash received from the sale) it should be retained as a long-term asset of the Society. A new lease was negotiated with Lloyds Chemists, who already rented the ground floor for a term of 6 years from June 1993 at a rent of £6,000 p.a. It was necessary to obtain the assent of the so-called Property Trustees who were in fact the Trustees appointed by the Society. They were the late Sir Richard Trehane, A Graham Enock, F Procter and Dr R J Macwalter, CBE. They were replaced by Dr David Armstrong, Mr Wilf Hipkins and Mr Peter Fleming (November 1998). The new lease provided a rent of £9,500 p.a. for a term of 6 years.

Crossley House, though it served the purpose, is an old and poorly designed building and repairs had to be made over the years. In a letter to Mrs Gale dated 19th October, 1989 Lines Commercial, Chartered Surveyor gave a brief description of the property:

"For the sake of the record I confirm that the property comprises a detached two-storey office building, the front section of which is of mid 19th century vintage, constructed of 9" solid brick work and under a slate roof, the rear section of which is of more recent construction, being of cavity brick work under a flat roof. The accommodation comprises three offices, totalling 373 square feet net with kitchen and two toilets on the first floor, and three offices, totalling 408 square feet net with toilet on the ground floor. Each floor is self-contained with its own entrance and services, including gas central heating and hot water heating. There is space at the side of the property for parking three cars with direct access to Ermine Street over a right of way and there is a small garden area at the rear, which we have assumed could be used for additional car parking or for extension of the building, subject to planning, sometime in the future. We have seen a plan showing a vehicular right of way between Ermine Street and Great Northern Street for the benefit of the property.

The ground floor suite of offices is let under a lease dated 9 May 1988 for a period of 21 years from 25 December 1987 at a rental of £3,250 per annum exclusive.

Under the terms of the lease the permitted user is 'offices'. The tenant must maintain the interior of the premises and pay a fair proportion of maintenance for the common parts and areas. The tenant has the right to assign or under-let the whole of the premises subject to the landlord's consent 'not to be unreasonably withheld'."

"Lines" valued the property at the time subject to the lease and other qualifications at £85,000.

There were also problems with access and the rights of adjoining property owners. These matters required time-consuming attention by the officers and staff of the Society.

Having considered alternatives (including a proposal from the RABDF and the IFST) the Council decided to contract the administration of the Society to the Plunkett Foundation (founded by Sir Horace Plunkett) located at Long Hanborough, Oxford and which plays a central role in the promotion of co-operatives in Great Britain and Ireland. The Plunkett Foundation took over the administration of the Society in October 1998 at an initial annual fee of £20,000. Mr Simon Rawlinson, Chief Executive, was a member of the Society and Dr David Armstrong was a long-standing Trustee of the Foundation.

Initially Plunkett coped well with the SDT work but strains emerged. Following the appointments of a new Chairman and Chief Executive, a review of the future strategic direction of the Plunkett Foundation was undertaken and it was apparent that the SDT work did not fit in comfortably with the new strategies. It was therefore decided by mutual agreement to terminate the contract. The Society decided to seek quotations to replace Plunkett and Dr Armstrong was entrusted with this operation. This was a time-consuming and difficult operation but in the end it was decided to award the contract to Agri-Food Consultancy, of which Dr. A. C. O'Sullivan, a Past President, is the Principal. Dr. O'Sullivan was obviously familiar with the administration of the Society and had also acted as Programme Planner. A formal contract was drawn up and A.C. O'Sullivan took over from Plunkett from 1 May 2001 as the Society's Executive Director.

At the same time the SDT had been reviewing the business plan for the organisation. It was decided that subscriptions, the main source of income for the Society would have to be increased, that payment should be made in future by variable direct debit in lieu of payment by cheque and standing orders which, in many cases, had not been updated. Deciding to raise subscriptions and introduce payment by direct debit at the same time was a tough decision but it was the correct one. The implementation of this scheme was admirably handled by Dr. O'Sullivan.

By 1947 the total membership of the Society was about 1200 and sections were flourishing in London, Scotland, Wales and the North West Midlands and West of England.

The hallmark of the Society, the Journal, was first published in October 1947 edited by Dr. Charles Crowther. Journal Vol. 1.1 No 1 presented papers given at a symposium on metals and milk in May 1947. The authors included Dr. A.T.R. Mattick, Professor E.L. Crossley, Professor E. Capstick and Mr D. B. Rogers. Several of these men made major contributions to the development of the dairy industry in the UK as well as the growth of the Society. The proceedings of the AGM held in London on 27th October 1947 were also recorded in this first edition of the Journal.

MILK INDUSTRIES OF THE UK AND IRELAND
The milk industry of the UK when the SDT was founded was essentially a British affair with a huge number of milk producers, many proprietary dairymen and a number of family founded, nationally known dairy companies.

ENGLAND AND WALES
The Milk Marketing Board for England and Wales, which had been set up in 1933 following an exhaustive inquiry into the depressed state of milk producers, acted as an agent for the Ministry of Food during the war years which controlled every aspect of milk and food production. The essential requirement was to feed the nation by maximizing home food production to replace imports severely disrupted by the war and specifically German submarine attacks on convoys.

*The MMB (England & Wales) Head Office at Thames Ditton, Surrey. (since demolished)*

*Mrs Parry, Peter Willis, Anne Hunter and Dr. David Armstrong at MMB E&W Brussels office.*

9

## SCOTLAND

*Sir Richard Trehane*
*An outstanding chairman of the*
*MMB and leader of the industry.*

In Scotland, as in Northern Ireland and in contrast with the absence of co-operatives in England and Wales, co-operative enterprises had developed in the first two decades of the 20th Century. Co-operation was supported by the formation in 1905 of the Scottish Agricultural Organisation Society (SAOS) and, by 1930, twenty-four co-operatives were operating in Scotland mainly in the South West of the country. But, as so often with British co-operatives, the reaction to market pressures was internecine competition which led to the establishment of the Scottish Milk Agency (S.M.A.) in November 1927. At that stage it was not possible to create an all Scotland approach and the Aberdeen and District Milk Agency was set up in 1928. At the start the S.M.A. had some 1800 producers and did some good work but it was flawed from the beginning by the fact that a number of influential co-ops stayed outside.

It was decided that Scotland should follow the example of England and take advantage of the Agricultural Marketing Act of 1931. The Scottish Milk Marketing Board started in 1933 but the north of Scotland was excluded from its remit. As Dr. Donald McQueen has noted in his article on "The Milk Marketing Boards in Scotland and their legacy," John Inglis of the Clydeside shipbuilding family was an enthusiastic supporter of the concept of Milk Marketing Boards and as a farmer in Somerset played a leading role in setting up the M.M.B. (England and Wales).

John Inglis (1904-1988) recalled in his private memoirs: 'It was a bitter fight (to win support for the concept of a Milk Marketing Board) with red faces all round and one of the bitterest against us was the man later to become Chairman of the Milk Marketing Board. For after all we narrowly won the day by 99 votes to 94 with over 100 abstaining. So great was the feeling aroused that the Union's President, a most distinguished man, and the Vice President walked out of the Union's Head Office that night and never came back". John Inglis came to the Scottish Board as Marketing Officer in 1951 from the MMB of England and Wales where he was Deputy to J L Davies (later General Manager of the English & Welsh Board.) He was appointed General Manager of the Scottish Board in 1958.

Having regard to subsequent events, it was significant that the N.F.U. (of England and Wales) was not wholehearted in its support of the concept of Milk Marketing Boards (an essential feature of which was that milk producers were required to sell their milk to the board.) The reasons for the N.F.U. reluctance were partly the lack of control of cheap imports which were depressing milk prices but probably more important that the N.F.U. perceived milk boards with their links to producers as a threat to their position as the sole representative of farmers. The N.F.U. also objected to the principle of compulsion.

Whereas the promoters of the Board in England and Wales were successful in setting up a national body, in Scotland regional rivalries and contrasting market factors restricted the Scottish board to the southern region. Ultimately, three boards were set up in Scotland: the main Scottish Board, The Aberdeen and District Board and the North of Scotland Board which was centred in Inverness. All subsequent attempts to form one Board for Scotland failed.

## NORTHERN IRELAND

In Northern Ireland during the war the Government of Northern Ireland through its Ministry of Agriculture controlled the milk industry. A feature of the dairy industry in Northern Ireland at the time was the part played by several relatively large co-operative societies which were involved in the manufacture of cheese and butter and the processing and sale of liquid milk. Nestlé and the Co-operative Wholesale Society (C.W.S.) had significant interests in milk manufacture.

The Ministry of Agriculture in Northern Ireland was active in supervising all aspects of the industry from milk production to manufacture and distribution

## IRELAND

The Society has two active regions in Ireland, North and South. The Northern Ireland section was established in 1950 and the Southern Ireland Section in 1966.

For many reasons the dairy industry in the Republic of Ireland took a very different development path from that in Northern Ireland, which was more influenced by the situation in the rest of the U.K.

In the 19th Century, Ireland, which was then governed as part of the U.K., was a land of very small tenant farmers, a large number of farm workers, and landlords with large estates. Generally holdings were small and cows were kept as a source of milk and butter for domestic use. The dependence on the potato as the main source of food for the rural population of peasants and labourers had disastrous consequences in famine and emigration.

The potato blight devastated the countryside for several decades but the potential assets of a climate and soil conducive to milk production were there awaiting development. This was facilitated by land reform, and especially the land purchase acts, which enabled farmers to acquire legitimately their own holdings. But what made subsequent growth of the milk industry in Ireland distinctive was the co-operation factor to which the industry owes its present shape. Though many individuals played a part all would agree that Horace Plunkett and his associates were the driving force. Plunkett had the vision to see that the best hope for Irish small farmers was self-help through co-operation.

He had the personality, the desire and the political nous to put his ideas into practice. On the Government level he was instrumental in setting up the Department of Agriculture and Technical Instruction for Ireland, which was a unique and pioneering Government department not only in the British Isles but also in Europe. He was the major force in promotion of enterprise, and innovation and organisation. His achievements were immense and all the more remarkable given the political turmoil of the time and the advent of the First World War.

*Sir Horace Curzon Plunkett (Knighted in 1903)*

Horace Plunkett, born into an aristocratic family, his father was Lord Dunsany, was educated at Eton and Oxford, and spent several years ranching in Canada. He was aware of the thinking of the pioneers of co-operation in England and especially Robert Owen, and was convinced that co-operation was the way forward summed up by his powerful message, as relevant today as it was in the 19th Century: *"Better Farming, Better Business, Better Living."* The pioneering days of the co-operative movement are recorded in the "Fruits of a Century" published by the Irish Co-operative Organisation Society (ICOS) to mark its centenary in 1994. The authors date the commencement of the Irish co-operative movement to 1889 with the opening of Plunkett's campaign and the foundation of the Belfast Co-operative Society (known by generations in Belfast simply and affectionately as the "Co-op" and more recently and sadly the victim of terrorist attack).

Plunkett had links with the Co-operative Union in Manchester and the C.W.S. (Co-operative Wholesale Society), as a major purchaser of Irish butter, had several depots in Ireland. The first co-operative was set up in Dromcollogher, Co Limerick with C.W.S. help.

Plunkett saw clearly that if Irish agriculture was to make progress, organisation had to start with the farmers and the model he chose was Denmark, which had successfully developed co-operative creameries, the butter from which was replacing poorly produced, badly packaged, badly presented Irish butter in the crucial British market. Plunkett also realised that he needed an energetic and enthusiastic organiser to travel round the countryside to promote the co-operative concept. He chose Mr R. A. Anderson, farmer's son, landlord's agent and Clerk of the Petty Sessions in Doneraile, Co. Cork. Plunkett and Anderson were received with hostility. In Anderson's words, "It was hard and thankless work. There was apathy of the people and the active opposition of the press and politicians. It would be hard to say now whether the abuse of the conservative Cork Constitution or the nationalist Eagle of Skibbereen, was the louder. We were 'killing the calves,' we were 'forcing young women to emigrate,' we were 'destroying the industry.' Mr Plunkett was described as a 'monster in human shape' and was abjured to 'cease his hellish work.' I was described as his 'Man Friday' and as 'Roughrider Anderson.' Once when I thought that I had planted a creamery within the precincts of the town of Rathkeale, my co-

11

operative applecart was upset by the local solicitor, who "gravely informed me that our programme would not suit Rathkeale…" 'Rathkeale' said he pompously 'is a nationalist town - nationalist to the backbone - and <u>every pound of butter made in this creamery must be made on nationalist principles, or it shan't be made at all.</u>' This sentiment was applauded loudly and the proceedings terminated."

Plunkett and Anderson were not deterred. They had the determination, a clear vision, and the social and political skills, and the stamina to present the case at innumerable meetings. By 1893 about 30 co-op creameries with a turnover of £140,000 were in business.

At this stage it became apparent that an organisation to bind together the individual co-op societies was required and the Irish Agricultural Organisation Society (IAOS) was set up in 1894 in the presence of everybody who was anybody. Plunkett knew how to mobilise the "A" list of the day. There was a dinner for the Press, which ensured the setting up of the new organisation got coverage. The gathering was also timed to take place during the Royal Dublin Society's Spring Show, a major social occasion.

Plunkett knew that education and technical instruction were essential ingredients in his co-operative concept and he also realised that self-help would not be enough, and that state involvement was necessary. As a result of skilful lobbying during the parliamentary recesses of 1895-6 the British Government set up in 1899 the Department of Agriculture and Technical Instruction for Ireland (DATI), and put Plunkett in charge as Vice President. The Department made an outstanding contribution to Irish agriculture and had it not been for the outbreak of the First World War in 1914 it could have been the model for other state economic and educational institutions, which could have gone a long way to satisfy the craving of the Irish for greater participation and control of their affairs.

In 1900 when Plunkett gave up his Presidency of IAOS to concentrate on the DATI, there were 840 affiliated societies of which some 350 were creameries. He had succeeded in creating a co-operative structure which was to last but his own standing was being undermined.

The ideas and thoughts incorporated in his book *"Ireland in the New Century"* (still interesting reading) managed to antagonise a wide range of people. As always, jealousy was a factor but Ireland was about to be mired in a series of upheavals: the Irish rebellion of 1916, its suppression, the activities of irregulars, and the civil war. Plunkett himself was a victim. His much loved home Kilteragh was burned own in January 1923 and with it Plunkett's hope for the future. He spent his remaining years in England where he set up the Plunkett Foundation for Co-op studies in Oxford.

PARTITION

The Government of Ireland Act 1920 set up two administrations which, in spite of many changes of style and upheavals including well-publicised terrorist activity, still exist today. One of the many unfortunate consequences of partition was the division of I.A.O.S. and the setting up of the Ulster Agricultural Organisation Society (U.A.O.S.) which still functions today.

Many of the co-operative creameries organised North and South in Plunkett's time are still there thought difficult to recognise as they have been re-branded, consolidated, integrated, amalgamated, taken over, and now become public liability companies (plcs) with businesses extending far beyond the shores of Ireland.

The decades following the setting up of an Irish Government in 1921 up to the outbreak of the second World War in 1945 were years of stagnation though there was much discussion and debate about how Irish agriculture and especially milk production

*Small country creamery accepting producer-delivered milk which is then separated with skim returned to producer (see skim return pipe line above iron railings).*

and the manufacture of milk products could be expanded. In the post-war years progress was made; milk production expanded and the co-operative creamery movement gained strength with

*As it used to be, self-delivery.*

*As it was! Self-delivery in 17 gallon churns by donkey cart.*

new creameries opened and the formation of larger units. Research facilities were also strengthened and worked closely with the management of the larger units now emerging.

THE KNAPP REPORT

Both the Irish and British Governments were watching the development of the Common Market in Europe but an application by Ireland and the UK was rejected in 1963. In that year Dr. J. Knapp, Administrator of the Farmer Co-op Service in the United States Department of Agriculture was invited by the Minister of Agriculture (Ireland) to review the agricultural co-op movement in Ireland. In the words of the authors of *"Fruits of a Century"* the Knapp Report marks 1964 as a watershed in the history of both the Co-op Movement in Ireland and the IAOS itself...." The principal recommendations made by Dr. Knapp were:

1.    IAOS as the central organisation of co-operatives should be reinvigorated with Government financed assistance and should be given responsibility for "working out general reorganisation plans for the dairy industry," and should be "encouraged to broaden its field of service to embrace more generally all types of agricultural co-operation."

2.    The Dairy Disposal Company (a state body managing a group of creameries) should be converted to a co-operative.

These recommendations were made in the context of the determination of the Irish Government to join the Common Market at the earliest opportunity. The Government correctly recognised that as a community of small farmers with natural advantage in milk production the Common Agricultural Policy was potentially a source of great riches while still retaining access to the all-important British market safeguarded in the Anglo Irish Trade agreement of December 1965.

Bilk-tank collection of milk began in Ireland in 1964 with this unit which was imported fully assembled by Lough Egish Co-op.

IAOS PROPOSALS FOR REORGANISATION

In 1966 IAOS published "Proposals for Reorganisation in the Dairy Industry." "The proposals for rationalisation were based on the establishment of 19 larger co-op creamery groups through the amalgamation of existing societies and through the transfer of the Dairy Disposal Company's creameries to co-operative control." These proposals were the precursor of the handful of large powerful hybrid PLC/Co-ops which now dominate and control not only the dairy industry in Ireland but also a significant part of the industry in the UK.

Attempts were made to improve the marketing of Irish dairy products, especially the main product butter the quality of which was variable and which had been sold traditionally on an individual contact basis using agents.

Throughout these years butter was by far the most important milk product produced in Ireland and the principal export. The United Kingdom was the principal market. The manufacture of chocolate crumb and milk powder had started and developed in association with British Companies. "During the period 1922-1960 the Irish dairy manufacturing industry was over-dependant on butter manufacture. It consisted of a large number of fragmented small units and it lacked the flexibility to respond to changes in world demand and markets…"

During this period the sale of Irish agricultural produce was disrupted by the Anglo Irish trade war and the outbreak of the Second World War in 1939.

Important points of policy were becoming clear: first if Irish agriculture was to expand, export markets had to be found and kept; second the quality of products had to be uniformly good; third there had to be discipline which in practice meant State enforcement; a coordinated professional approach to marketing was essential and prices had to be competitive.

## MARKETING ACT 1961: AN BORD BAINNE

These key points were reflected in the Dairy Produce Marketing Act of 1961, which empowered the establishment of An Bord Bainne, the Irish Dairy Products Export Board. The objectives of the new board were stated to be "to promote, facilitate, encourage, assist, coordinate and develop the exportation of milk and milk products."

The functions of the Butter Marketing Committee were taken over by the new body which became responsible for the export of almost all the principal products manufactured. Exceptions were the export of chocolate crumb, milk powder, non-creamery butter and some special cheeses.

At the outset the Government paid the Board a substantial annual grant and it was authorised to collect levies from producers.

There were nine members of the Board broadly representative of all sections of the milk manufacturing industry and including the Minister for Agriculture. Mr Patrick J. Power, representing manufacturers of milk powder was elected Chairman, a position which he held until his death in May 1971. The most colourful member of the new Board was Mr Martin J. Mulally who represented the Dairy Disposal Company. Martin was a powerful personality who had been actively involved in the fight for Irish independence, but he did not allow that to sour his relationships and he threw all his experience, energy and charm into making the new Board a success.

The appointment of the charismatic Ireland and British Lions rugby player, Tony O'Reilly, as General Manager of the new organisation was a stroke of genius.

Though now long forgotten there was a political and trade battle over access to the British butter market, the largest in the world, and, in spite of ups and downs, the most profitable. New Zealand was fiercely protective of its privileged position (that has not changed) as were the other long-term suppliers. Without recording all the details, the UK imposed a penal duty on imports of butter from Ireland and trade ceased from November 1961 until February 1962. So the new Board started business facing a crisis in the export of butter. Prices had collapsed because of over-supply. There were many consultations and meetings (GATT was also involved) and ultimately it was reluctantly agreed that supplies to the British market should be restricted and Ireland was allocated a quota of 2,000 tonnes (compared with 159,000 tonnes for New Zealand and 63,000 tonnes for Australia).

At this time cheese production in Ireland was small but there were also problems of over-supply in the UK market and in this case there was a voluntary understanding to restrict supplies. As in the case of butter, the main driver for a restriction of supplies was New Zealand. Informal meetings of the leading cheese suppliers were held in the Savoy Hotel in London led by the formidable Bill Rodden. Curiously and sensibly the question of restrictive trade practices never arose, though the Authorities were aware of the activities of the "cheese club."

It is interesting and instructive that half the quota of 12,000 tonnes was sold in Northern Ireland. It was not unknown in the past for butter sold to agents in Britain to find its way back to Northern Ireland to the benefit of the agents. The clear message for An Bord Bainne and its inexperienced but energetic management was to find markets for Irish butter outside the UK and, most importantly, sell it under a distinctive brand. In those days as General Manager of the Milk Marketing Board for Northern Ireland, I had many informal talks with Mr Tony O'Reilly and urged him to press ahead with the development of a brand which would stop the traffic in bulk butter and ensure Irish butter was recognised by consumers in retail outlets. The result was **'Kerrygold'**, a name derided by some at the time but quickly established as an outstanding success.

As Tony O'Reilly was linked with the success of Kerrygold, his standing and reputation was enhanced. He was now in demand. In 1966 he was appointed Managing Director of Comhlucht Siúcra Erin (Irish Sugar Company) & Erin Foods Ltd, a staging post which led to Heinz and an immensely successful career. In four years he had put Kerrygold and An Bord Bainne on the world map. His successor, Mr Joe McGough continued the good work.

In 1969 the value of dairy exports hit a record £31 million compared with £7 million in 1961. The Board was now exporting a wide range of dairy products to 49 countries.

*Joint meeting of Northern Ireland MMB and An Bord Bainne CA 1962.*

| | |
|---|---|
| *Centre* | *J.K. Lynn, Chairman MMBNI* |
| | *P.J. Power, Chairman An Bord Bainne* |
| *3rd from right front row* | *Tony O'Reilly, G.M. An Bord Bainne* |
| *3rd from left front row* | *David Armstrong, G.M. MMBNI* |
| *2nd from left front row* | *Martin J. Mullally* |
| *Back* | *Joe McGough, Tony O'Reilly's successor* |

EUROPEAN ECONOMIC COMMUNITY
The next major step for the Irish dairy industry was entry into the European Economic Community (EEC) which was done in conjunction with the United Kingdom. Meantime the expansion and improvement of agriculture and especially the production of milk continued with the aid of substantial state support. Among the beneficiaries was An Foras Taluntais (The Agricultural Institute), to which funds were made available under the American Counterpart Fund instead of the non-applicable Marshall Plan. This national agricultural research organisation, established in 1958 when the Republic of Ireland was coming out of a deep recession, received a State grant in 1972-73 of about £2 million. The functions of the Institute were to "review, facilitate, encourage, coordinate, promote and undertake agricultural research." The Institute at the time employed some 220 scientists and 360 technicians. The main centre for dairy activity was established at Moorepark, Fermoy, County Cork which benefited from the expertise of the New Zealand dairy industry.

COOKE SPRAGUE REPORT
The Dairy Science Faculty of University College, Cork was closely tied to An Foras Taluntis and conducted research projects in co-operation with the Dairy Products Research Centre, Moorepark and An Bord Bainne. Some of the projects were jointly financed by the Irish National Dairy Council, An Foras Taluntais and Bord Bainne. This relationship between Moorepark and An Bord Bainne provided the industry with valuable research as well as the day-to-day maintenance of quality standards. The Co-operatives had new multi-purpose plants treating milk as a raw material for manufacture of the most profitable products. Further impetus towards rationalisation and larger scale manufacture was given by the Cooke and Sprague investigation and report in 1968. Hugh L. Cooke was a dairy expert at University of Wisconsin and Dr. Gordon W. Sprague was the economist for Land O Lakes Creameries Inc of Minnesota. Among the bodies consulted by Cooke & Sprague were the Milk Marketing Boards of Britain and Northern Ireland. Irish agriculture was at this time heavily subsidized by the State. "Total subsidies in 1971/2 on 538 million gallons of manufactured milk were about £30 million (£28 million on all products in milk price subsidies, £2 million in export subsidies.) These represent

about 44% of milk price..."

The burden on the Irish Exchequer of these comparatively massive subsidies could not be sustained and ironically the solution was to join the EEC and shift the burden onto the Common Agricultural Policy to which the UK was to become a major contributor. In effect, the burden on the Irish Exchequer was, transferred to the British Exchequer!

The involvement of the State in An Bord Bainne made it unacceptable under the Common Market's rules. The solution was to transform An Bord Bainne into a voluntary co-operative organisation; and An Bord Bainne Co-operative Ltd was registered on September 14th, 1972. Thus the Irish managed to retain central marketing of Irish dairy products. In the UK special regulations were drawn up to enable the Milk Marketing Boards to continue in operation, but these were flawed from the start and the consequence was the ultimate disappearance of the Milk Marketing Boards with a adverse impact on all aspects of milk production, manufacture, and marketing of milk and milk products in the UK.

An essential element in the transition from the statutory An Bord Bainne to the co-operative was financial support from the Irish Government in the form of a guarantee of borrowings up to £20 million by the co-operative in 1974, which increased to £90 millions in 1979. Both the UK and Ireland became members of the EEC in 1973.

The consequences for Irish Agriculture were enormously beneficial; the consequences for the UK dairy farmer were the opposite, though this was not immediately apparent. Even before entry was completed, An Bord Bainne and the major co-operative groups had started to acquire businesses in the UK dairy industry and further afield, especially in the USA. The first significant purchase was Adams Butter Ltd (later Adams Foods). Brian Joyce, who was Chief Accountant of the Board, was appointed Managing Director; and later, he succeeded Mr Joe McGough in charge of the parent co-operative.

Other purchases of UK businesses were made and this was followed by the establishment of trading relationships in Continental Europe and further afield.

GROWTH OF IRISH DAIRY INDUSTRY

In recognition of the international development of the business, An Bord Bainne became the Irish Dairy Board. The growth of the Irish dairy industry following entry into the EEC (now the EU) was spectacular. From 1960 to 2002, sales of Irish dairy products overseas grew from £7.5 million to £1.9 billion. The Irish Dairy Board had become Ireland's largest food exporter.

According to Sir Anthony O'Reilly, first General Manager, now President of Independant News and Media: "The Irish Dairy Board has developed an effective track record as a marketing agency. Ireland's 28,000 dairy farmers have had forty years of experience of the Irish Dairy Board, they know it, believe in it, trust in it and appear to take a certain vicarious pride in it. It appears to have developed an excellent and effective staff dedicated to the principles and objectives both of the Board and of Irish agriculture. In summary the Irish Dairy Board has a brand name, Kerrygold, a selling and trading capability, a unity of purpose, a dedication to marketing and a history of success."

"In 1960 there were 97,500 milk suppliers producing 280 million gallons of milk from 669,000 creamery cows, giving an average herd size of 7 cows per supplier and average yield of 418 gallons per cow. The average price per gallon of milk was about 8p..." "In 1981 there were 64,626 suppliers producing 820 million gallons from 1,268,613 creamery cows, an average yield of 656 gallons per cow. The average price per gallon was about 60p. In the new millennium 28,000 dairy farmers (note the drop from 97,500 suppliers in 1960) tend over 1.3 million cows with an average yield of 4,408 Kg resulting in over 5,338 million tonnes of milk a year." The drop in the number of milk suppliers was matched by concentration of the processing side into four large groups, which though co-operative in name are in reality large, commercial, profit-driven PLCs with businesses in the UK and throughout the world. Further amalgamation can be expected. Though we are concerned mainly with milk, the commercialisation of the general side of co-operative business went ahead in parallel and the Irish Agricultural Wholesale Society (IAWS) is now a large and expanding multinational business with a range of products and activities from catching fish to making bread. These developments, in conjunction with industrial and commercial growth, have brought unparalleled prosperity to Ireland though the strains and stresses of rapid growth are now appearing in what has been described as the "Celtic Tiger" economy.

CONSEQUENCES OF PARTITION

The partition of Ireland in 1921 had many consequences and one of them was that the development of agriculture in Northern Ireland veered away from that in the Irish Free State. Farmer's organisations had to adapt to the new political framework. The Ulster Agricultural Organisation Society hived off from the Irish organisation and involved itself in the problems of the co-operative organisations in Northern Ireland and the Ulster Farmers Union took up the cudgels on behalf of Ulster (Northern Ireland) farmers. Though Northern Ireland was given its own Parliament, Government primary legislation emanated from Westminster. In the agricultural sphere the most relevant and important legislation was the Marketing Acts of 1931, which was mirrored in Northern Ireland by the Agricultural Marketing Act of 1933 (see Young, The Importance of Agriculture in N.I.; Harkness, Progress and Trends in the Dairy Industry in N.I.)

Though a Pigs Marketing Board was set up in Northern Ireland in 1933, it proved impossible to set up a milk-marketing scheme as in England and Scotland. Instead, the Parliament and Government of Northern Ireland enacted the Milk and Milk Products Act (N.I.) 1934 which provided a framework for the development of liquid milk and milk manufacture and the improvement of milk quality. An element of price support was also introduced. Late in 1936 the Government set up the Butter and Cream Marketing Board with the object of improving the marketing (and export) of cream and butter.

WAR YEARS IN NORTHERN IRELAND

During the war years (1939-1947) the Government of Northern Ireland, acting as agent for the Ministry of Food, took control of agricultural production with the primary objective of expanding the production of food. On 1st November 1940 the Ministry of Agriculture (Northern Ireland) became the purchaser of all milk sold off farms in Northern Ireland - a bold step which facilitated the establishment of the Milk Marketing Board for Northern Ireland after the war. Other measures were taken of a nature only possible under wartime regulation. As a result, by the end of the war, the dairy industry in Northern Ireland had been "transformed." Total sales of milk off farms increased by 50% and there was a four to fivefold increase in annual payments to producers. Sales of liquid milk doubled and in 1945 Belfast became the first city in the UK to be scheduled as a "safe milk area." Exports of liquid milk to the UK had grown significantly and the manufacturing industry had been diversified.

MILK MARKETING BOARD FOR NORTHERN IRELAND

With the ending of food control and the restoration of marketing powers in 1954 to the Milk Marketing Boards in GB, the time had come to set up a milk marketing scheme in Northern Ireland. As the newly appointed economist of the Ulster Farmers Union I was involved in the discussions and negotiations which resulted in the establishment of the Milk Marketing Board for Northern Ireland (MMBNI) in 1955. As a historical footnote consideration was given to extending the operations of the Milk Marketing Board (England & Wales) to Northern Ireland, the Province being treated as an additional region of the Board. Dealing with the comparatively insignificant milk production of Northern Ireland would not have been a physical problem for the Milk Marketing Board E&W but it was felt that the Northern Ireland region would not receive the attention that was required and this option was not pursued.

In his contribution to the SDT publication ***"The Milk Marketing Boards - A Concise History"*** Dr. George Chambers has given a lucid and comprehensive account of the start-up of the Board in Northern Ireland and its subsequent successful development. Between April 1955, when the Board commenced operations, and 1984, when the European Community introduced milk quotas on disadvantageous terms for the UK, milk sales off farms increased from 96MGAL (436 M L) to 312 MGAL (1,418 ML) and annual payments had increased from £14 million to £215 million gross. The growth in production was accompanied by a dramatic reduction in the number of milk producers from 23,000 to 8,300 but an increase in herd size and yield per cow. Creameries had been updated and expanded and the product range diversified. Payment on quality was successfully introduced and indeed every aspect of the business of production, processing and manufacture of milk had been improved with immense benefit not only to milk producers, but also to all involved in the industry. The confidence of the Board and its producers was reflected in the move from the dilapidated nissen huts in Castlehill Road to new purpose-built offices and central laboratory on the Antrim Road, Belfast which were officially opened by The Duchess of Gloucester in 1962. Dr Armstrong at that time was General Manager of the Board.

But there was not time to bask in success. As in the case of the rest of the UK, membership of the European Community generated problems for the Board. The introduction of quotas, the complexity of European regulations compounded by the vexatious way in which they were applied by UK Governments, ingenuity and greed by particular groups, and growing political discontent, which quickly developed into ferocious and brutal violence, weakened the Board's position. However, it still commanded the support of most milk producers and in the reorganisation forced on the industry by the combined efforts of the EU and the British Government, United Dairy Farmers and its processing arm, Dromona Quality Foods (now Dale Farm Ltd), was successfully launched. The Board's existence was terminated formally on 25th February 1995 and the successor body, a voluntary co-op in structure, started operations the next day.

In the intense negotiations which led up to the termination of the Northern Ireland Board, it was able significantly to retain control of its processing and manufacturing activities as a wholly owned subsidiary.

DEREGULATION

The sorry tale of "deregulation", a misnomer if ever there was one, must now be told. The documentation is massive. The Milk Marketing Boards in Scotland, England and Wales and Northern Ireland were required to submit reorganisation proposals. Lawyers were called in to add to the confusion. Though there were important variations in the ultimate results of reorganisation, certain key points were common. The Boards ceased to exist as statutory bodies under the Agricultural Marketing Acts as modified by a series of European Regulations, and were replaced by voluntary co-operatives or, in some cases, companies; producers of milk were no longer required to sell their milk to the new organisation; the commercial operations of the Boards were hived off (with the exception of Northern Ireland and the two small Boards in Scotland); the technical services of the Boards (with some exceptions) withered; research was damaged; and much capital investment made over the years was wasted. At the same time ownership of a large and important segment of the UK dairy industry passed into foreign hands, including ironically the integrated co-operatives of Ireland and continental Europe - a process which is still ongoing.

Milk Marque, the co-operative successor to the Milk Marketing Board for England & Wales, though launched in ringing terms as a commercial version of its parent has now collapsed, and has been replaced by a number of co-op ventures the future of which is shrouded in uncertainty. It has to be noted that the Government by its refusal to countenance vertical integration, commonplace in Ireland and the continent of Europe, was a major factor in the collapse of Milk Marque.

One positive outcome in England is that Dairy Crest, floated as a public company, is commercially successful and has ensured the survival to date of the long established trademark ***"Dairy Crest."***

Concurrently with the abolition of organised milk marketing, Governments, both Conservative and Labour, withdrew largely from support for dairy research and the provision of advisory services to farmers.

MILK DEVELOPMENT COUNCIL AND DAIRY COUNCIL

One of the consequences of de-regulation was the disruption of funding of the National Dairy Council through the Joint Committee of the Milk Marketing Board and the Dairy Trade. In place of that arrangement the Government facilitated the establishment of the Milk Development Council funded by a levy on milk producers, ironically a statutory levy! (producing an annual budget of about £7m.)

The Milk Development Council (MDC) is a levy-funded Organisation, which was established on 7 February 1995 by Statutory Order under the industrial Organisation and Development Act 1947.

The MDC is classified as an Executive Non-Departmental Public Body sponsored by the Department of the Environment, Food and Rural Affairs (DEFRA), with support from the Scottish Parliament and the Welsh Assembly.

MDC's core activities include market development, market information, research and knowledge transfer.

Funding of MDC activity is obtained from a statutorily imposed levy on all milk produced and sold in Great Britain. The maximum rate of levy collected from milk producers is determined each year and must receive Ministerial approval.

Through these functions the MDC works to improve the competitiveness and profitability of British dairy farmers.

NATIONAL DAIRY COUNCIL (NOW THE DAIRY COUNCIL)

In a leaflet dated 1998 the National Dairy Council (now The Dairy Council since early 2001) described the origins of the Council and outlined its activities as follows:

"For over seventy years the dairy farming and dairy trade have come together to provide an information programme and to undertake promotional activities in order to strengthen the market for milk and dairy products. The organisation which they formed for this purpose was the National Dairy Council (NDC).

When it was founded in 1920, as the National Milk Publicity Council, it was supported by voluntary funds from the industry. At its first meeting it passed a resolution to "make known in the interests of the nation, the value of pure milk" and began a nationwide propaganda campaign to raise levels of milk consumption.

When the Milk Marketing Scheme ended in 1994, it was generally agreed within the industry that the NDC should continue as an independent body. However a new method of funding needed to be found, as the system whereby the publicity levy was collected through the Milk Marketing Board could no longer operate.

The core areas of the NDC, such as Information Services and Market Research activities, are now funded by dairy farmers through the Milk Development Council and by processors, manufacturers and distributors via the Dairy Industry Federation (now the Dairy Industry Association Ltd, or DIAL).

It has two main objectives, agreed by both sides of the industry,
- to maintain the good image of milk, its products and the industry as a whole in the eyes of key influencers and the consumer
- to increase consumption of dairy products.

The Dairy Council has operated through the following departments:
- Information Services
- Marketing and Research

INFORMATION SERVICES DEPARTMENT

This Department offers an expert resource for the dairy industry.

Each year it answers approximately 40,000 enquiries and distributes over 1 million items of literature, promoting the industry and its products to consumers, health professionals, school teachers and children, and to the media.

EDUCATION PROGRAMME FOR SCHOOLS

Young people are major consumers of milk. TDC's education programme is designed to build their loyalty to milk and dairy products through a strong presence in schools.

HEALTH AND NUTRITION PROGRAMME FOR CONSUMERS AND HEALTH PROFESSIONALS

Health professionals, as well as consumers, are swamped with advice and misinformation often through the media and via pressure groups. Even Government initiatives such as the Health of the Nation with its emphasis on avoiding 'fatty' foods, can undermine the usage of "natural products such as milk and dairy products". TDC runs a comprehensive Nutrition Information

Service to keep key positive messages about the health benefits of dairy products at the forefront. It produces literature on the role of milk in a healthy, balanced diet and organises seminars and exhibitions for health professionals to help put the facts across and combat one-sided information. It also works closely with health promotion agencies, organisations representing the health professions and independent health experts, adding to the industry's credibility and authority.

### NUTRITION MONITOR

To monitor the continual flow of new scientific research on nutrition and health, a database was established in 1986 at the Milk Marketing Board. This was transferred to the NDC in 1995. By scanning the medical and scientific literature, it examines the main issues associated with milk and dairy products, producing critical appraisals of pertinent research and comprehensive reviews of current hypotheses relating to diet.

Currently there are 13 Dairy Councils or their equivalents in other countries subscribing to the database. Long term, the intention is to ensure that this activity becomes self-funding by marketing the service to the dairy industry internationally.

### ISSUES MANAGEMENT FOR THE DAIRY INDUSTRY

Food and farming industries are particularly vulnerable to criticism. TDC continuously monitors the media and pressure groups, as well as the activities of companies marketing competitive products. It addresses, where appropriate, the misrepresentation of milk and dairy products and works to ensure that damaging 'scare' stories are investigated and accurate information supplied to counter them.

In the event of an incident or issue arising which could threaten public confidence in dairy products (e.g. contamination, food scare, disease link, pressure group activity), it co-ordinates industry action and response drawing on the advice of independent as well as industry experts. It provides a common source of information and comments aimed at reassuring the consumer.

### MARKETING

This department ran a variety of marketing programmes on behalf of the dairy industry. However, following the end of the Milk Marketing Scheme in November 1994, funding for national generic advertising from the industry in England and Wales ceased for a number of years. But, a joint promotion of milk, as the *"White Stuff"* campaign, was funded by both sides of the dairy industry and ran from June 2000 to December 2001. Because the dairy trade sector did not agree to continue this joint funding, TDC's Marketing Department was closed down in late 2002.

However, TDC is currently running a series of advertorials in women's press, promoting milk to mothers of children aged between five and ten. This activity is funded by the European Commission. Part of this EU funding is also used to sponsor the junior section of the British Triathlon Association. The Milk Triathlon series features events for eight to fifteen year olds where competitors take part in three disciplines consecutively - swimming, cycling and running.

### RESEARCH

As part of its agreed role, The Dairy Council has an obligation to monitor the effectiveness of it activities as well as to meet the information needs of the industry.

In order to do this, TDC manages the purchase, analysis and dissemination of AGB Superpanel data on behalf of the dairy industry. This research consists of a panel of households who record all their food purchases and transmit this data to AGB, who collate and analyse it. It enables TDC to measure the size of the household market, and break it down by source of purchase, type of milk, type of packaging etc.

Much of this research, together with other relevant information is brought together in a quarterly publication "Milk Monitor".

The Research Department is also responsible for the production of ***Dairy Facts and Figures***. As the sister publication, ***EC Dairy Facts and Figures***, is no longer being produced, UK Dairy Facts and Figures is supplemented by a section containing key statistics on the dairy industries in other European countries.

# CHAPTER TWO

## SECTIONS, REGIONS AND BRANCHES

SECTIONS, REGIONS AND BRANCHES

Within a few days of the first general meeting of the Society, the Midlands and South Wales Sections were established. Professor Kay remarked in his paper on the early history of the Society that Mr John Lewis at the time the "President - elect" had much to do with this speedy and enthusiastic move" (JSDT Vol. 11 No. 1 1958)

The development of Sections and Branches in Great Britain is recorded in the Journals of the SDT.

NORTH WEST SECTION OF THE SOCIETY OF DAIRY TECHNOLOGY (MR CHRIS ASKEW, CHAIRMAN)

"The earliest records that remain on file for the Section are of the Committee meeting held at Dobson's Dairies, Levenshulme, November 1952 when the minutes of the 29th August 1952 meeting were approved. Even at this early time it was proposed to hold a joint meeting with the Midlands Section at the Buxton Hotel the following April followed by a visit to the local A.I Centre. Sadly, of the 68 people who attended, 20 were from the North West and only 3 from the Midlands. Travelling large distances to events was the main stumbling block, a trait that still persists today. It was not, however, until February 1953 that recruitment of new members was put firmly on the agenda."

"Throughout its history the Section has always sought to offer a mix of technical, practical and social events in its calendar of proceedings. It was therefore fitting that one of the first such courses was an *'Instruction in Lancashire Cheesemaking'* arranged with Major Brockholes at the Lancashire Institute of Agriculture, Hutton. Over the years, the Section has also maintained strong links with the other 'educational' centres within the region e.g. Reaseheath, Cheshire (to encourage future generations of dairy technologist to become active members). Ron Lawton was the Head of Section for many years as well as a staunch member of both the Section and Society and received the 'Award for outstanding efforts in regard to the Society' in 1980. Another stalwart was Wilf Hipkins who joined the Society in 1949, became Section Secretary in October 1957, Chairman in 1961 and President of the Society in 1982/83. In December 1958, the Constitution and Rules for the Preston Sub-Branch, to be known as the Fylde & Furness Branch of the North West Section, were proposed by Mr. P. J. Kearns and seconded by Mr. J. F. Heald."

"It was fascinating to observe from a trawl through the early Minutes that 'attendance at our meetings might be affected,' not for the most common reasons we hear today of there being very few dairy people in the Region or pressures of working longer hours, but because meeting dates coincided with local shows (e.g. Chester & Altrincham). Sadly there are very few of these remaining 50 years on, with probably the best-known survivors being the Cheshire Show at Tatton and the Nantwich Cheese Show."

"To accommodate everybody in a very large geographical area, meetings have been held in Liverpool, Manchester, Chester and Blackburn. In later years we have even tried holding 'joint' meetings with Lakeland and North East Sections."

"What goes around comes around with even the Society as recently as 2000 asking its members in a survey what they wanted from meetings and the reason for poor attendance. The same approach was adopted in 1959 by the Section in order to address falling numbers and also included other suggestions e.g. meetings with NFU, encourage farmers, and even a summer social event and coach party trips e.g. to Glaxo Laboratories and the MMB, Kendal."

"Pressure on finances has always remained a key focus but an unusual precedent was set in September 1962 when the Committee agreed that individual members would be responsible for the cost of their own meals at functions held within the Section. The Section would bear the cost of meals for any guests as approved by Committee but happily this draconian measure did not apply for long."

"In 1967 the Section organised the Spring Conference in Blackpool. This included an evening function and engaging a civil dignitary to attend one session of the Conference. Following this, the Section sponsored expenses for two delegates to attend the Annual Conference until it next hosted the Spring Conference at Chester in 1976."

"In the dairy industry the potential scourge of Foot & Mouth is always present and precautions and preventative measures have to be taken to stop its spread. Therefore, it was with great reluctance that the Section took the decision in February 1968 to cease all activities for a period of 6 months."

"The annual dinner has always been the highlight of any Section and the North West is no exception. In the early years the Barton Grange Hotel in Preston was a very popular venue as was the Park Royal, Charnock Richards. In later years more adventurous venues were sought including a Medieval Banquette and a tour of Manchester's Granada Studios when, after a 'virtual reality ride', we assembled for a well-deserved meal. Theme nights at Smithill's Coaching House were also popular alternatives and these included a Bavarian Evening with an Oompah Band, a 60s Disco with a competition 'Crackerjack' evening. During the '80s, more formal dinner dances were (alternatively) arranged with the North East Section and these illustrious occasions rivalled London and the South East Section's claim to being the prestigious social event in the Society's calendar."

"Guest speakers were generally arranged by a member of the Committee. In March 1973 one speaker at Reaseheath who was bound to be 'an enormous' attraction and guaranteed to keep everyone on the edge of the seats was Dr. Magnus Pike with his talk on 'Synthetic Milk'."

"To many people, acquisitions, mergers and closures only really affected the dairy industry since the late '90s. A proposal was carried in April 1975 that the Fylde & Furness Branch be officially dissolved due to recent poor attendances as they had not held meetings for some considerable time or even formulated a programme for the forthcoming year. Furthermore they had experienced great difficulty in recruiting Committee members to replace those who had resigned or left the area. This has also been a longstanding problem in recent years where, despite numerous 'pleas' and personal invitations, the North West Section also suffered the same fate. The Committee nevertheless soldiered on when sometimes it seemed they were putting together programmes where they were the only people to attend, along with a few other resolute members."

"One of the main reasons for falling membership and attendance at Society events is that dairy companies now appear to be run by 'men in grey suits'. In past years, Factory Managers used to be present at meetings and even supported their staff attending, even to the extent that the dairy often paid the membership fees, but this no longer applies. It is, therefore, a sad reflection that very few, if any, 'captains of industry' support the Society or its role in education and training of dairy people. This is borne out by the fact that the budget for the Society for 1975/76 reported a drop of membership from 2,600 to 2,000 due to increased subscriptions - remember this was the era of high inflation."

"The first recorded Section 'tour' was to have been the D.L.G Exhibition in Frankfurt on the 6th - 10th May 1980. Unfortunately, due to a disappointing response, this trip had to be cancelled. However, David Booth was on the committee at that time. Not long afterwards, in 1982/83, he together with David Rimell and Margaret Dawson of the North East Section, started to arrange annual Study Tours to various parts of Europe. Visits were arranged to Austria, Belgium, Channel Isles, Denmark, France, Germany, Holland, Ireland, Isle of Man, Scotland, Spain and Wales."

"On each occasion an average of 35-40 SDT members enjoyed a combination of study/enjoyment excursions over a period of some 12 years, with the North West and North East Sections alternately taking responsibility for the organisation. The generous sponsorship from many dairy equipment supply companies throughout this period should not go unrecognised."

"By 1994 a general depression in the dairy industry reduced the number of members resulting in the final 'joint' study tour. However, until they retired or left the industry, a nucleus of several of the dairy industry and previous tours continued to get together once a year and travel into Europe. The final tour took place in 1999 to Roscoff."

| YEAR | STUDY TOUR | SECTION | ORGANISERS |
|------|-----------|---------|-----------|
| 1983 | Denmark | North East | D. Rimell & M. Dawson |
| 1984 | Holland | North West | D. S. Booth |
| 1985 | Germany (North) | North East | R. Thompson |
| 1986 | Ireland | North West | D. S. Booth |
| 1987 | Belgium | North East | D. Rimell |
| 1988 | Channel Isles | North West | D. S. Booth |
| 1989 | Germany (Middle) | North East | R. Thompson & J. Hogg |
| 1990 | Scotland | North West | J. Coates |
| 1991 | France (Normandy) | North East | J. Hogg |
| 1992 | Germany (South) | North West | D. S. Booth |
| 1993 | Ireland | North East | P. Moody |
| 1994 | France | North West | D. S. Booth |
| 1995 | England (Cumbria) | North West | J. Coates |
| 1996 | Austria | North West | J. Mitchell |
| 1997 | Spain | North West | D. S. Booth |
| 1998 | Isle of Man | North West | D. S. Booth |
| 1999 | France | North West | D. S. Booth |

PARTICIPANTS IN THE 1989 STUDY TOUR TO GERMANY

"One notable visit in recent times was a nostalgic look back to the 1939-45 era with a tour of the underground world of Stockport. Local inhabitants took refuge from the might of the German air force in a labyrinth of caves situated beneath the centre of town that is now preserved as a museum. This event proved very popular, especially with the younger element, and brought home the reality of living in wartime Britain. An unlikely, but extremely well liked, attraction was a visit behind the scenes of Manchester Airport."

"The North West Section no longer exists, as we once knew it, because the Society's AGM in November 2000 agreed to a number of reforms - one of which was the current thirteen Sections would be consolidated into seven Regions. The new Region, known as the North & Midlands Region, now embraces the former Northern, North East, North West, East Midlands and West Midland Sections. The 'broad' geographical spread of the Region now covers Carlisle & Northumberland to the North, Northampton to King's Lynn in the South and as far West as Chester to Chepstow."

"In coming to the end of this brief stroll through memory lane, it is regrettable to have to report that the Section has not managed to stage any meetings/events since June 2000. This was primarily due to a dearth of willing volunteers to serve on Committee (or 'special interest groups' on a more 'localised basis') but, more importantly, remaining members being dedicated enough to attend meetings."

"If the Region and the Society are to reaffirm their presence as a strong, vibrant and influential force, then it is imperative that 'active' support is provided by the dairy companies, especially the senior management teams. It is appreciated that it is an uphill battle to attract the younger members who need to spend valuable time with their families."

"On a final note I would like to apologise to all those people, too many to mention individually by name, for all their sterling service. Thank you to everyone who has given so freely of their time, whilst serving in various capacities as Committee members, for the benefit of both the Section and the Society."

## THE NORTH EASTERN SECTION
JSDT Vol. 1 No. 3 July 1964

"The 29th June 1951, saw the birth of the North Eastern Section at the inaugural meeting, which was held at the Royal Station Hotel, York, when 50 members and 43 visitors were present."

"Miss M. Morrison then addressed the meeting, explaining that there had been a suggestion by the Chairman of the North Western Section that the time might be appropriate for the formation of a North Eastern Section. The Council of the Society had been approached and approved, which would entail the transfer of members in Yorkshire and the North East. A meeting had then been called of a small number of interested members who had explored the possibility, and the well-attended meeting of the 29th June was a most satisfactory result of their initiative and enthusiasm."

"A business meeting was then suggested by Miss Morrison, which entailed the adoption of a local constitution and rules and the election of a Committee. Mr. C. H. Westwater, who had been proposed as Chairman, was unanimously elected and then took the Chair. Miss M. Morrison to the office of Honorary Secretary, Mr. D. E. Ralph to that of Honorary Treasurer and the rest of the Committee were all nominations receiving unanimous approval."

"Dr. A. T. R. Mattick had been asked to address the meeting but an apology for absence had been received from him, and consequently Mr. W. W. Cuthbert read Dr. Mattick's paper, 'Looking Round the Dairy Industry,' which was followed by a general discussion and tea."

AREA

"Geographically the North Eastern Section is comprised of the following counties: Yorkshire, Northumberland and Durham. However, as one may well imagine, the density of the Section's membership tends to develop in the industrial and mining areas of the West Riding and the heavy engineering cities and towns of Newcastle, Sunderland and Middlesborough. The majority of the members in these areas are mainly involved in liquid milk processing and distribution. Milk manufacture is to be found in the Section with a demand from the cocoa and chocolate industry of York. Dairy engineering establishments are very well represented in the area, also ancillary products such as chemicals, glass bottles and cartons, churns, etc."

"Early 1964 saw the formation of the new Northern Section, which covered the area of Cumberland, Westmorland, Northumberland, Durham and the northern part of the North Riding of Yorkshire, leaves the East, West and part of the North Riding of Yorkshire within the North-Eastern Section."

MEETINGS

"In general, meeting have been held each year to satisfy the demands that might be expected from a Section whose membership is spread over all three membership groups. The main centres have been York, Newcastle, Harrogate and Leeds. During the past 10 years joint meetings have been arranged with the National Farmers' Union, Food Manufacturers' Federation and the East Midlands Branch of the Midlands Section."

"The Society honoured the Section by inviting it to act as host on the occasion of the 1958 Annual Spring Conference. This was held at the Cairn Hydro, Harrogate, from 27th April to 2nd May 1958. The weather was indeed made to order, which allowed the 150 delegates to enjoy the beauty of the broad acres and dales of Yorkshire when visiting dairying establishments, farms and other places of interest."

EDUCATION

"Following the inaugural meeting of the North Eastern Section in June 1951, the newly formed Section Committee met on the 17th September 1951, and the subject of education within the dairy industry was the first item on the agenda. Two sub-committees were formed to meet the representatives of the C.M.D.C. in the Northern Region and the West Riding, respectively, to discuss the possibility of arranging further courses."

"Miss M. Morrison, Honorary Secretary, attended a meeting of the North Eastern Division of the N.D.A. in the March of 1952, where she outlined the position. Contact was also established with the Yorkshire Council for Further Education. As a result of this work, courses were implemented at various centres in the Section's area to prepare for the Milk Pasteurisation and Distribution examination of the City and Guilds of London Institute."

"The Section has provided lecturers for Section 159 and 160 of the City and Guilds syllabuses and has also arranged for suitable members to act as local examiners whenever any request has been made."

"The Newcastle Branch, which was formed in March 1961, showed great enterprise on this matter following a talk given to the Branch by the Vice-Principal of Kirkley Hall Farm Institute. Sufficient interest was aroused in the possibilities of arranging courses for employees in the dairy industry on a similar basis to those offered in agriculture, and a Sub-Committee was formed to which the Vice-Principal of Kirkley Hall was co-opted."

"It was decided in the first instance to aim at a day-release course suitable for inside dairy employees, broadly based without too much detail and limited to a short period in order to whet the appetite of the employees so that their enthusiasm might be transmitted to their fellow workmen."

"It was arranged to centre the course on the Houghall County School of Agriculture, where facilities existed for teaching the basic principles of dairying. The response was excellent and 16 students were selected for the first Course. A second 10-week course was held early in 1963. In view of the success of the two first 10-week courses, a 20-week course was commenced in October 1963, treating the subject in greater detail."

"In addition to these activities, Section members participated in a Careers Convention for the National Union of Teachers which was held at the James Graham College, Leeds, from February 23rd to 25th, 1962. This convention was staged for careers teachers who took full advantage of the information available from the careers stand of the National Milk Publicity Council, which was supervised by Miss M. Corbridge, Education Executive of the N.M.P.C."

BRANCH FORMATION

"The first branch of the North Eastern Section was formed at the Royal Station Hotel, Newcastle-on-Tyne, 29th March 1961."

"The Chairman (Mr. G. A. Barley) welcomed the members and guests with special reference to the presence of the then President, Dr. R. J. Macwalter, and the visiting speaker, Professor E. L. Crossley."

"After an address by the President (during which he indicated his pleasure at being present at the inaugural meeting of the Newcastle Branch), Professor Crossley presented a paper on 'Education for the Dairy Industry', which was published subsequently in the Society's Journal."

"The following nominations for Branch officers had been received and were unanimously accepted by the meeting.

| | |
|---|---|
| Chairman: | Mr. T. Reed. |
| Hon. Treasurer: | Mr. E. Lee. |
| Hon. Sec: | Mr. R. G. Hallett." |

"The Branch was formed with a membership of 28 and, as the strength has more than doubled itself in three years, the Branch Committee felt sufficiently confident to apply for Section status. The Committee of the North-Eastern Section were wholeheartedly in favour of this proposal and at a subsequent Council Meeting approval was granted in early 1964, culminating in an Extraordinary General Meeting of the new Northern Section for the appointment of officers and Committee members in February 1964."

SOCIAL

"The first annual dinner-dance was held at the Griffin Hotel, Leeds, in February 1959, which was attended by over 100 guests. This function was sufficiently successful for the Committee to stage the second dinner-dance at the same venue in 1960."

"On this occasion, the highlight of the evening was the presentation by the Section Chairman, Mr. D. B. Bell to Mr. and Mrs. W. A. Cuthbert, who were leaving the Section owing to the appointment of Mr. Cuthbert as head of the Bacteriology Department at the National Institute for Research in Dairying. The membership of the Section had contributed towards the gift as an appreciation of the sterling work done during the previous 10 years by Mr. and Mrs. Cuthbert on behalf of the Section. Miss Marjorie Watson, National Dairy Queen, 1960, who resided at Goole, was the guest of honour on this occasion, which was so highly successful that it confirmed the annual dinner-dance as permanency in the Section's programme. The Newcastle Branch followed suit in arranging dinner-dances in its own area."

*Northern Section Summer Ball*
*From left Mr John Westwell, Northern Section Chairman, Mrs Elizabeth Westwell, Mrs Marilyn Lee, Mr Grahame Lee, President of the Society of Dairy Technology.*

MEMBERSHIP

"The total membership is now 126, with the majority being found in the West Riding of Yorkshire."

ACKNOWLEDGMENTS

"In conclusion, it is fitting that a most sincere appreciation should be recorded of the various members who have served as Section officers and Committee members and the various organisations which have rendered close co-operation in assisting at the meetings over the years. The Section Secretary feels that special mention should be made of the facilities which are always made available by the management of Express Dairy Company (Northern) Ltd. and the Milk Marketing Board, Harrogate, as and when accommodation may be required for a Committee or General Meeting. This whole-hearted support is most encouraging and creates a feeling of confidence when considering the future of the North Eastern Section."

NORTHERN SECTION

(John Bird (Past President))

"The Section was broadly based on the Northern Dairies (now Arla Foods) site at Newcastle and supplemented by members from across the border from Aspatria and Appleby. A number of employees of smaller dairies in the area were also committed members of the Society. The driving forces behind the resurrection of the Northern Section at that time were John Crombie and John Westwell, and they carried out most of the legwork in organising the re-launch of the Section. My part as President (John Bird) was to bring the resources of Council to bear and assist the local members in their endeavours."

"The re-launch of the Northern Section took place at the beginning of 1992 at the Metro Hotel in Gateshead, and many active and retired members of the Society attended the evening. In fact, many members travelled great distances to support the evening. Robert Hamilton, the Honorary Secretary, travelled from Northern Ireland and Peter Fleming, Vice President, came down from Glasgow as well as many other members from the midlands and adjoining Sections. A turnout well in excess of fifty enjoyed the course of the evening, and the event ended with a buffet supper supplemented by a delicious cheese-board provided by one of the members of the section. The section returned to strength and is still active today."

## LAKELAND BRANCH (NORTHERN SECTION)
### (Maurice Walton President)

"The Lakeland Branch was part of the Northern Section for many years and had around 25-30 members at most. Meetings were held throughout Cumbria mainly, with infrequent excursions into Lancashire and the North East. Regular Committee meetings were held at the Dairy Crest site at Kendal, which had a reasonable number of members. Other sites in existence at the time which provided membership and support included: Dairy Crest Aspatria, Longridge and Carlisle; Express Appleby; Nestlé Dalston; Farleys and Lakeland Creamery, both at Kendal. Supporting suppliers at the time included Alfa Laval, Tetra Pak, Diversey, Reddish Savilles and Oakite."

"Numerous factory visits were undertaken which included all of the aforementioned sites along with a memorable trip to the now defunct Hartleys Brewery in Ulverston - the Brewing industry has undergone the same transition (fate) as the dairy industry!"

"The highlight in the life of the Lakeland branch was, undoubtedly, the study tour of Scotland. Around 40 members of the Society joined a coach at the Dairy Crest, Shaddongate Creamery in Carlisle, first stop was the Cheese factory at Stranraer which is now the Caledonian Cheese Company. At that time they produced mainly Cheddar cheese in bulk and also had an ion exchange whey demineralisation plant, which was of particular interest to members of the party from Express at Appleby which also had a similar installation. In addition to this, the site was starting to produce Feta-style cheese for canning and export."

"The next stage of the tour was to the Scottish College of Agriculture at Auchincruive where the delegates were given a tour of the site and facilities. It was then on to the hotel for an evening meal, a few drams and even a dip in the sea for the more resilient (although I believe that some of the resilience could be attributed to the malt)."

"The second leg of the tour took us to the Scottish Farm Dairies site in Glasgow, followed by a visit to Tennants Caledonian Brewery to investigate the similarities in technology between the dairy and brewing Industries. It was then on to the Auchentoshan Distillery on the outskirts of the city en-route to a very pleasant overnight stay in Historic Sterling where traditional Scottish fare was consumed in one of Sterling's foremost 'curry houses'."

"The scenic route back via Edinburgh and the A7 including a delightful high tea near Galashiels as the conclusion to a successful and highly enjoyable study tour."

## THE MIDLANDS SECTION
### (JSDT Vol. 14 No. 4 1961)

"At a meeting of representatives of all aspects of the dairy trade held at the Midland Counties Dairy Ltd., Birmingham, on 28 April, 1943 under the Chairmanship of Mr. Edwin White, the aims of a Society of Dairy Technology, which had recently been formed, were explained by Mr. Frank Procter, and it was unanimously agreed that a regional section of that Society should be formed in the Midlands. A small *ad hoc* Committee was immediately elected, with Mr. F Procter as Secretary, and the first general meeting of the Midlands Section of the Society of Dairy Technology took place on the 3rd November 1943 in Birmingham, when a paper dealing with heat treatment of milk was given by Mr. Harold Smith of Leicester. "

"The first Chairman of the Section was Mr. Edwin White, and the first Honorary Secretary, Mr. William Bailey, whilst Mr. Harold Gurden undertook the duties of Honorary Treasurer."

"The following members have also served in the capacity of Chairman of the Section up to the present time:

| | |
|---|---|
| F. Procter | R. J. Frazier |
| H. J. Smith | T. C. D. Atkinson |
| J. White | R. Scott |
| H. Gurden | W. W. Ritchie |
| C. A. Scarlett | M. S. Munro |
| A. T. Ridout | J. Freedland |
| J. O. Fowler" | |

"In 1944 Mr. Procter was elected President of the Society. That high office has since been held by another member who was also very active in the formation of the Midlands Section - Dr. A. L. Provan."

"The Section has taken a keen interest in dairy educational matters and in 1947 a Sub-Committee co-operated with the Birmingham and District branch of the National Dairymen's Association and the Local Education Authority with a view to organising suitable part-time courses of instruction for dairy personnel. This Sub-Committee later merged with that of the local N.D.A. to form a Joint Education Committee with Mr. Scarlett as Chairman, (that position is now held by Mr. R. Scott), and Mr. K. J. Curtis of the N.D.A. as Secretary of that Committee; this resulted in very successful City and Guilds courses in *'Milk Pasteurisation and Distribution'* and *'Milk Processing and Control'* being organised and run in Birmingham continuously during the past twelve years, the principal lecturer being R.J. Frazier, the present Honorary Secretary of the Section. This latter post has also been held by Dr. L. F. L. Clegg, Mr. E. J. Nokes and Mr. A. T. Ridout."

"In 1952 consideration was given to the formation of sub-sections or branches of the Society and the Midlands Section was probably the first section of the Society to form such a sub-section. The inaugural meeting of the East Midlands Branch - Derby/Leicester area - was held at Derby on 25th February 1953, Mr. Brindley being elected Chairman and Mr. Dilworth, Honorary Secretary of the Branch. The meeting was afterwards addressed by Professor Kay on *'The Functions and Work of the National Institute for Research in Dairying.'* A second branch of the section, covering the North Staffordshire area, was formed in 1955, the first Chairman being Mr. R. A. Coulthurst and Honorary Secretary, Mr. Adkins. Both branches are still very actively carrying out the functions of the Society."

"The Midlands has been the venue of a number of General Meetings of the Society over the years, one of which, held at Messrs. Cadbury Brothers, Ltd., Bournville, in March, 1956, being an outstanding success."

"An account of the Midlands Section would be incomplete without mentioning the Spring Conference of the Society which was held within the area of the Section at Buxton this year. This proved to be one of the best functions of this type so far, due in no small measure to the work and enthusiasm of the Section Chairman, Jack Freeland, and other members of the Midlands Section Conference Sub-Committee - notably A. G. Wyman and A. T. Ridout."

THE WESTERN SECTION 1945/62

(JSDT Vol. 16 No. 2 1963)

FORMATION

"The 20th February 1945 was an important day for dairying in the West of England, for, at a Council Meeting of the Society, it was agreed to act on a proposal from Mr. J. W. Egdell that a local Section be formed to serve the interests of the Society in that part of England. An *ad hoc* Committee was formed under the chairmanship of Mr. E. L. Crossley, with the result that on the 27th September 1945, at Royal Fort, Bristol University, the inaugural meeting of the Western Section was held, when Mr. E. L. Crossley was elected as the first chairman. The meeting was addressed by the President of the Society, Mr. F. Procter, on *'Some Aspects of Management in the Dairy.'* This sparked off a most lively discussion, setting the tempo for what has ever since been characteristic of the Western Section. Forty members and twenty-five guests were present."

"The Section quickly grew in numbers and enthusiasm and in one year its membership was trebled to 160. Its second meeting, held in Exeter on December 12th, 1945, with an attendance of 108, proved unique in the history of the Section. It was held in the Courtroom at the Castle, and the Chairman and Speaker had the unusual experience of standing in the dock and witness box respectively. The third meeting, held in Taunton on 15th March 1946, gave further evidence of the popularity and success of the Section, the attendance being a little over 200. Mr. J. W. Egdell's confidence had indeed been fully justified."

"The Western Section covers a wide area, and includes the counties of Herefordshire, Gloucestershire, Wiltshire, Somerset, Dorset, Devonshire and Cornwall. With such a large territory, involving long distances to travel within its boundaries, meetings have been so arranged that each part of the Section has been served, the most popular choice of venues being Truro, Plymouth, Exeter, Taunton, Bristol, Bath, Gloucester and Cirencester."

"Early in its history the Section did much pioneering work in connection with broadcasting, which resulted in a series of three programmes under the general title of *'A Pint of Milk'* being broadcast by the B.B.C. in its Western Region. This series was sponsored by the Section, and was designed to increase popular appreciation of the dairy industry in the West of England. A B.B.C. observer and recording van visited a milk producer, a milk pasteurising and bottling plant, and a large manufacturing creamery on three successive weeks. The series proved to be a great success and an appreciation of the work done by Mr. J. W. Egdell to make it possible must be recorded."

"The Section has always been keen to foster interest in dairy education and in its early days was fully conscious of the lack of facilities. With this in mind, a survey was undertaken in 1947 to ascertain the facilities for education in dairying within the boundaries of the Section. The Education Committees of the seven counties were consulted, and the report on the survey drew everyone's attention to the unsatisfactory situation, which the Section soon helped to remedy. The Bristol branch in 1949/50 arranged a course of ten lectures in milk technology. The Gloucester Branch later conducted a similar course; and nearly all the lecturers were members of the Society. As a result, the Bristol College of Technology agreed to run a course in *'Milk Handling and Distribution'* to prepare students for the City and Guilds examination. This was followed by a request for advice from the Principal of the Technical College at Chippenham, who was anxious to organise a course at Melksham, which was later arranged. Assistance was readily given to other educational authorities, and it is pleasing to record that a series of lectures for courses leading to the City and Guilds of London examination was held in Bristol, Frome, Taunton, Gloucester, Plymouth, Wincanton, Yeovil, Exeter, Melksham and Barnstable. It is not claimed that the Section was solely responsible for all these courses, but advice and assistance were readily and willingly given to help bring them to a successful conclusion."

"One outstanding achievement of 1949 was the formation of the Bristol Branch. Under the Chairmanship of Mr. J. W. Egdell, and inspired by his enthusiasm, a Committee of seven with Miss E. R. Bird as Hon. Treasurer, and Miss J. M. Phillips as Hon. Secretary, saw the foundations well and truly laid. The Branch quickly progressed and became firmly established. The formation of the Gloucester Branch followed the next year. In November 1950, Mr. F. C. Kingscote became its first Chairman, Miss J. M. C. Faill, Hon. Secretary and Miss B. A. Whittal, Hon. Treasurer. The southern part of the region now came up for consideration, Exeter being considered a suitable centre. Miss J. M. C. Faill, who worked so well to establish the Gloucester Branch, offered her services, and with the help of Miss J. Smithson, the Exeter Branch was duly formed on December 13th, 1951. Mr. K. Dowding was its first Chairman, Miss J. M. C. Faill, Hon. Secretary and Miss J. M. Smithson, Hon. Treasurer. The Section has been well served by its three Branches."

CONFERENCES AND COMBINED MEETINGS

"The Society held its first Summer Conference at the Royal Agricultural College, Cirencester, in 1950. The Section was indeed honoured to play host and the Conference has been generally considered one of the best ever held. The Society's Spring Conference of 1955 was also held in the Section at Torquay, which again proved an outstanding success, all who attended it taking back happy memories of the West Country."

"A combined meeting with the Sanitary Inspectors' Association was held at Bristol on 25th September 1951. Dr. Betty C. Hobbs (Central Public Health Laboratory) addressed the meeting, which was presided over by the President, Dr. J. A. B. Smith."

"Encouraged by the success of this meeting, a second combined meeting was arranged, this time with the National Dairymen's Association at Bournemouth on 29th May 1953. The President, Mr. E. L. Dobson, presided at the morning session, the afternoon session being in the hands of the N.D.A.'s President, Mr. J. S. McQuillin. This meeting was held on the last day of the N.D.A Conference. Three papers were presented by Mr. F. Procter, Mr. J. W. Egdell, and Mr. A. W. Marsden. It was felt this joint meeting served a useful purpose in promoting co-operation between the two organisations."

"The Section has enjoyed several joint meetings with the Midlands Section, alternating each year in each Section…."

"The social life of the Section has throughout its history played an important part in its progress."

"1960 saw the first Annual Dinner at Bristol. The President, Mr. J. Ridley Rowling, with the Secretary, Mr. M. Sonn, graced the West Country with their presence, making it a never-to-be-forgotten occasion."

"The Branches have continued to arrange varied programmes of social event, including socials, skittle evenings and cheese and wine parties. Annual cricket matches have been very popular, the Egdell Cup creating keen rivalry between Bristol and Exeter."

"The total membership of the Western Section is now 332, a healthy figure that clearly indicates its continued progress, reflects the sterling work of the Branches and augers well for the future."

"Early in 1960, the Section invited for the first time a speaker from overseas to address a General Meeting. This proved most successful and, as a result, Mr. Gustav Fredrikson of Vadstena, Sweden, who addressed that meeting, invited the Section to make a tour of Sweden to study the progress made in the dairy industry. A small party left Tilbury on July 7th, 1962, and thoroughly enjoyed a memorable visit. One thousand miles were covered, embracing many aspects of the industry, over a period of ten days…."

"The Section owes a great debt of gratitude to the Officers who have worked so hard to ensure it success. The following is a list of those who have served the Section:

*Chairmen:*

| | |
|---|---|
| 1945/47 Prof. E. L. Crossley | 1956/57 Mr. E. Steane |
| 1947/48 Mr. L. J. Walker | 1957/58 Mr. A. C. Page |
| 1948/49 Mr. J. W. Egdell | 1958/59 Mr. F. C. Kingscote |
| 1949/50 Mr. S. A. Wilcox | 1959/60 Mr. A. P. Stacey |
| 1950/52 Mr. E. Steane | 1960/61 Mr. P. W. Hart |
| 1952/53 Mr. S. A. Wilcox | 1961/62 Mr. C. L. Silverstone |
| 1953/55 Mr. F. C. Kingscote | 1962/63 Mr. L. C. Couch |
| 1955/56 Mr. J. W. Egdell | |

*Hon. Secretaries*

| | |
|---|---|
| 1945/58 Mr. W. A. Johnson | 1958/- Mr. S. W. Stilton |

*Hon. Treasurer*

1945/- Mr. R. W. S. Warren"

"It is of interest to record that since the formation of the Western Section, there have been only two Hon. Secretaries and Mr. R. W. S. Warren is still its Hon. Treasurer."

THE SCOTTISH SECTION 1944-1961

(JSDT Vol. 15 No. 3 1962)

FORMATION

"Of the two hundred and thirty nine members who founded the Society of Dairy Technology in March 1943 only eight (3 per cent) lived north of the Border but, by June 1944, when the Society's total membership had risen to six hundred, those from Scotland numbered forty-two (7 per cent)."

"The long distance from the Society's centre of activity, and the absence of a Journal at that time to report the Society's proceedings made the formation of a Scottish Section imperative, and an inaugural meeting was convened by two of the founder members of the Society, Dr. N. C. Wright and Professor D. M. Smillie. More than one hundred representatives of Scotland's dairy industry and related organisations met at the Central Station Hotel, Glasgow, on 28th September, 1944, and, after supporting the formation of the Section, heard Dr. A. T. R. Mattick discuss '*Heat Treatment of Milk with Special Reference to H.T.S.T. Pasteurisation*'. In recognition of the part he had played in the Section's formation, Dr. Wright was elected Chairman, but his duties took him out of Scotland for much of the first year and Mr. A. McBride made a very popular deputy."

"Five meetings were held in that first year and the success of the venture enabled Mr. T. W. Gibson, the Honorary Secretary of the Section, to report to the Section's Annual General Meeting that Scotland could now claim one hundred and seventy members out of the Society's total of eight hundred and sixty eight (20 per cent)."

ACTIVITIES

"During the ensuing years the Section settled down to a fairly regular pattern of five afternoon meeting between the months of September and March. Speakers from all parts of the United Kingdom, from many branches of the dairy and allied industries, addressed audiences averaging between sixty and seventy in number. The topics ranged from milk production through milk processing and manufacture, to dairy engineering, water and electricity supplies, planning and research."

"In 1947 a summer outing to the West of Scotland Agricultural College, Auchincruive, proved so popular an innovation that similar outings to various other places of interest have since take place almost every year. Apart from the meal which precedes the Annual General Meeting, this outing is the only social occasion enjoyed by the Section. To encourage a slightly less formal atmosphere, and to provide an opportunity for the attendance of junior staff who could not easily get away in the afternoon, evening meetings were first suggested in 1952. They became a reality in 1955 when the first was held in October of that year. Although the response from the group for whom they were intended was not overwhelming, attendances were good enough to merit the holding of two evening meetings in every succeeding session."

"In January 1947, a General Meeting of the Society was held in the Merchants' House, Glasgow, when Dr. A. Nicholls, Messrs W. Alexander, L. J. Meanwell and F. C. White addressed an audience which included members from other Sections. This meeting was so popular that the suggestion to hold a Spring Conference in the Scottish Section area in 1951 was heartily welcomed. The Marine Hotel, Troon, was selected as the base for visits to local institutions and for paper reading sessions while the weather co-operated in making the occasion a pleasurable one for all who attended. The last day of the Conference was spent in 'Edinburgh. Troon appeared to find favour with the visitors because it was again selected by the Council as the centre for the 1960 Conference and the response evoked was highly gratifying to the speakers and organisers. Once again the various establishments visited extended generous hospitality and the four days passed all too quickly. During the latter Conference the seeds were sown which led to the formation of two Scottish Branches."

BRANCHES

"The choice of Glasgow as the main centre for meetings of the Section was convenient for residents no further away than Edinburgh or Ayr but Southwest and Northeast Scotland, both with important dairying activities, were poorly served. Glasgow is 142 miles from Aberdeen, 169 from Inverness and 84 from Stranraer and it was not surprising that a survey of the Scottish Section membership in 1946 showed that 60 per cent came from Glasgow and North Ayrshire."

"In June 1948 the Section Committee discussed the formation of an East of Scotland Branch and in each of the years 1949, 1950 and 1951 one meeting was held in Edinburgh. Support was disappointing, however, and no further action was taken."

"As mentioned earlier, it was the second Conference at Troon which provided the spur to the formation of two Scottish Branches. A paper presented at the Conference by Mr. John Smith emphasised the concentration of creameries in the south of the Section area in Dumfriesshire, Kirkcudbrightshire and Wigtownshire. The Society's visit to Sandhead and Dunragit creameries found Mr. W. McLean and Mr. T. E. Hayter, the respective managers, enthusiastic about the formation of a local branch. At a meeting in the Masonic Arms, Gatehouse of Fleet, on Thursday, 25 August, 1960, a Branch Committee was formed and, with Mr. R. Prentice as Chairman and Miss M. K. Russell as Honorary Secretary, the Dumfries and Galloway Branch was an immediate success. Its present meeting place, the Murray Arms, Gatehouse of Fleet, is an old hotel set in a very attractive village which, though small, has many historical associations. It is the most central spot in the Branch area although some of the regular attendees have still to travel sixty miles."

"Mr. M. Boyle and Mr. T. Foley of the Aberdeen and District Milk Marketing Board returned to Aberdeen from the 1960 Troon Conference having agreed to try to form a branch in their part of the country. They sought assistance from Mr. D. McKenzie of the North of Scotland College of Agriculture who subsequently became first Chairman of the North Eastern Branch. This was formed on the 30th January 1961, and the opportunity thus offered to hold local meetings brought an influx of new members. In a review of the first few months of the Branch's operation, Mr. H. Malloy, the Honorary Secretary, reported a membership of twenty-five with a current target of forty."

"The total membership of the Scottish Section which had risen to 206 by 1947 but dwindled to 150 by 1958, received an added boost from the Branch activities and now stands at the healthy figure of 240."

EDUCATION

"Unlike some other Sections of the Society, the Scottish Section has never assumed responsibility for courses of education for workers in the dairy industry. It was, however, largely due to the efforts of Professor Smillie, who had long advocated some instruction in dairy technology, and to Mr. James Kirkwood and their colleagues that various courses, including one of two years instruction for a City and Guilds of London Certificate, were started at the West of Scotland Agricultural College. Dairy education is constantly under review by the Section and discussions have taken place from time to time with a view to providing facilities for a part-time, non-diploma course for dairy employees in the distributive trade. It has not been possible to initiate this yet, possibly because of the nature of the industry in Scotland where the number of large establishments is small and labour scarce. Under these circumstances the Section and Branch meetings fulfil, to a large degree, the educational functions of the Society by passing on new knowledge to its members."

"In order to arouse more interest among milk producers, the Scottish Milk Marketing Board publicises Section activities in its Bulletin and the Section Committee recently sent a letter to dairymen pointing out the advantages to be gained by encouraging their staffs to attend meetings of the Society whenever possible."

"No history would be complete without a list of dates and it is appropriate to pay tribute to past officers of the Section by publishing the following table:

*Chairmen*

| | |
|---|---|
| 1944-1945 | Dr. N. C. Wright (Mr. A. McBride deputised) |
| 1945-1946 | Dr. N. C. Wright |
| 1946-1947 | Mr. W. D. Smith |
| 1947-1948 | Prof. D. M. Smillie |
| 1948-1949 | Mr. T. W. Gibson |
| 1949-1950 | Dr. J. A. B. Smith |
| 1950-1951 | Mr. C. H. Chalmers |
| 1951-1952 | Mr. A. B. M. Tulloch |
| 1952-1953 | Mr. A. McMillan |
| 1953-1954 | Mr. C. E. D. Colquhoun |
| 1954-1955 | Mr. James Kirkwood |
| 1955-1957 | Dr. R. Waite |
| 1957-1959 | Dr. J. F. Malcolm |
| 1959-1961 | Mr. W. B. Sampson |
| 1961-1962 | Mr. J. Crawford |

*Hon. Secretaries*

| | |
|---|---|
| 1944-1946 | Mr. T. W. Gibson |
| 1946-1952 | Mr. C. E. D. Colquhoun |

| 1952-1959 | Mr. T. A. Blair |
| 1959- | Mr. J. Abbot |

*Hon. Treasurers*

| 1945-1953 | Mr. J. B. Rice |
| 1953-1959 | Mr. I. A. McAlpine |
| 1959- | Mr. J. Hamilton" |

"Any success achieved by the Section, however, must be equally attributed to the many committed and other members whose names do not appear and to all the speakers who have readily shared their knowledge. The many commercial organisations, large and small, which willingly allow members of their staffs to discuss matters which at one time might have been jealously guarded as hard-won trade secrets, have also given staunch support to the activities of the Society in Scotland."

"With pride the Section can record that two of its members have held the office of President of Society, Dr. J. A. B. Smith in 1950-51 and R. Waite in 1961-62, both past Chairmen of Section and very active in furthering the aims of the Society."

SUBSEQUENT EVENTS (DR. R. M. CRAWFORD, PAST PRESIDENT).
"Despite falling membership one and two-day symposia were successful with topical subjects such as *'Milk Utilisation'*, the latter organised by the Dumfries and Galloway Branch for 75 members and guests in 1989. The Annual Jim Robinson memorial lecture arranged jointly by the Scottish Section and the Scottish Branch of the Institute of Food Science and Technology generally attracted a good attendance of members of both organisations; the 1989 meeting on *'Listeria in the Dairy Industry'* attracted 68 members and guests."

"The highlight of the Dumfries and Galloway Branch from 1984 until 1999 was the Annual Cheese Show held at Gatehouse - of - Fleet as the last event of the Branch's winter programme. The show attracted entries from cheesemakers in Scotland and England and provided an excellent technical and social event for up to eighty or more members and guests. Unfortunately this popular event was cancelled in 2001 because of the outbreak of Foot and Mouth disease in South-west Scotland and the Borders. At the Society's AGM in November 1995 the Honorary Secretary reported that the Scottish section had run into very serious difficulty in getting people to turn out."

"As reported, the decision was taken to unite the Scottish Section with the Dumfries and Galloway Branch, eventually representing the Scotland Region."

"The North East (Aberdeen) Branch ceased to function in the mid-nineties. It had endured several losses of popularity. In 1978-79 the Branch Committee proposed the closing down of activities. However, after various discussions, Mr. Michael Boyle was persuaded to form a new committee and arrange a programme for 1978-80. This initiative was successful and activities continued for more than a decade. The attendance at the branch's AGM in 1990 was boosted to hear the Scottish Section's Honorary secretary Mr. Gordon Paterson give an important lecture on *'Preventing Pathogens gaining access to Dairy Products'."*

EDUCATION FOR WORKERS IN THE DAIRY INDUSTRY
"The earlier report in 1962 on the Scottish Section of the Society referred to the continuing reviews of the need for provision of dairy education for workers in the various sectors of the dairy industry."

"In 1962 the Section Committee convened a meeting of representatives from the dairy industry, the West of Scotland Agricultural College, Elmwood College, Fife, the Milk Marketing Boards in Scotland, the Department of Agriculture, the Department of Health and local authorities in Scotland. It was hoped that the various recommendations and suggestions from the meeting would lead to improved facilities for the education of all grades of workers in the Scottish dairy industry."

"At the Society's AGM in October 1964 it was announced that it was the intention to launch two elementary courses on dairy technology one in Fife and the other in Glasgow. These courses were held in the winter of 1965 and the majority of the students passed the examination. ".

"Successful students from the elementary courses could progress to more advanced courses organised by Elmwood college, Fife and the West of Scotland Agricultural College with the possibility of sitting the examination for the City and Guilds of London Certificate in milk Processing. Society of Dairy Technology certificates were awarded to students successful in the elementary dairy technology examinations held after courses and examinations in Fife, Glasgow, Stranraer and Dumfries from 1967 to 1969."

"The courses were discontinued after 1969 because of the implications of the impending Industrial Training legislation for dairy companies. It was felt that the Society's role in devising and promoting the courses in elementary dairy technology had been a success."

## THE WELSH SECTION
(JSDT Vol. 15 No. 1 1962)
John Lewis

"It is a matter of both pride and pleasure to many that the formation of what was to be the Welsh Section took place in the same year as the parent Society held its first general meeting. Evidently some of those once referred to by the Society's first President, D. Kay, as the 'founding fathers' of the Society had strong Welsh connections and the dairying interests of the Principality much at heart!"

FOUNDATION

"Following meetings of an *ad hoc* Committee in March and September 1943 to discuss preliminary arrangements, what was then known as the Mid and South Wales Section held its inaugural meeting in Carmarthen on the 25th November, 1943. It was very appropriate that on this occasion those present should receive guidance and inspiration from addresses delivered by Dr. H. D. Kay and Mr. W. G. Sutton, Founder President and first Joint Honorary Secretary respectively of the parent Society. The minutes of the day's proceedings as entered in the Section's first minute book record that D. Kay gave an *'illusive'* address to the meeting (no doubt a slip of the pen!) and that Mr. Sutton after eulogising the value of local Sections expressed his hope later on 'to publish a Journal of the Society'. Not many years passed before this cherished ambition was realised."

TERRITORY

"As its title implied, the first Section of the Society in Wales was intended to cover seven counties which, rather surprisingly, did not include either Glamorgan or Monmouth. However, this omission was rectified early in 1944. Following an effort to extend its territory further, it was announced at the Section's Annual General Meeting in November 1945 that 'subject to ratification by the Council of the Society, Wales shall henceforth have its own Section of the Society of Dairy Technology, which will be referred to as the Welsh Section'. Confirmation of this proposal by Council was reported to the Section Committee in February 1946. It seems possible that the rumours then current concerning the possible annexation of the North Wales counties to the newly formed North Western Section of the Society may have stimulated Wales to take united action, thus breaking down the traditional barrier between north and south. In more recent years questions have been asked concerning the precise boundaries of the Welsh Section area, especially in regard to Monmouthshire. However, it has now been satisfactorily established that the area comprises the thirteen counties of Wales - a feat not yet accomplished by some other societies and organisations."

BRANCHES

"Despite this act of union by the Welsh counties, the difficulties of travel and communication within the Section area were known to militate against attendance at general meetings, especially under adverse weather conditions during winter. These circumstances led, in October 1945, to the establishment of sub-sections at Carmarthen and Cardiff. About five years later and mainly due to the departure of a few local enthusiasts, the activities of the Cardiff sub-section lapsed. This left the Carmarthen centre to operate a Branch for members in the South Wales area. In

October 1952, a vigorous North Wales Branch was inaugurated at Wrexham. These two branches continue to flourish and the activities of the Society in Wales owe much to the sustained efforts of members resident in their vicinity."

ACTIVITIES

"The activities of the Section and its Branches over the years as recorded in the minute books and published papers presented at meetings make interesting reading. Addresses have been given by persons who are specialists in one or other of several branches of the dairy industry and the contributions of both senior and junior workers in many fields have always been well received. It is also apparent from the records that members of the Welsh Section have been responsible for some of the Society's pioneering ventures. One of these was the visit to Southern Ireland undertaken by a party of forty-six members and friends in April 1946. Another innovation which preceded the publication of the first volume of the Society's Journal, was the compiling of a bound volume of the nineteen papers presented at Section meetings over the period March 1944 - July 1947. This volume, which is adorned with a crest designed for the Section by one of its staunch supporters, is now deposited in the departmental library of the University College of Wales, Dairy Buildings, Aberystwyth, as a permanent record for reference by members of the Society, dairy students and others interested. Still another venture of interest in the annals of the Welsh Section was the staging of a small exhibition of objects connected with the history of milk production and dairying in the area. This attracted an assortment of exhibits including early textbooks on dairying and a varied collection of farm dairy equipment. Some of these, although worm-eaten or bearing evidence of extensive usage, aroused keen interest and considerable speculation."

"In order to preserve due balance between the serious and no-too-serious aspects of the Society's activities in the Principality, social functions including dinner-dances, whist drives, debates and summer outing, have been held from time to time at both Section and branch level. It has been found that they are not the least effective method of drawing the crowds together or of affording individual members of the opportunity of displaying their talents! The more informal nature of Branch meetings has resulted in social activities possibly figuring more prominently in their calendar of events that at Section meetings."

VENUES OF MEETINGS

"For years past it has been customary for the Welsh Section to hold its Annual General Meeting in or near Aberystwyth. It is one of the most accessible points for members from the north, mid- and south areas to get together and as the reputed 'agricultural centre of Wales,' the town and district always ensures well-attended meetings. The Dairy Buildings of the University College have so often been the venue of both general and Committee meetings that it has come to be regarded as the spiritual home of the Welsh Section."

"For other Section meetings, an effort is made to alternate the venue between north and south of the area. Most of the sizeable towns ranging between Bangor in the north and Cardiff in the south have by now been visited on at least one occasion. Annual joint meetings with the Midlands Section have given members from Wales the added pleasure of being invited over Offa's Dyke at biennial intervals. Most of the creameries in Wales have also on appropriate occasions received parties from the Section and its Branches. The warmth of the welcome and the hospitality received at these centres testify to the high regard in which the Society is held by creamery proprietors, management and staffs in the area."

EDUCATION

"One of the declared objects of the Society is to encourage technical education for persons engaged in the dairy industry. In order to further these aims the Section Committee has assisted in promoting courses in dairy technology at centres in both north and south Wales. Help in this has also been forthcoming from the Branch Committees, which are closely linked geographically with the technical colleges at Carmarthen and Wrexham. Since the inauguration of dairy courses at both these centres in November 1952 and September 1953 respectively, members of the Society in the area have done much to ensure continuity of attendance and to promote their general usefulness to workers in the dairy industry. The Section Committee also delegates some of its functions in regard to technical training to two Education Sub-Committees, one to collaborate with each training centre and their reports form a regular feature on the Agendas of Committee and Annual General Meetings of both the Section and its Branches."

MEMBERSHIP

"One of the criteria by which the success or otherwise of the Society's Section in Wales can be measured is its strength in terms of membership and the support which is given to its various meetings. It is, therefore, pleasing to record an almost continuous increase in the Section's membership from the time of its first meeting at Carmarthen when there were reported to be sixty people present, but of whom probably fewer than twenty were actual members of the Society. Members who are affiliated to the Welsh Section now number approximately 150, which is the highest figure so far achieved but which, it must be admitted, falls considerably short of the potential among Welsh dairying interests. Attendance at each of the general meetings of the Section is in the region of fifty members and friends, and at the annual General Meeting an attendance of about twice that number is usual."

LINKS WITH PARENT SOCIETY

"This historical review of the Welsh Section would be woefully incomplete without reference to the pleasure which its members derived from official visits which the parent Society has made to Wales for two of its meetings. The first occasion was in January 1946 when members of the Society assembled in Cardiff for a general meeting. This, incidentally, must have been one of the first occasions on which the society departed from its previous practice of holding meetings in London. The second visit on a more ambitious scale and of longer duration was for the 1952 Spring Conference, centred at the Craigside Hydro Hotel in Llandudno. Mention should also be made of the pleasure derived by members of the Society resident in Wales from the annual visitations which successive Presidents of the Society have made to attend one or more Section meetings."

"The main purpose of this contribution is to present in outline a history of the Welsh Section. It has not been an easy task to try to condense nearly four hundred pages of recorded minutes into less than two thousand words. There may be many important omissions, but the one the writer is most conscious of is the names of those Section and branch officers who have helped to make the society's activities in Wales a modest success. This omission is intentional, as the list would be a long one. However, in a tribute to the services rendered by two ex-Officers of the Section who had left the area, there is recorded in the Welsh Section's first minute book a verse which may serve to inspire gratitude for the past, high hope for the future and a fitting conclusion to this review.

> 'Lives of great men all remind us,
> We should make our lives sublime
> And departing leave behind us
> Footprints in the sands of time.'"

COMMENTS ON WELSH SECTION BY MISS E RICHARDS (PAST PRESIDENT) AND DR S WALKER (COUNCIL AND TECHNICAL & EDITORIAL BOARD)

"The inaugural meeting of what became the Welsh Section of the Society of Dairy Technology was held in Carmarthen in November 1943. Since that time it has played an important role within the Welsh dairy industry. Because of the difficulties of travel within Wales, the Branches became the focus of SDT activity. In 1945 branches were established at Carmarthen and Cardiff and in 1952 a vigorous group in North Wales established a branch at Wrexham."

"The Cardiff Branch foundered at an early age. The other two were very active especially during the 1950s and 1960s. The Carmarthen Branch held speaker meetings almost every month with subjects such as 'milk utilization,' 'creamery management,' 'safety on the farm' etc. The list of subjects is indicative of local interests and membership. One active member was dairy farmer W. J. Hinds, who was chairman of the Section in 1964/65 and an elected member of the England

& Wales Milk Marketing Board. The Branch also held the occasional social event, which allowed it to be almost self-funding. Although this was frowned upon by some of the senior members of the Society, it did not deter the Carmarthen members. Members of the North Wales Branch were far more serious minded, with regular meetings of an educational nature, again indicative of local interests."

"In the North, encouraged by the management of Mr Charlie Hares at Marchwiel and latterly Mr Mike Harvey at Llandyrnog, meetings where also held on a regular basis. The numbers were further boosted by colleagues living in the Border counties of Cheshire & Shropshire and students sent by Mr Ron Lawton at Reaseheath and Mr Fred Cunningham at Llysfasi. With this in mind several of the meetings were of an educational nature; indeed the co-author of this summary can remember being told to present a 20 minute paper on the *'Fractionation of Butter-Oil'* when turning up to work an afternoon shift, with about 4 hours to prepare!"

"With the closure of the dairy at Aberystwyth, the Section meetings tended to move around the Principality. In latter years these were very poorly attended and the last one held at Buith Wells in the mid-90s reflected, as with the Society in general, just how difficult it is to maintain interest in what might be deemed a specialist subject."

"The Section annual general meeting was held at Aberystwyth until the closure of the dairy department at the college there. John Lewis, Director of what was then known as the University College of Wales at Aberystwyth, was President of the Society in 1957/58. It was his goal to achieve a membership of over 3,000 - a goal he just failed to achieve."

"What is obvious. when reading of the early years of the Society in Wales, is how it encompassed all parts of the industry. It was all-inclusive; there was no education or qualification bar. The opportunity was given to any interested person to be actively involved. Thus the members included dairy farmers, advisors, processors - both management and staff - research scientists, college staff and students and people from companies producing a range of ancillary products."

"There were what some have called 'pioneering ventures'. Among these was a tour of Ireland in 1946. A party of 46 members and friends spent four days experiencing Irish hospitality and gaining an insight into the dairying activities surrounding Cork. The last of these tours took place in 1988 when Mr Robert Willis took a party over to Yorkshire and Humberside to visit fellow members and their places of work in that region. A Crest designed for the Section by P. F. Carter is among the many items now deposited at the National Library of Wales for safekeeping. The Welsh Section has also been involved with sponsorship of events, for several years supporting the Royal Welsh Show by sponsoring the prizes in the Butter Section and providing speakers for evening meetings for many of the people who stay for the three-day duration of the event."

"The terrain of Wales does not make for easy travel and many of the Section meetings were one-day events. If held at Aberystwyth, it was expected that students and staff attended the meetings, providing a readymade sizeable audience for the speakers. This early introduction to the Society was an invaluable source of information and contacts at a later date."

"There have been three Welsh Presidents. The first was John Lewis (1957/58), followed by Idwal Jones (1969/70) and Eurwen Richards (1988/89). These three represent the various strands of the Society - academia, advisory services and retail/manufacturing."

"The first Spring Conference was held in 1952 at Llandudno, when Charles Brissenden was President, in 1964 at Llandrindod Wells, back to Llandudno in 1982 and Cardiff in 1992. The abiding memories of these early Conferences, especially that of 1964, was the opportunity younger members of the industry had to meet and talk with the 'captains of industry.' In later years these 'captains' have not been as involved and in some cases have actually hindered membership of the Society."

"To mark the official 50th Anniversary of the Society, a special event was held in mid-Wales hosted by Llangadog Creamery, where special commemorative plates were presented to those within the Society in Wales. Sadly this event would be the last organised by the mainstay of the Welsh Section in recent years, Hazel Rowlands. It is perhaps a sad reflection on the Society in the Principality that, despite Hazels' efforts in her last few years, she did not receive the support and backing of successive Section Chairmen and the Section in Wales finds itself in its current position."

"Changes in the structure of the dairy industry, fewer milk producers, the closure of creameries, of dairy departments in colleges, reduction of advisory staff and of research establishments both in Wales and wider afield, has affected the membership. Many resigned on having to seek other employment, and increasingly over the past decade there has been a dearth of potential members."

"Much of the milk produced in South Wales is now shipped across the border to England. In the north of the country the situation is not as dire, with the remaining creameries being rewarded for their entrepreneurship. There has been a dramatic increase of farmhouse cheese-makers as well as of small-scale manufacturers of yoghurt and of ice cream. The Specialist Cheese-makers Association, which came into being as a result of the Listeria scare, and the British Sheep Dairy Association are examples of organisations that have developed in recent years to answer the needs of the small-scale producer."

"The vision of the early members of the SDT as the educative and informative force of the dairy industry has changed. The journal is now more orientated towards an international market, leaving the more practical, less academic literature to others. The nature and subject matter of Society meetings are unlikely to attract many of the small-scale producers. However, there remains a core of people in Wales committed to the Society, with the will to see it continue to a more favourable climate for the industry as a whole. We are indebted to these enthusiasts for the future of the Society of Dairy Technology."

"Messrs John Lewis and Fred Cunningham and Miss Rowlands ensured that all the paperwork and other items of interest relating to the Welsh Section were deposited in the National Library of Wales in Aberystwyth. Of particular note, amongst the artefacts held there is a special crest designed for the Section by P. F. Carter. Perhaps this along with the commemorative plates may turn up on an Antiques Roadshow one day!"

WELSH SECTION
JSDT Vol. 45 No. 4 Nov 1992

"The first Chairman of the Welsh section was Mr. Percy Carter manager of Wilts United Dairies Creamery in Carmarthen. He did the pen and ink drawing of the Section Crest which he presented to the section in 1947 and is now kept in the National Library of Wales in Aberystwyth."

EVOLUTION AND DECLINE OF THE LONDON, SOUTH AND EASTERN SECTION
(Dr G. Chambers & Alan Stack)

"At the 14th AGM of the Society on 21 Oct 1957 there was no verbal or written report from London & the South East [JSDT Vol. 11 No. 1 (1958), pp 3 -14]; nor was there any reference to London and the South East in a statement of officers and council members of the Society for 1957/58 which appeared at the beginning of the volume [*ibid*, p1]. However in Section Notes in the same volume [*ibid*, p47-8] there was a detailed report from the 'London Branch' covering the period 1 Jul 1956 - 30 Jun 1957. The average attendance at their monthly meetings had improved and they had recruited 20 new members of the Society. At their <u>branch</u> AGM in May 1957 I. A. Howard had succeeded Charles Brissenden as branch chairman. Also in a further statement of section and branch officers for 1957-58 for the whole of the UK (which appeared in the same 'edition' of Section Notes) the South East is represented only by 'the London Branch'. i.e. there is no sign of a section. In a further edition of Section Notes later in the year [ibid 16] it was reported that J. F. Hunter had become chairman of <u>the London Branch</u>. And <u>the branch</u> featured yet again in the next Section Notes, with details of its winter programme for 1958/59. But significant developments occurred in 1959; and were reported in JSDT Vol. 12 (1959). First, on p1 of that volume, where the names of office-bearers and Council members for 1958/59 are again set out, there is still no reference to a London & South Eastern <u>Section</u> - but the name 'J. F. Hunter (London Branch)' appears at the end of the list of 'Chairmen of Sections'; second, in his report to the Society's AGM on 20 October 1958 the President, John Lewis, said: '...in the revised Constitution and Rules there is, for the first time, provision that the Chairman of the

38

London Branch shall *ex - officio* be a member of Council. The same privilege is already available to all the Chairmen of Sections' [*ibid* p 4]; third, at the end of the series of reports of sections to the AGM, there was a report from J F Hunter in respect of the London Branch [*ibid* pp 13-14]. Thus the London Branch had in effect been accorded the status of a section, in all but name. The branch was also the subject of a most significant report in the next edition of Section Notes in the Journal, which began: 'The London Branch moves from strength to strength. Founded in 1954, it now has a mailing list of 180 members, with average attendance at monthly meetings of around 100' [JSDT Vol. 12 (1959), p 1291]. At the AGM in October 1959 the new status of the branch was again recognised; and the branch chairman (A. Rowlands) reported that the branch had enrolled 57 new members in the course of the preceding year [JSDT, Vol. 13 (1960), pp 13-14]. The branch was again treated as a section at the Society's AGM in Oct 1960: C. S. Miles was now branch chairman and another 31 new members had been enrolled [JSDT, Vol. 14 (1961), p 15]. But by the next AGM, on 23 October 1961, there was another significant development: this was announced by the retiring President, Dr. R. J. Macwalter, in his presidential report. After referring to the formation of three new branches during his year of office (Dumfries & Galloway and North Eastern in Scottish Section and Newcastle-upon-Tyne in the North Eastern Section) Macwalter said: 'The London Branch has now been granted Section status, covering a very wide area' [JSDT, Vol. 15 (1962), p 4]. And later in the meeting H. Burley presented what he said would be the last report of the London Branch: it was for 1960/61 [*ibid*, p 13]. The first identification of the grouping as the 'London, South, and Eastern Section' occurs in a tabulation of sections and their branches and memberships of their committees for 1961/62: in this the London Branch has disappeared and is replaced by the 'London, South, and Eastern Section' with John Glover of CWS as section chairman and Vic Cottle as section secretary [*ibid*, p 62]. This brought the number of sections in the society to nine, and the process of evolution was capped at the 19th AGM in London on 21 Oct 1962, when John Glover gave a report on the first year's activities of the new section. He began by referring to 'the elevation in status of the old London Branch' to that of a section: and gave the geographic compass of the new grouping as Berkshire, Buckinghamshire, Cambridgeshire, Essex, Hampshire, Hertfordshire, Holland (Lincolnshire), Huntingdonshire, Kent, Middlesex, Norfolk, Oxfordshire, Suffolk, Surrey, and Sussex - a very large parish indeed. Thus the old London Branch which from 1954 had been the main - and for most of the time the only-devolved focus for members' living in the capital and in the South and East of England passed into history. This research has not shown what became of another branch in the South of England, formed at the same time as the London Branch: this was the Brighton & District Branch. The two branches were formed in 1954 within days of each other, but unlike the London Branch, the Brighton & District Branch had ceased to feature by 1958 and possibly earlier. It should be noted, however, that within two years of the establishment of the London, South, and Eastern Section its committee was discussing the desirability of forming branches in Reading and Cambridge - which was not surprising since the membership of the Section was then 596 [JSDT Vol. 17 No. 1 (1964), p 48-9]. The outcome was the formation on 7 November 1963 of the Eastern Counties Branch based on Cambridge and Chelmsford; and on 5 March 1964 of the Reading and Southern Counties Branch based on Reading, with H. J. Brooks and Dr. James Rothwell as the respective foundation chairmen [S C H Suggate, Section Chairman, at the Society's AGM in London, 26 October 1964 - JSDT Vol. 18 No. 1 (1965), p 25]."

Some thoughts and comments on the run down of this Section.

"The most active supporters of the section were to be found in the independent sector of the dairy industry suppliers - notably APV, Alfa Laval and Tetra Pak."

"The Reading Branch was principally supported by Cliffords in Bracknell and the various organisations attached to the University of Reading and NIRD."

"Once Cliffords closed and the NIRD was wound up, then there was very little support for the Branch and its activities more or less stopped in 1997/98."

"We have tried on at least two occasions to rekindle the society from this base but the support for the traditional type of evening meeting in the Reading area is sparse."

"The main section also lost several independent dairies in the early 90s with the result that meetings became very difficult. I recall going to an evening meeting at MD Foods at Oakthorpe where there were 4 Dairy Crest people, Oakthorpe's plant manager and some committee members. It was really disappointing for the host, the organisers and the speaker who had travelled some distance to speak to us."

"In the late 80s and early 90s our most successful meetings were held at the Alfa Tower in Brentford with up to 60 people involved and standing room only in the Conference room. These meetings were often in the form of a late afternoon, early evening meeting with 3 or 4 papers covering relevant subjects. We repeated this style of event at the Tetra Pak offices in Stockley Park with some initial success but found it very hard to keep the momentum with the correct range of subjects and speakers."

"The Spring Conference of 1998, which we had to cancel, was very traumatic for those us involved in the organisation. We spoke to many people in the Industry only to find that many of them felt they couldn't afford the time or money to attend a two-day Conference. We believed this to be symptomatic of the time and cost pressures the industry faces."

"The section also ran a Dinner Dance in early November. The event was thought of as the major dairy industry social event and most of the major suppliers would take tables for their customers. This event also became more difficult to organise as the industry rationalised. The number of possible guests and the number of suppliers prepared to act as hosts reduced. Consequently, when numbers dropped by about 40%, we took the conscious decision to stop this event before it turned into a financial disaster. At the time the Gateshead Summer Ball was growing and many felt that this was a more cost-effective event as it was not held in a West London Airport hotel."

"In a nutshell, this section has many members and they are spread over a wide area. Distances may be comparatively short, but the traffic, and working hours, make the organisation of the traditional evening meeting difficult with little chance of success."

"Many have commented that they miss the social events. These were often the only opportunity individuals had of meeting with old friends and colleagues in other companies where they could discuss in wide issues."

"I guess the section could be described like some skills use it or lose it, and I am afraid that it has been lost through a lack of use by the dairy industry itself."

CAPITATION GRANTS

"The finance of sections was aided by capitation grants referred to below:
(JSDT Vol. 36 No. 1 January 1983)."

"The state of the Sections is one of the best mirrors of the health and success of the Society. For a further year, the Sections overall have been largely self-supporting financially, raising income from various local events and activities, which almost totally covers local administration: net expenses of the sections were only £226 this year and again Section Committees and treasurers are to be congratulated. This has meant that the capitation grants have been untouched and are carried forward in the Section bank accounts."

"In view of this healthy situation and to encourage further Society activities at Section level, Council approved in August some revised guidelines for the administration of the capitation grant, which was itself increased for the year 1982-83. These changes are designed to encourage a wider range of speakers to local events by covering travelling costs which otherwise might be inhibiting."

The revised guidelines for capitation grants were set out in the Journal (JSDT Vol. 37 No. 2 April 1984):

"The question of Capitation grants to Sections has in the past created some confusion because of a complicated system devised about 12 years ago, which took into account the number of members in each section, the distance from London to the centre of each section and several other factors, resulting in a complicated equation to determine the amount due to each Section."

"Consequently, it was felt that the whole question of Capitation grants required a review to produce, if possible, some new guidelines. These revised guidelines were formulated by a sub-committee on 19 May 1982, further amended on 25 June 1982, approved by Council in August

1982, and reviewed by Council in August 1983.

1. A new flat rate will be introduced applicable to all fully paid up Society members as at 1 July each year, including those of the London, South and East section (who previously received considerably less than other sections because of the mileage factor which has now been abandoned). The new rates will be determined by Council at its August meeting each year (agreed at £1 per member at August 1982 meeting with a recommendation at the August 1983 meeting that the rate remain the same for the coming year) and reviewed annually thereafter. All future increases in the rate will normally be related to increases in the membership subscription rate.

2. The expenses of speakers at Section or Branch meetings will be met from the main funds of the Society, for a maximum of two speakers per year from within the UK and Ireland for both each section or branch, such expenses being normally limited to a total of £85 (increased from £75 at August 1985 meeting) per Section in any one financial year.

3. Capitation grants based on the above principles will not be paid out automatically. After the Capitation rate for the year has been reviewed by Council in August, the funds will be held in the Society's main account, but will be available to Section Committees, to be requisitioned by Treasurers only when the amounts held in local section bank accounts (including Branches) fall below a total equal to £3 per head of fully paid-up Section membership. The balance should be maintained at or above the bank minimum requirement so as not to incur bank charges.

(a) The purpose of the Capitation grant is to permit Sections to operate their own programme of meetings.

(b) Where necessary, the travelling expenses of the Section Chairmen or their representatives to attend Council meetings will be paid by the Society's offices from the main funds of the Society.

(c) The cost of postage continues to represent the main expense of Sections' administrative work and all Sections are urged to send as many local notices as possible to the Society's office for inclusion in main circulations from there, or have them included in the supplement to the journal."

There is no continuous record of these grants but reference is made to them from time to time in the Journals and accounts (JSDT Vol. 42 No. 1 February 1989). Normally decisions on the grants were taken at the August meeting of Council."

SECTIONS IN IRELAND

The formation of Sections in Ireland came later, stimulated by the diversification and growth of the dairy industry in both Northern and Southern Ireland.

The section in Northern Ireland was founded on 28th April 1950, at a meeting in Belfast as recorded in the following extract from the Journal (JSDT Vol. 4 No. 1 24-29 (1950)

"The Society had 18 members in Northern Ireland before any move was made to start a section there. N.R. Knowles, a senior colleague of John Murrays was the first Hon Secretary. The prime mover in setting up the section was Peter Clerkin, a microbiologist in the faculty of agriculture at Queens University, Belfast."

"The idea of having a Northern Ireland Section of the Society, which has been under consideration for some time with the benevolent support of the Council of the Society, received an emphatic local endorsement at a preliminary meeting held in the Agriculture Building, Elmwood Avenue, Belfast, on Friday, 28th April, 1950 which was attended by some 85 persons representing all branches of the Northern Ireland dairying industry. Those present included Mr. D. A. E. Harkness, C.B.E., Permanent Secretary of the Northern Ireland Ministry of Agriculture, who, after the Chairman's opening remarks, delivered the valuable and interesting address on "Progress and Trends in the Dairying Industry in Northern Ireland, 1924-49 (see 'Papers of Special Significance, Chapter 8.)"

"In opening the meeting the Chairman (Mr. P. Clerkin, Queen's University, Belfast) reviewed the aims and objects of the Society of Dairy Technology and its rapid growth from the small band of enthusiasts who brought it to life into the vigorous and steadily growing organisation of over 1,400 members of today. As a member of relatively long standing, he was satisfied that the

Society gave him good value for his money and assured prospective new members that they would find it a very live and helpful organisation."

"He stressed the difficulty experienced by members who lived far away from London in deriving the full benefit that the Society could provide, but could testify from seeing a Section at work how greatly and successfully that difficulty had been minimised in those parts of Great Britain where Sections had been organised."

"Turning to the programme of the meeting, he expressed the pleasure of all present that Mr. Harkness had found it possible to implement his sympathy with their aims not only by his presence but by the important contribution that he was about to make to their proceedings, to which they all looked forward with great interest."

"Mr. Harkness might not pose as an expert dairy technologist but he was certainly the expert on the set-up of the dairy industry in Northern Ireland. Although no longer directly responsible for the Milk Section of the Ministry, his personal interest in this field clearly remained unabated. It was now many years since his first published work on Milk Reorganisation in Northern Ireland came out, and his mass of research carried out in the late 'twenties and early thirties' furnished the basis of the price-fixing structure (as used by the Joint Milk Council) which saved the industry of those days. No one was better qualified to give an outline of what the industry has come through since those early days and the great advances that have been made nor, if his official responsibilities would permit, to give a shrewder forecast of the probabilities of the future in the address he would now give."

The Northern Ireland Section flourished with support from the processors and manufacturers, the Milk Marketing Board for Northern Ireland, the Northern Ireland Milk Alliance (representing the trade), the co-operative movement (several important creameries in Northern Ireland were co-operatives), the Ministry of Agriculture for Northern Ireland, the Universities, the centres of research and the general public.

Dr. G. Chambers gives this account of the growth of the N. Ireland section:

"Nowhere was the growth of the Society in the first two decades of its existence more dramatic than in Northern Ireland. The result was that by the middle of the 1960s membership in the Region had climbed from around eighty to almost three hundred - with every branch of the industry represented, and with business meetings and social events regularly attracting attendances of a hundred to a hundred-and-fifty including a number of very welcome enthusiasts from the Republic of Ireland. In the first instance the main thrust of expansion both in numbers and in influence had come from the Ministry of Agriculture for Northern Ireland, which was still in complete control of the local industry. Charged with maximising the contribution of Northern Ireland agriculture to the national larder during World War II and the subsequent period of recovery and continued food-control, the Ministry had enforced rationalisation on a massive scale, established or developed six new creameries of its own, introduced a number of national and international processors to the scene, and built up a whole new infrastructure of regulatory, advisory, quality control, research, and educational support, provided in the main by technical and scientific staff. This was fertile ground for Society membership, especially when a strong personal lead had been given by the heads of the various segments of the technical-support arrangements. The result was that a high proportion of the staff involved elected to join the Society and to encourage their contacts in the industry to do likewise. Those who were particularly influential in that regard included Percy Bell of the dairying inspectorate, O'Hara O'Neill of the farm advisory service, and Professor John Houston, Peter Clerkin, and Dr John Murray, senior scientists in the Ministry's research, teaching, and quality control organisation based in Queen's University Belfast - and all five of these worthy advocates of Society membership duly fulfilled the role of chairman of the Northern Ireland Section, Dr Murray going on to become national President in 1966/67."

"A second boost to the growth of the Society in the province came from the establishment of the Milk Marketing Board for Northern Ireland in 1955. Modelled on the organisations that had been created in Britain in the 1930s, the Board took over responsibility for the assembly and marketing of ex-farm milk and for the operation of the creameries established by the Ministry during the war. Except for industrial employees only a small proportion of those who had been involved in these operations under Ministry control transferred to the Board. Hence many of the Board's administrative, marketing, and technical staff were new to the scene - and became potential members of the Society. It was in these circumstances that the author and editor of the present work, Dr David Armstrong, became a member of the Society. He was the foundation administrator of the Board and became its general manager less than three years later. At that time the role of the Society in personal development and in the general advancement of the industry was also being promoted within the Board's organisation by two key members of staff recruited from outside the Province: these were the senior technical officer, Harry Rostern, and his assistant, George Grier, who had been respectively manager of the English Board's creamery at Wem in Shropshire and assistant manager of the Scottish Milk Powder Company's plant at Kirkcudbright. From the beginning it was not only technical staff of the new organisation that joined the Society: senior staff in other disciplines soon recognised that the Society provided a forum in which for them to learn more about the industry in other parts of the United Kingdom and in the Irish Republic and to become acquainted with a wide range of individuals from the Ministry and from private, co-operative, and public companies with which the Board did business, either as suppliers of milk or as customers for dairy equipment. Dr Armstrong moved from Northern Ireland to the Milk Marketing Board of England and Wales in 1963 and was succeeded in Northern Ireland by Dr George Chambers who had been chief chemist of the Board from 1958 and secretary of the Northern Ireland Section of the Society from 1961. Under his leadership more members of staff from a range of disciplines and at all levels were encouraged to join the Society, and a number of Board members also became actively involved. In the fullness of time Harry Rostern, Dr Chambers, and George Grier all chaired the Northern Ireland Section - and Dr Chambers and George McL Grier went on to become national Presidents of the Society in 1975/76 and 1986/87 respectively."

"Important as were the contributions of the Ministry and the Milk Board to the expansion of Society membership in the formative years, they were at least equalled by the combined effect of two further developments at the end of the 1950s and the beginning of the next decade. The first was the formation in 1958 of the Section's Western Branch based in Omagh; and the second was the institution in 1961 of dairy diploma courses at the Ministry's agricultural college at Loughry near Cookstown, under the inspirational leadership of J A Young as principal and with the enthusiastic support of the Society. The formation of the branch in Omagh brought in new members from creameries in the traditional dairying counties of Fermanagh and Tyrone; and the dairy diploma courses at Loughry became a prolific source of new student members, the majority of whom graduated to full membership as they took up posts in the still-expanding industry. In this connection it has been said that the first two heads of the dairy department at Loughry - Scotsman Bob Renwick and his successor and fellow Scot, George Lang - did not regard a dairying student's education at the college to be complete until he or she had joined the Society! As a result of these developments and of coincidental expansion and diversification both of product manufacture and of farm services, the strength of the Northern Ireland Section increased from around 140 in 1958 to a peak of 290 in 1966, making it the third largest section in the Society at that time. Over the next three or four years the strength eased back to around 260, one factor in that reduction being the loss of members to the new Section in the Republic of Ireland formed in 1966. Thereafter the strength of the Northern Ireland Section remained at around 250 for many years, and the malaise of drastically reduced membership and disappointing attendances at meetings did not appear in Northern Ireland until some five years after it had become a matter of deep concern in Great Britain."

"In the long period of stability the Section had developed and maintained annual programmes that included a number of novel features, appealed to members, and commanded good attendances. The highlight of every year was the annual general meeting held at Loughry College, a venue that was equally suitable to members from all parts of the province. Invariably the occasion included an excellent meal provided by the college and an address by a speaker of national or international standing, dealing with a topic of wide general interest. It was therefore not unusual for one of those gatherings to attract an audience approaching 200, including the students involved in the dairy diploma courses. Likewise the annual dinner dance, generously sponsored by firms in the industry and held in a good hotel, was the social highlight of the year and helped the section to build up some independent funds. However the core of Section and Branch activity was the winter programme of technical and marketing lectures, staged mainly at the School of Agriculture and Food Science at Newforge Lane in Belfast, and often repeated on the following night at the Nestlè factory near Omagh for the Western Branch. A high proportion of the speakers at these meetings came from Great Britain or the Irish Republic; but every annual programme also featured a young members' night where three 'young' (or not so young!) members from Northern Ireland addressed their colleagues on aspects of their work in the industry."

"Another novel feature was a series of annual meetings aimed specifically at dairy farmers. Staged at Greenmount Agricultural College near Antrim and attracting large audiences 'recruited' by the Ministry's advisory staff, these events certainly raised the profile of the Society amongst milk producers and communicated the latest thinking on important topics - but they did not succeed in attracting more than a handful of producers to membership of the Society. On the other hand the annual spring outing, occasionally arranged on a joint basis with the Southern Irish Section, was invariably successful in its aim of a group of members and their guests spending a pleasant day together in an industry context - an uncomplicated precursor of the expensive 'bonding' and 'team-building' exercises that are *de rigueur* in many forms of human activity today. Valuable too was the unfailing involvement of Northern Ireland representatives in national seminars, conferences, and study tours organised by the Society: these invariably paid dividends in new ideas, improvements in efficiency, modernisation of equipment, and growth in the confidence and professionalism of the individuals involved. But perhaps the best indicator of the sustained vitality of the section over a period of almost fifty years was its ability to host the Society's Spring Conference on five occasions - 1954, 1963, 1977, 1987, and 1997 - and to attract a sizeable number of delegates from Great Britain and the Republic of Ireland on each of these occasions. Like the 1987 conference the last of these events was also notable for its taking place during the national presidency of a Northern Ireland member: this was Robert Hamilton of Pritchitt's Foods, who after his presidency in 1996/98 made another significant contribution to the Society through his editorship of the Members' Newsletter for a number of years. During that period, however, the consequences of de-regulation and of the dismemberment of the whole fabric of a cohesive technology-based industry were beginning to be apparent; and the challenge of keeping the Society alive in Northern Ireland in the face of economic pressures and technological apathy has been a formidable one. But three individuals have succeeded in doing so and they deserve the warmest commendation for their efforts: they are Dr John Crawford and Dr Ian McDougall who between them have sustained the chairmanship of the section for the past five years, aided by the untiring and ever-optimistic secretary, Miss Joy Alexander who has been honorary secretary for the whole of that time. A measure of their success was the attendance from inside and outside Northern Ireland at the section's residential conference on 'CIP - *For a Clean Future'* held at Loughry in the period 10 - 12 September 2000. But it has not been easy and the prospects for the future look somewhat bleak."

SUMMER MEETING IN NORTHERN IRELAND 1954 (JSDT VOL. 7 1953/54)
The notice of the meeting was as follows:

"At the invitation of the Northern Ireland Section the 1954 Summer Meeting of the Society will be held in Northern Ireland from Sunday, 2nd May, to Friday, 7th May 1954, inclusive."

"The party will be accommodated at the Northern Counties Hotel, Portrush, and, if necessary, at other hotels in the vicinity. Portrush is situated on the north coast of County Antrim and is surrounded on three sides by the Atlantic Ocean. It offers every amenity for enjoyment including golf (championship course), tennis, bathing, fishing and boating and those who decide to stay on for a holiday after the termination of the Conference will be indeed fortunate."

"This must have been one of the most significant conferences in the whole series in terms both of the eminence, quality and knowledge-bank of the battery of speakers assembled and of its timing, in the first year of decontrol in GB and the last year of Ministry marketing of milk in NI. The President was Professor E L Crossley, and the chairman of the NI Section and therefore of the organising committee was a future president, Mr John G Murray, who would become Dr J G Murray (Northern Ireland's 'JG') in the following year. The first speaker was J I Magowan, then permanent secretary of the Ministry of Agriculture for Northern Ireland. His paper was entitled *'Dairying in Northern Ireland'* and it concluded with an important and perceptive look forward into the consequences of decontrol and new marketing arrangements in the Province. This was followed by a masterly survey of the development of dairy husbandry in the local scene by Professor James Morrison, in the run up to an official lunch hosted by the then Minister of Agriculture for Northern Ireland, Rev Robert R Moore MP. Next morning saw two more most distinguished speakers on the rostrum: Professor Hubert G Lamont, head of veterinary research in Northern Ireland, who spoke on *'Animal Health and Milk Production'*, and Professor Estyn Evans, professor of geography in the Queen's University of Belfast, whose paper was entitled *'Dairying in Ireland through the ages'*. And the final paper-reading session, on the Thursday morning, was devoted to current milk utilisation, with J Guild of the dairy department of the Belfast Co-operative Society and vice-chairman of the Northern Ireland Section, speaking on the liquid milk market and W A Manahan of the Nestle Company telling the story of the remarkable diversification of the manufacturing milk market in the province between 1938 and 1952. These papers plus a comprehensive programme of tours and social activities made this into one of the most significant dairy-industry events ever held in Northern Ireland."

SUBSEQUENT SPRING CONFERENCES IN NORTHERN IRELAND, 1963

"The Society's second conference in Northern Ireland - the first of the Northern Ireland events to be rightly called 'a Spring Conference' - was based in the Slieve Donard Hotel Newcastle 21-26 April 1963. This conference had many notable features. According to President Walter Ritchie's summary 'the audience was the 'largest ever recorded at a spring Conference'. Notable too was the substantial involvement of the local Minister of Agriculture, Harry West. He turned up on the Sunday evening of the delegates' arrival, formally opened the conference on the Monday morning, stayed for the first session, and hosted a Government dinner at Stormont on the Tuesday evening. It was also a time when the Northern Ireland industry was recording two 'firsts' of great <u>significance: first in the UK with a scheme for the payment for milk on broadly in</u> accord with the recommendations of the Cook Committee; and first too with arrangements for the eradication of brucellosis in the cattle population. Other developments that interested the delegates included the still relatively new head office and central laboratory of the MMBNI on the Antrim Road in Belfast and the institution of a dairying diploma course at Loughry Agricultural College at Cookstown in County Tyrone. The latter provided an excellent example of the influence of the Society on dairy education in the UK, in that the then principal of the college told delegates that the new course was introduced in 1961 'mainly as a result of representations from the local section of the Society of Dairy Technology'. Dr Armstrong's paper on milk marketing in the province was a masterpiece and Dr. Chambers used it as a model on at least two subsequent occasions and Dr Chambers was secretary of the organising committee of the 1963 conference and Gerald Malcolm of the Ministry was chairman [JSDT Vol. 16 (1963)]"

"Because of the outbreak of civil unrest and politically - motivated violence in Northern Ireland in 1969 and Council's perception of consequential risks there was not to be another Spring Conference in Northern Ireland until 1977, in the presidency of Vic Cottle. Like the 1954 summer meeting this took place in the Northern Counties Hotel Portrush, 24-29 April 1977. The attendance included strong representation from Great Britain and the Irish Republic, together with one visitor from New Zealand (Dr Wayne Sanderson), a 'speaker from the European Community, and a large turnout of the local membership (which, incidentally, had by then grown to 233). The section chairman at the time of the conference was John Campbell, Secretary of MMBNI [JSDT Vol. 30 (1977) & Vol. 31]"

"The fourth conference in Northern Ireland was based in the same venue as the second, the Slieve Donard Hotel Newcastle 26-29 April 1987, during the presidency of George Grier, director of creameries of MMBNI. The section chairman at the time was Robert Agnew of the Department of Agriculture, and the secretary was Elizabeth Bailey of MMBNI, The conference attracted an attendance of around sixty delegates from overseas (GB and further afield), together with about twenty from the Republic of Ireland and a high proportion of the local membership. It is notable that the conference was a four rather than six-day event, in reflection of increased economic pressures in the industry and a parallel and short-sighted diminution in appreciation of the contribution that dairy science and technology could make to its future [References: Section and branch activities supplement to issue no 3 of Vol. 40 August 1987.]"

"The fifth conference in Northern Ireland was also held in the Slieve Donard Hotel Newcastle 20-22 April 1997. In accord with changing times and changing emphasis the theme was *'Food Safety: Protection and Enhancing your Business'* and the duration was further reduced from four days to three. The event was attended by eighty delegates from Great Britain, the Republic of Ireland, and Northern Ireland. The conference dinner on the evening of 21 April was graced by the attendance of the direct-rule Minister, Baroness Denton, and I had the privilege of responding to her address and talking about the history of our society. The chairman of the section at the time was David Abernethy of Pritchitt's Foods (News Supplements of the Journal in May and November 1997) and two of the papers presented at the conference were published in the Journal, Vol. 50 No. 1 (1997), pp 89-95 and 96-98."

'CIP - FOR A CLEAN FUTURE': RESIDENTIAL CONFERENCE NORTHERN IRELAND 2000

"This event devoted to the very practical subject of cleaning-in-place (CIP) was staged at Loughry College 10-12 September 2000. It attracted speakers and delegates from all parts of the UK and from the Republic of Ireland, together with one or two from Continental Europe. The conference was addressed by Mrs Brid Rogers MLA, the Minister of Agriculture and Rural Development in the local administration; and I had the privilege of replying on behalf of the Society, making a plea for the preservation of the integrated system of technical and scientific support of the agri-food industry in Northern Ireland which had served it well for almost eighty years."

*Northern Ireland Annual Dinner Dance 1996*
*Left to right: Robert Hamilton (Vice-President), Marilyn Lee, Grahame Lee (President), Roberta Cargan, Alistair Gault (Section Chairman)*

*Martin McEvoy, IFST Chairman, Norah Whittaker, (Speaker) Pritchitt Foods, Andrew White, Chairman of the SDT Northern Ireland Section and Michael Rowe (speaker) Dept of Agriculture for Northern Ireland at a joint meeting of the SDT Northern Ireland Section and the IFST local Branch in February 1995.*

OFFICE BEARERS N. TRELAND SECTION AND AGM SUBJECTS OVER 53 YEARS
(Contributed by Dr. I. A. McDougall)

| Year ending Sept | Chairman | Secretary | Treasurer | AGM Speaker & Subject |
|---|---|---|---|---|
| Inaug 1950 | P Clerkin, Min Agr. | | | 28 Apr, Elmwood, 85 persons; DAE Harkness: Progress &Trends NI 1924 -1949 |
| 1950 | P Clerkin | N R Knowles | C Damaglou | J White: Origins of the Society |
| 1951 | P Clerkin | N R Knowles | C Damaglou | CB Benwell: Instrumentation |
| 1952 | WD Thomson | N R Knowles | C Damaglou | JA Young:Problems of Milk Producers |
| 1953 | H Simmons | N R Knowles | C Damaglou | Brains Trust |
| 1954 | J G Murray | N R Knowles | C Damaglou | Robinson: Dairying in retrospect in NI |

| 1955 | J Guild | E Alexander | J Robinson | AE Muskett: panel discussion |
|------|---------|-------------|------------|------------------------------|
| 1956 | JG Young | E Alexander | J Robinson | Morley Parry: 'not to be quoted' |
| 1957 | E Wilson | (E Alexander co-opted) | (J Robinson | MMBNI Panel, incl. JK Lynn D Armstrong |
| 1958 | C Damaglou | J Patton | (J Robinson co-opted) | Mr Hamill: Farmer 'made good' in NZ |
| 1959 | EPW Hill | J Patton | (J Robinson | S Shaw, Farmer, Ford Foundation Scholar USA. |
| 1960 | FP Bell | J Patton | SW Buller | AM Raven, Kellogg Foundation Scholar: Dairy products in USA |
| 1961 | Prof Houston | G Chambers | SW Buller | Panel, A Lovesy and others |
| 1962 | H Rostern | G Chambers | SW Buller | CH Chalmers: Progress in Dairy Hygiene over past 50 years |
| 1963 | WG Malcolm | G Chambers | JC Hood | WR Trehane: R&D in the Dairying Industry |
| 1964 | SW Buller | JE Lamb | JC Hood | AJF O'Reilly: An Bord Bainne, The Irish Dairy Board |
| 1965 | SD Gibson | JE Lamb | JC Hood | P O'Neill: Some Problems in Distributive Industry |
| 1966 | H O'H O'Neill | JE Lamb | JC Hood | Dr Jean Graham: Pastoral Farming in Ireland in the Past |
| 1967 | G Chambers | JE Lamb | RJ Hyde | D Armstrong: The MMB as a Commercial Enterprise |
| 1968 | JC Hood | TJ Noble | MM Blackburn | JR Rowling: The Promotion of Milk & Milk Products |
| 1969 | RJ Hyde | | AW Acheson | JAB Smith: Milk Production Progress in past 25 years & Future |
| 1970 | JR Meaklim | | AW Acheson | WM Drummond, RD Rolston: Aspects of Industrial Training, incl. The Food & Drink Training Board. |
| 1971 | WJ Simpson | | (Mr Gilliland) | JA Young: Dairying in NI Agriculture |
| 1972 | G McL Grier | | Mr Gilliland | F Kearns: Entry to Europe and its Implications for NI Dairying |
| 1973 | GIA Lang | | F McWha | P Newbould: Conservation and the NI Countryside |
| 1974 | D McGivern | | F McWha | A McClelland: History & Dev of Agric. Industry in NI |
| 1975 | JJ Jack | J Campbell | | J Simpson, QUB: The Role of Agriculture in Economic Growth |
| 1976 | B Duncan | J Campbell | | J Johnston, BBC: Ulster Farm |
| 1977 | J Campbell | R Workman | RP Smvth | P O'Keefe: Future for Dairying |
| 1978 | H Martin | R Workman | RP Smyth | Dr Love: The Myth of Diet and Coronary Heart Disease |
| 1979 | AW Acheson | B Maybin | A Gilmour | JE Seydel: Planning for Change and Business Dev |
| 1980 | RP Smyth | B Maybin | A Gilmour | HC Bang, Cow & Gate: International Marketing of Dairy products, incl. Baby Foods. |
| 1981 | D Stewart | B Maybin | MT Rowe | D Armstrong: The EC Community |
| 1982 | J Hamilton | J Logan | MT Rowe | JA Young: The Importance of Agriculture in NI Economy |
| 1983 | T J Noble | J Logan | JDM Neill | HS Tate, IDB: The Economic Recession - Problems & Opportunities |
| 1984 | A Gilmour | J Logan | JDM Neill | WM Dijkstra: World Dairy Situation |
| 1985 | J McClelland | J Logan | JDM Neill | Prof JA Rook (Hannah RI): Dairying |

| 1986 | F McWha | Miss E Bailey | JDM Neill | Dr Barbara Pickard: |
| 1987 | W R Agnew | Miss E Bailey | C Paul | F Harding: Future Dev in Milkfat Utilisation. |
| 1988 | R Hamilton | Miss E Bailey | C Paul | B Gardiner, Agra-Europe: [European Dairy Scene] |
| 1989 | Miss M Gilliland | Miss E Bailey | | P Lee: Experiences before and After the SDT |
| 1990 | J Cooke | Miss E Bailey | | Lord Skelmersdale, Minister for Agr. NI: [The NI Dairy Industry] |
| 1991 | Dr M Rowe | Miss E Bailey | | N Shaw: The Future of Dairy Farming in NI |
| 1992 | I J Kettyle | Miss E Bailey | S Nixon | R Wright, BBC: Any Publicity is Good Publicity |
| 1993 | B Moffatt | Miss G McFarlane | S Nixon | J Megaw, Stewarts Supermarkets: [Marketing] |
| 1994 | W Weatherup | Miss J Alexander | | [C Gibson: The Challenge of Change - Threat or Opportunity] |
| 1995 | A A White | Miss F Dickson | R. Smyth | Rankin, Ulster Farmers Union: |
| 1996 | Alister Gault | Miss F Dickson | R. Smyth | [Spring Tour to Killeshandra/Town of Monaghan] |
| 1997 | David Abernethy | G. Stevenson | R. Smyth | 2 Speakers, Tesco. Policy & Developments for NI |
| 1998 | J C Crawford | Oswald Mason | | No AGM |
| 1999 | LC Crawford | Miss J Alexander | | Rachel Rowlands: Organic Processing |
| 2000 | I A McDougall | Miss J Alexander | | M Drayne Mrs B Rodgers: Keynote speaker at Conference: CIP: For a clean Future |
| 2001 | I A McDougall | Miss J Alexander | | M Drayne Dr T Quigley, Food Safety Promotion Board: Food Safety in Focus; Mark Neale, EU Working Group: E. Coli 0157 |
| 2002 | I A McDougall | Miss J Alexander | | M Drayne, M Megarry, M Tynan: IPPC (Environmental Mgt) |
| 2003 | I A McDougall | Miss J Alexander | | M Drayne Miss N Childs, Prof G Rechkemmer: "Health Claims in USA and EU" |

*Pictured at the Northern Ireland Section AGM - Joy Alexander, acting Honary Secretary, Andrew White- incoming Section Chairman, Christopher Gibson and Wilf Weatherup, outgoing Section Chairman.*

## FORMATION AND DEVELOPMENT OF THE SOUTHERN IRELAND SECTION

The late Martin Mullally and I were good friends and in the course of many discussions about the future organisation and development of the markets for milk and milk products in both Northern Ireland and Southern Ireland we talked about extending the activities of the SDT to the Irish Republic. These informal talks bore fruit.

Dr. G. Chambers recalls:

"The credit for the spread of the Society to the Republic of Ireland must go to the legendary Martin J. Mullally, then chairman of the Dublin & District Milk Board and of the comparable body in Cork - and member also of An Bord Bainne (the Irish Dairy Board). From the establishment of Bord Bainne in 1962 Mullally had been promoting cross-border contact with the industry in Northern Ireland and also with milk-publicity interests in Great Britain. In this context he became a member of the Northern Ireland Section of the Society and frequently chugged up from Dublin in his huge diesel-powered Mercedes to attend evening meetings in

Belfast or Cookstown, which involved a round trip of over two hundred miles. Soon a few kindred spirits in the Republic followed his example, and this was warmly welcomed at the annual general meeting of the Society in London on 25 October 1965 when reported by the then chairman-designate of the Northern Ireland Section, Henry O'Hara O'Neill, head of the farm advisory service of the Ministry of Agriculture for Northern Ireland and former British Lions prop-forward. Throughout 1966 Martin Mullally, aided by Eammon P McCormick of An Foras Taluntais (the Irish agricultural institute) and other key figures in the industry, laid the foundations for the establishment of a section in the Republic, in close consultation with prominent members of the Society in Northern Ireland; and the decision to proceed was taken at a meeting in Cork in October that year, when the Society was represented by serving President John Glover, President-elect Dr. John Murray, and Secretary Michael Sonn. Deciding on a name for the section proved difficult: if it were to have been called the 'Irish Section' that - in the political climate of the time - might have been embarrassing to the well-established Northern Ireland Section; but if it were not to be called the 'Irish Section' that could have proved off-putting to many potential members in the Republic. In the end Martin Mullally conjured up a solution: envisaging that the section about to be formed would draw its membership mainly from the great dairying counties in the deep south of the Republic - Cork, Kerry, Tipperary, and Limerick - he proposed that it be called the *'Southern Ireland Section'*, leaving the terms 'Eastern Ireland' and 'Western Ireland' for possible future sections based respectively in Dublin and Galway. This proved acceptable all round, and so the inaugural meeting of the *'Southern Ireland Section'* took place in the Silver Springs Hotel in Cork on the evening of 24 January 1967, involving almost a hundred members of the Society. The meeting was addressed by Dr. John Murray of Queen's University Belfast, who was then President of the Society; and by Dr George Chambers, then general manager of the Northern Ireland MMB and chairman of the Northern Ireland Section of the Society. Dr. Chambers in association with Mr James McDowell also delivered a paper on the quality control of ex-farm milk in Northern Ireland, after which the provisional committee, headed by Martin J. Mullally as chairman and Eamonn P. McCormick as honorary secretary, was confirmed. By September 1967 when the first annual general meeting of the section was held at Cahir in County Tipperary and Dr. John Foley of University College Cork became vice-chairman, the membership had grown to around 200; and it was already clear that the initiative of Martin Mullally and Eamonn McCormick would turn out to be one of the most important developments in the whole history of the Society."

Dr. P M Kelly, Moorepark continues the story:

THE FORMATIVE YEARS (1966 - 1980)

"The formation of the Southern Ireland Section was announced to the assembled gathering of the 23rd Annual General Meeting of the Society in London in October 1966. At that meeting, Mr. C. H. Brissenden, speaking on behalf of the then President, Mr. J. Glover stated that there was also "that very interesting development across the Irish Sea." Furthermore, he quoted Mr. Glover as saying that amongst his highlights of the past year "of course, we have organized the Republic of Ireland"...... He also said that he hoped that "they would bear this in mind, as I have no doubt it will, the newly formed and very welcome Section of the Society in the Republic of Ireland goes through the teething troubles which all new Sections are bound to have". At the same AGM, the report by the Chairman from the Northern Ireland Section, Mr. H. 0 H. O'Neill, noted "the progress of the proposed new Section in the Republic of Ireland, and assure all our friends there of our support and encouragement in their venture." A year previously (1965), Mr. O'Neill's report noted that a number (of the Northern Ireland Section's 268 members) "live in the Republic of Ireland" and that "Martin Mullaly, Chairman of the Dublin and District Milk Board, deserves congratulations" for encouraging interest in membership of the SDT. Negotiations leading up to the formation of the Southern Ireland Section during 1965/66 peaked with the visit to Cork in October 1966 by the Society's top delegation consisting of the President, Mr. J. Glover; General Secretary, Mr. M. Sonn and President-elect, former Northern Ireland Chairman, Dr. J. G. Murray, that saw the formation of an ad hoc committee to draw up Section Rules and "appoint a committee to deal with the business and conduct of the Section." At local level, the roles of the "present" members, Mr. M. J. Mullaly and Mr. E. P. McCormick, Dairy Liaison

Officer of the Agricultural Institute, in convening the meeting, circulating all members, and leading a recruitment drive were singled out in particular for acknowledgement. In addition, a letter of encouragement from the Minister of Agriculture in the Republic was sent to the new Section organisers. Anecdotal evidence indicates that there was much discussion during negotiations regarding the form of association that SDT would operate within the Republic. Operating as a branch of the N.I. Section was considered an option but, with a relatively high concentration of potential members in the Munster region, the formation of a 'Southern Ireland Section' affiliated directly with SDT head office offered the best solution in terms of regional balance. Thus, on the 24th January 1967, the inaugural meeting in Cork of the Southern Ireland Section was addressed by the new President, Dr J. G. Murray and heard a presentation by Dr. G. Chambers, Chairman, Northern Ireland Section on *'Quality Control of Ex-farm Milk in Northern Ireland'*. Officers and committee elected at the inauguration of the Southern Ireland Section included Chairman: Mr. M. J. Mullaly; Vice-Chairman: Dr. J. Foley; Secretary/Treasurer: Mr. E. P. McCormick; Committee: Mr. B. E. Booth; Mr. D. Corkery; Dr. W. K. Downey; Mr. J. J. Myers; Mr. D. O'Connor; Mr. D. F. O'Driscoll; Mr. C. O'Leary. Ex-officio: Mr. M. J. Walshe. Later, meetings were held in Cork and Dublin where a presentation on *'A Camera View of the United States Dairy Industry'* was given by Professor R. C. March, Cornell University. Society support was again evident for the fledgling new Section member during September of that year when the new President-elect, Dr. R. J M. Crawford, West of Scotland Agricultural College, presented a lecture on *'Recent Developments in Cheese Manufacture'*. It is worth turning to the report of the Northern Ireland Section to Society's AGM in October 1967 to capture the excitement of the preceding year's activities - Dr. G. Chambers, N. I. Section Chairman, remarked that it was 'one of the most notable in the history of the Society' in that the Section provided a President of the SDT, Dr. J. G. Murray, who was also on hand to guide the setting up of the Southern Ireland Section. In equal measure was the 'appreciation' expressed in the first S. I. Section report by its Chairman, Mr. M. Mullally, for the 'help and guidance' received from the Northern Ireland Committee as well as the 'unstinted assistance' of the President and General Secretary in getting the S. I. Section effectively under way. It was also recognised that 'with a view to the general advancement of the Society's interests, meetings at each side of the Border were enjoying the attendance of members from the other side'."

"By the following year (1968/69), the calendar of events for the Southern Ireland Section included an evening meeting during the month of January in Limerick on the subject of *'Continuous Buttermaking'*, featuring H. Rostern and M. F. Murphy as the speakers. A two-day event followed in May at Kilcoran Lodge Hotel, Cahir (transferred from its original listing at the National Dairy Research Centre, Moorepark due to size of audience) on the subject of *'Ultra-High-Temperature Processing of Dairy Products'*. This event was organised by Dr. A. C. O'Sullivan, the first Head of Moorepark's Dairy Technology Department, ably assisted by Dr. R. F. Murphy of the Dairy Chemistry Department. The speakers included Dr. T. R. Ashton, Dr. H. Burton, D. Shore, Dr. A. C. O'Sullivan, Dr. W. K. Downey and J. Kenny. As one of the first major symposia to be hosted by the newly-opened dairy research laboratories at Moorepark, SDT was to play a significant role in attracting the participation of a wide circle of specialists in this field. It also appears to have been the first time that SDT devoted a major conference event to discuss UHT as an emerging technology. Furthermore, the proceedings of this event were subsequently published in booklet form by SDT under the joint editorship of Dr. T. R. Ashton, Professor E. L. Crossley, Mr. F. O'Connor, and Dr. A. C. O'Sullivan."

"In reports to the Annual General Meeting in 1968, the Society's programme all over Ireland was severely curtailed due to precautions being taken as a result of a foot and mouth epidemic in the UK. The N. I. Section did not appear to be too concerned at the 'gap in our numbers which followed the formation of the Southern Ireland Section' as their membership appeared to be growing very slightly and stood at 266. Meanwhile, membership in the Southern Ireland Section continued to grow and was approaching the 200 mark. It was also noted that the Section was entrusted with working out details for the Society's Study Tour to Southern Ireland during that year. Judging by the following year's AGM reports (President and S.I. Section), it was obvious that the Annual Study Tour to the Section was a tremendous success and that 'generous hospitality' was 'quite a feature of the tour.' The outgoing President, Dr. R. L. M. Crawford, in his address to the AGM, drew attention to the coat of arms of An Bord Bainne (The Irish Dairy Board) that was presented as a memento of the visit to him in Dublin on the opening evening of their tour. In addition, he singled out several Section members: Basil Booth, Joe Dunne, Martin Mullally and Eamonn McCormick 'who helped to make the tour such a success'. With over 60 members of the Society participating, the Section decided to coincide their Annual General Meeting with a visit by the Study Tour participants and an address by the then President, Dr. R. J. M. Crawford. In addition, Dr. J. G. Davis was among the visiting party, and presented a stimulating paper to the meeting on *'Milk Products of the Future'*. The Section report goes on to mention the co-option of Mr. B. Cahill to the Committee for the year 1968-69. Mr. Cahill, who headed up Carbery Milk Products from an early stage, was later to go on to play prominent executive roles in Express Dairies, and in particular, the 'grooming' of young graduates for future leadership position in the dairy industry and other areas of business. A keen supporter of SDT over the years, no doubt Mr. Cahill's presence on the Committee was instrumental in facilitating the organisation of the First Annual Spring Outing to Carbery Milk Products, where the visitors had an opportunity to view 'their magnificent new factory at Ballineen'. The success of the first Spring outing prompted the Committee to organise another event in the following year with a visit to the cheese manufacturing operations of Wexford Creamery Ltd., followed by an afternoon visit to the Soils Research Division of the Agricultural Institute in Johnstown Castle. Attendance at this outing included 20 final year dairy science degree students from UCC accompanied by their professors. Membership of the Section had now reached 240. One of the most well attended meetings during 1969/70 period was on the topic of *'Milk collection in the '70s'*. With many dairy farmers among the attendance of 150, the Section decided for the future to 'open its meetings to dairy farmers when the subject of the lecture was of particular interest to them'."

"By 1970/71 Ireland's accession to membership of the European Community was being hotly debated, and the theme chosen for the Annual General Meeting held in Mallow was *'Prospects for Irish Agriculture, Especially Dairying in the EEC'*. The speakers at that meeting included Dr. Garret Fitzgerald, Economics Department, UCD, and then member (TD) of the Irish Parliament, (Dáil), who was to become Prime Minister (Taoiseach) of the Republic within the space of a few years. With many farmers attracted by the title of the event, numbers reached approximately five hundred and the report to the Annual General Meeting of the Society that year mentions the presence of an anti-EEC picket. The then President of SDT, Mr. Idwal Jones, attended the meeting and at the end of his address thanked the audience in Welsh, English and Irish, for which it is noted that he received 'an enthusiastic applause'."

"Membership within the Section was concentrated largely in the very southern part of the country with smaller pockets in Dublin and along the border counties of Cavan and Monaghan. In order to make up for the weaker membership base in the Dublin area, the Section teamed up with the Agricultural Economics Society of Ireland in jointly sponsoring one meeting which attracted an attendance of 50."

"There were signs of fatigue in the payment of membership fees according to the Chairman's report, and a relatively small number of members in arrears out of a total membership of 250 were threatened with having their names removed from the membership list."

"The Annual Spring Outing that year toured dairy plants (Town of Monaghan Co-op, Killeshandra Co-op and Bailieboro Co-op) along the border counties, and it was noted with great satisfaction that the party included 20 members of the Northern Ireland section and as such marked the first official shared function between the North and the South."

"At a meeting in Cork (November 1971) on the subject of *'The Organisation of Labour and Facilities in the Manufacture of Dairy Products'* it was noted that 'a good meeting' was enhanced 'by the attendance of a lively bunch of dairy science students (amongst whom was the writer) from University College Cork.' By 1972 the visiting President, Mr. P. A. Hoare, and his wife took some extra time out following visits to dairy plants to view the West Cork scenery under the guidance of the then section Chairman, Mr. J. Lyons, himself a native of that part of the country. EEC issues continued to dominate the agenda again that year - *'Increasing milk production to prepare for the advantages of membership of EEC'* was such a success when it was held in Limerick that Avonmore Creameries requested a repeat in Kilkenny during February, 1972 with an astonishing attendance of 500 under the Chairmanship of Mr. J. Lyons. Later, the focus moved from milk production to one of *'Markets for Irish dairy products in the EEC.'* The Annual Spring Outing that year suffered a minor setback due to industrial action experienced by the national electricity supplier (ESB). However, it eventually took place a month later to Hughes Dairy Ltd and HB Ice Cream in Dublin."

"The visit of the President, Dr. J. G. Davis, to the Section in 1972 took place at an event where another eminent dairy scientist, Dr. T. C. A. McGann, National Dairy Research Centre, Moorepark, gave a talk following the business part of the AGM on the subject of *'Centralised Milk Laboratories for Payment of Milk, Breeding Programmes and Quality Control'.*"

"Now that Ireland had signed up to EEC membership, expansion of milk production followed and to some extent the agenda or the topics of discussion at the Section meetings now had a new emphasis. *'Financing Co-operatives'* was the topic of one meeting, while at another event *'Milk Production and Assembly'* was addressed by Prof. T. Raftery (University College Cork) and Mr. M. Fleming and Dr. T. O'Dwyer (Agricultural Institute). In relation to the latter, it was observed that the intensity of audience participation was so great that the Chairman had to bring the meeting to a halt after four hours when the volume of questions showed no sign of abating."

"Visiting researchers and academics to the country in 1973 were targeted as potential speakers and contributors to Section-hosted events. For example, Dr. P. Robinson, New Zealand Research Institute, spoke on *'Developments in Cheese Mechanisation,'* while a Prof. Padberg, Cornell University, USA addressed a joint meeting with the Agricultural Economics Society of Ireland on *'Food Marketing in the Future'.* On the other hand, Section members were also travelling abroad, for example three members from part of an Irish group which went on a study tour of New Zealand in 1972 (Prof. T. F. Raftery, UCC, Mr. J. Lyons, Past-Chairman and Mr. D. Lynch, Dairy Farmer). Their report formed the substance of an evening meeting in Charleville on *'Can Irish Dairying Learn Anything from New Zealand'*?"

"More visiting Dairy Technologists to the country figured on the Section's programmes during 1974. Mr. L. L. Muller, CSIRO, Australia, presented a paper on *'Utilisation of Dairy Products in Food',* while another Australian, Dr. P. M. Linklater, University of New South Wales, presented a talk on *'The Cheese Industry in Australia'.* The Spring Outing in 1994 took place to Avongate Milk Products Ltd, a joint venture processing company owned by Avonmore Creameries and Unigate. The visiting party was welcomed by the Chief Executive, Mr. R. Brennan, and the Unigate Director in Ireland, Mr. R. Cooper. The group was joined by a twelve-member party from the Northern Ireland Section led by their Chairman, Mr. D. McGivern. Later that year, the Chairman (Mr. P. Dowling) and Secretary (Mr. E. P. McCormick) reciprocated by joining the Northern Ireland Section for their Spring Outing. Northern Ireland also figured on the programme of a meeting held in Virginia, Co. Cavan, in March when Dr. D. Stewart, Agricultural Bacteriology Division, Queen's University, Belfast, presented a talk on *'Quality Control and Standards for Milk in Northern Ireland'.* In October 1974, the Section extended its activities to education, when it joined with the Agricultural Institute and the Irish Creamery Managers Association to organise a two-day course on the *'Mechanisation of Cheddar Cheese Manufacture'.* The highly successful event included speakers from Holland, Sweden, England and Scotland and attracted approximately 75 participants."

"The Annual Spring Outing went to Scotland, where it visited Galloway Creamery, Stranraer, and Liquid Packaging Ltd, Dumfries. The group was joined by members of the Northern Ireland Section also. The AGM of the Section featured a paper on *'Dairying in the EEC'* by Dr. T. O'Dwyer, Director of Dairy Products Division, EEC. Dr. O'Dwyer was a member of the Section's Committee shortly after start up, when he was a member of staff of the Agricultural Institute. Topics that were prominent on the programme of the Section during that year included a campaign for increased milk production, pathogens in dairy products and automated methods for bacteriological analysis of dairy products and milk assembly. The highlight in 1976 was the attendance of Dr. George Chambers, both as President of SDT and as Leader of the joint Northern and Southern Ireland Sections Annual Spring Outing to Ballyclough Co-operative

Creamery Ltd. The day's event was rounded off by an enjoyable reception and dinner hosted in the Central Hotel, Mallow, by the Chairman and Management of Ballyclough Co-op."

"In 1977, the intensive range of activities included a paper on *'Automated Methods of Microbiological Analysis'* by Dr. E. D. Campbell, US Food & Drugs Administration. Another visiting technologist in the country at the time, Mr. T. Cheeseman, NIRD, was invited to participate in a seminar on Milk Quality and Metering."

"In the report to the AGM in 1978, it was noted that the Section's AGM was attended by the then President, Mr. S. G. Coton, and that the subject *'A Management Renaissance'* resulted in a 'lively discussion.' The Annual Spring Outing that year went north of the border with a two-day tour, which included Armaghdown Creamery, Banbridge, and Metal Box Ltd, Portadown. A certain cultural hunger seems to have been satisfied on the part of the Section Secretary, Mr. Eamon McCormick, when he reported that a visit to Clanwellan Forest Park helped him to discover the location of the historic 'Dolly's Brae.' Mr. Coton was back in Ireland during the following year as Immediate Past President when he presented the request from the Programme Planning Committee for the Section to host the Society's Spring Conference in Killarney in 1981. The Section's programme also included a visit that year by former President, Prof. J. G. Murray, Queen's University, Belfast, and a central figure in the establishment of the Section some ten years previously. This was his final official visit in advance of his retirement and he took the opportunity to present a paper on *'Dairy Research, Education and Training in Northern Ireland.'* The Annual Spring Tour to England that year included visits to Express Dairies, Minsterley, MMB Research & Development Centre, Crudgington, and Adams Foods Ltd, Leek."

THE PERIOD OF CONSOLIDATION (1981 - 2003)

"The annual programme of the S.I. Section began to follow a typical pattern of seminars/workshops and factory visits. A visit to the Section by the Society's President in office in any year was usually tied in with one of the programme events. The so-called 'farmers' meeting' usually kicked off the winter programme. A willing dairy cooperative usually agreed to host the event and the agenda featured issues of topical interest to farmers. Needless to say, the European Commission's Common Agricultural Policy and its various states of implementation and revision over the years were never too far off the agenda. Gradually, as the industry became more consolidated and commercial sensitivity dominated, factory visits became fewer."

"The 'Students' meeting was, and still is, another regular feature of the annual programme. A jointly organised event with the student-based, Dairy and Food Science Society, University College Cork, the focus is usually on career opportunities in the industry and gaining an 'inside track' from practitioners engaged in their fields of work. Speakers, drawn from different sectors of the industry, usually gave willingly of their time to relate a profile of their own career, and experience for a brief moment what life as a professor in front of student audience might be. Many an anecdote was related as a 'confession' to the attentive audience by an invited speaker about their 'wayward behaviour' during their student days, which contrasted starkly with their prevailing image later as leaders of business and stalwarts of society. After all that effort, generous hospitality was afforded over many years by the local Beamish & Crawford brewery. These 'Students' meetings were always used to promote SDT membership and students were encouraged to join at the earliest opportunity following graduation."

"Seminars were also organised jointly on a number of occasions with other professional bodies, e.g., with the Irish branch of The Institution of Chemical Engineers (I. Chem. E) on *'Powder Handling'* (Cork, December 1986); with Dairy Executive Association/AFT on *'Rationalisation of Milk Sampling and Testing'* (Moorepark and Cavan, February 1987)."

"For the most part, the S. I. Section was singularly responsible for the organisation of seminars and symposia on topics of major significance to the industry. The decline in butter consumption during the early 80s gave rise to industry concerns. A seminar on *'Consumer Attitudes to Dairy Products'* wrestled with issues affecting the market. A keynote address on 'Dietary Fat and Disease' was given by Dr. R. C. Cottrell, British Nutrition Foundation, who also provided a synthesis of the critically important UK NACNE and COMA reports. The steadily evolving structure of the Irish dairy industry figured at various events from time to time, but was given an airing in its own right during a seminar on *'Dairy Co-operatives - Their future'*? (October 1986). EEC milk quotas were beginning to restrict business growth, and Kerry Co-operative had already taken the unique initiative of forming a public liability company (plc). Mr. Michael Dempsey, Deputy Chief Executive and Financial Controller, Waterford Co-operative, gave a good insight into the funding options facing co-ops that wished to expand their businesses. Other panellists on that occasion included Mr. Jim Maloney, Director General, Irish Cooperative Organisation Society (ICOS), and Mr. Joe Rea, the then President, Irish Farmers Association. The Chernobyl nuclear disaster led to small increases in radiation levels even in the outermost regions of Western Europe and an SDT seminar held in October 1987 discussed *'Radiation Levels in Dairy Products due to Nuclear Fallout'* with the help of several national experts (Dr. J. Cunningham, Nuclear Energy Board; Dr. P. l. Mitchell, University College Dublin; Dr. W. Reville, University College Cork; Mr. Michael McElroy, Irish Dairy Board)."

"The Section hosted the Annual Spring Conference on three occasions - Killarney (1981), Limerick (1991) and Kinsale (2000) - the latter doubling as the Society's Millennium Conference which the S. I. Section was uniquely honoured to host. Kinsale, renowned internationally for its 'Good Food Circle' restaurants, proved a success with visitors when the programme allowed much flexibility as regards choice of venue for evening dining. To finish off in style, the weekend-long event concluded with a 'working breakfast' address by Dr. George Chambers on the final (Tuesday) morning before the participants dispersed. Hosting of major events such as the Spring Conference concentrated the minds of the Section, and 'prominent' Section members were sought to lead the respective organising committees. The Chairman of 1981 event, Mr. Eamonn McCormick, took over that position after relinquishing the role of secretary to Dr. Joe Phelan a short time previously. Committee member and former Section Chairman, Mr. Michael Lane, took responsibility at the Killarney event to organise traditional Irish music and dance for the entertainment of guests on the night of the conference dinner. Well-known traditional Irish singer, Mr. Seán O'Sé, was secured as the leading artiste on the night."

"SDT member and dairy equipment business entrepreneur, Mr. Paul Moody, joined the organising committees of both the 1991 and 2000 events as a result of relocating to Ireland and helped with innovative ideas and valuable contacts. It gave Professor John Foley, Chairman of the 1991 Spring Conference Organising Committee, extra pleasure to host the event in his native Limerick City on the 800th anniversary of the city's establishment. An 'old hand' on SDT matters, Professor Foley was Vice-Chairman of the Section at the time of its inauguration in 1967/68 and succeeded Martin Mullally subsequently as Chairman. The theme *'Strategies for Competitiveness'* focussed on how to compete within Europe. Mr. Andrew Dare, Managing Director, St. Ivel, was a keynote speaker who explored a wide range of issues concerning milk marketing in the UK. His anticipation of EC pressures bringing about an end to the role of the UK Milk Marketing Board in 1992 and the postulation of alternative structures gave rise to considerable audience interest and debate. Another topic on the programme which grabbed attention was a talk by Mr. J. Gill, an Equity Analyst with Solomons Stockbrokers, who laid it on the line to those dairy companies planning a stock market flotation that investor expectations were high and only companies like UK-based Northern Foods with its 16% earnings growth should be regarded as a benchmark of what the market expected. Today, the subject of competitiveness is still prominent in SDT programmes, but this time taking in a much wider global context. The social side of the programme included a medieval banquet in Knappogue Castle where guests were invited to 'behave' like the natives of several centuries ago. Diners were entertained with the ceremonial dispatch of a high profile guest, framed for stealing cows, to the castle's dungeon for a little while in between courses. Meanwhile, management of the Limerick Inn Hotel, the Spring Conference venue, and members of the organising committee were struggling to maintain stand-by electrical power supply to the hotel amidst the chaos caused by an industrial dispute occurring at the national electricity supplier (ESB). Tension mounted in advance of the conference dinner as the scarcely adequate generator coped with the demands imposed by the hotel's kitchens. Thankfully, much to the relief of everybody, mains electricity became available for the duration of the dining period. The formalities at that conference dinner included the awarding of honorary life membership of SDT to Eamonn McCormick, having retired a short time previously from his post as Dairy Liaison Officer with the Agricultural Institute (AFT), in recognition of his untiring commitment to the Society both before and since the inauguration of its Southern Ireland Section. While on his feet to express his appreciation, Eamonn very shortly had his audience in raptures while listening to his repertoire of anecdotes and short stories - many of them being very humorous embellishments of personal experiences encountered in his early days as creamery manager in remote parts of rural Ireland. Regrettably, few of these were ever committed to paper. Fortunately, the commemorative issue of the Society's Journal at the time of the 50th anniversary in 1993 did entrap one such tale. In recounting Eamonn's recollection of the 1970 Study Tour of Scotland, the story (p.27) goes that "although the technology is fading from his mind, he well remembers the whiskey increasing in volume and strength as they progressed from the English border up through the Highlands. He remembers sharing a room with a food inspector from Glasgow, who after a particularly heavy night, swung his legs out of bed, held his head in his hands and muttered 'By God, Eamonn, I will be glad to get back to ma porridge,' to which Eamonn replied 'Amen, to that' for he was not feeling the best himself'."

"After the 1991 Annual Spring Conference, it was only with great reluctance that the Section's Committee allowed him to withdraw gradually from its routine business. Finally, the moment of 'letting go' was reached on the 11th November 1994 when a special function was held the 'Two Mile Inn' Hotel, Limerick to mark his retirement from the Committee of the Southern Ireland Section. A measure of the esteem with which Eamonn was held was evidenced by the attendance of 16 of the Section's 20 former chairmen, including some no longer involved in the industry. Chairman, Michael Hickey gave a brief résumé of Eamonn's career, followed by a presentation to him and his wife, Anne, by Prof. John Foley, the senior surviving Section chairman. Prof. Foley was well able to jog memories with his recollections and close association with the Section since its formation. Again, Eamonn did not disappoint with his response on the night as now he was, for once, amongst 'his own' and the protocols and etiquette of formal functions could be put to one side. Musical talent was also in plentiful supply amongst the collection of past Chairmen with the result the function ran into the early hours of the following morning as guests renewed acquaintances and contributed to the party atmosphere."

"Sadly, the almost 10-year gap between S. I. Section-hosted spring conferences was to take its human toll, so that by the time the Millennium Spring Conference (2000) came around, the Organising Committee were proud to remember their departed colleague by inaugurating the **Eamonn P. McCormick Memorial Medal** as a mark of respect for the late Eamonn's service to SDT. The medal was presented by his widow, Anne, to the postgraduate student winner (David O'Sullivan) of the best-presented research poster displayed at the conference. An innovative feature was that the poster competition was confined to dairy and food science postgraduate research students throughout the UK and Ireland that were also SDT members. The Kinsale Conference also had its social moments especially when former, retired SDT Secretary, Mr. Peter H. F. Lee, decided to pay a courtesy call on the event. This was significant because Peter had been closely involved with the local organising committee during the 1981 event, and also in a 'handing over' capacity in the lead-up to his successor, Mrs. Ros Gale, taking control in 1991. Peter held the Conference Dinner audience captive with his descriptive account of participating and winning in the TV programme 'How to Become a Millionaire'."

"Some significant seminars/workshops and conferences were held during the 1980s and 90s:

1980s
'Rationalisation of the Dairy Industry' (Kilcoran Lodge Hotel, 1982), 'Developments in Evaporation and Drying' (Nenagh, 1982) 'Demineralisation of whey' (Fermoy, 1983), 'Export markets for Irish Food' (UCC, 1983), 'Marketing of Dairy Products' (Kilcoran Lodge Hotel, 1983), 'The Yoplait Story' (UCC, 1983), 'The effect of Extended Storage on the Suitability of Milk for Processing' (Fermoy, 1984), 'Quality in the Marketing Mix' (UCC, 1985), 'Consumer Attitudes to Dairy Products' (Fermoy, 1985), 'Equipment Cleaning in the Dairy Industry'

(Portlaoise, 1986), *'Dairy Co-operatives: their Future?'* (Kilcoran Lodge Hotel, 1986), *'Powder Handling Systems'* (with I. Chem E. in Hotel Blarney, 1986), *'Radiation levels in Dairy Products due to Nuclear Fall-out'* (Kilcoran Lodge Hotel, 1987), *'Rationalisation of Milk Sampling and Testing'* (Moorepark and Cavan, 1987), *'Eradication of Animal Disease and Bovine Tuberculosis'* (Adare, 1989)

1990s

*'Marketing Irish Food'* (Cork, 1993), *'The GATT Agreement: Implications for Irish Agriculture'* (Newcastlewest, 1994), *'Fat Replacement in Foods'* (Kilcoran Lodge Hotel, 1994), *'Dairying towards the Year 2000 - European, Irish and U.K. Perspectives'* Mallow and Cootehill, 1994, *'Emerging Technologies in Dairy and Food Processing'* (Fermoy, 1995), *'The Implications on the Irish Dairy Industry of the EC Directive on Packaging and Packaging Waste'* (Portlaoise, 1996), *'Information Technology and the Food Business'* (Kilcoran Lodge Hotel, 1996), Protein Standardisation (Kilcoran Lodge Hotel, 1997), *'EU Dairy and Meat Policy - The Future'*, (Kilkenny 1997), *'Developments in Dairy Policy and Regulations'* (Clonakilty, 1998), *'The Cost of Quality'* (Waterford Institute of Technology, 1998), *'Advances in Food Safety'* (University of Limerick, 1998), *'Computer-aided Design & Analysis of Food Ingredients'* (UCC, 1998), *'Outlook for the Dairy Sector'* (Newmarket, 1998), *'Handling and Disposal of Effluent Sludge'* (Waterford, 1999), *'Advances in Milk Pasteurisation and ESI, Technology'* (Cavan, 1999),

"The programme content of the Section's seminars during the 1990s also reflected a new level of technological evolution impacting on the dairy industry. Genetically-engineered chymosin was commercially realised for the first time and an SDT seminar *'Coagulants for Cheese Manufacture'* (Kilcoran Lodge Hotel, February, 1992) afforded cheesemakers a public forum to hear presentations by its manufacturer, Pfizer Inc. and the experience of an US end-user representative, Mr. E. McLoughlin, Golden Cheese Co., Corona, CA. Cheese was also the subject of further innovation when later that year, a Mr. T. Hori, Snow Brand Co. Japan, presented a talk on the *'Development of the Hot-Wire Method for the Detection of Coagulation during Cheesemaking'*. Developments in new analytical techniques were also captured in a seminar entitled *'Applications of NMR & Immunocapture in the Food Industry'* (Moorepark, May 1992). Nuclear Magnetic Resonance (NMR) spectroscopy, hitherto a research tool confined to research and specialist laboratories, was now a user-friendly technique that could be applied in manufacturing operations for the measurement of solid-fat content and other properties. Fat was also to feature as a theme of another seminar *'Fat Replacement in Foods'* (November 1994) - this time with an emphasis on product formulation demands associated with its substitution. A new generation diagnostic system could detect the presence of over 300 strains of Salmonella within 24h. With emerging pathogens like *Listeria monocytogenes* dominating the headlines in the early 90s, the organisation of a joint seminar with the Institute of Food Hygiene was opportune to review the operation of food hygiene and safety protocols. This event was also supported by trade exhibitors who displayed a range of rapid microbial diagnostic tests."

"A desire on the part of Prof. John Foley to bring to the fore the outputs of academic research in dairy technology conducted at University College Cork to industry led to the planning of a seminar on *'Trends in the Processing, Modification and Usage of Milk Fat and Milk Proteins'* (Blarney Park Hotel, November 1995). Prof. Foley took the opportunity to relate and link the outcomes of research undertaken by many of his recent postgraduate students in a paper entitled 'A Review of Protein Functionality and Methods of Measuring Functional Properties'. Staying within academic circles, the Section accomplished another novel undertaking by organising a seminar on *'Opportunities & Challenges for the Food Industry'* at the Institute of Technology, Tralee to coincide with the nationally organised Science Week in November 2000. This event was groundbreaking also in that SDT awareness was achieved at the Tralee-based Institute now that it had recently commenced training of Food Science students to degree-level qualification."

"Throughout the years, members of the Section participated eagerly in social programmes and tours especially when planned by others, especially the N. I. Section. Travel by boat on a Spring tour to English dairy plants in 1979 was an "outstanding technical and social success" according to Eamonn McCormick's report: "the B & I run from Dublin to Liverpool will never be quite the same again after the revelry generated by 'diluting' Southern Comfort and Paddy Whiskey with Cork Gin, a practice, incidentally, only to be recommended when B & I are doing the driving." The Section's archives, however, do reveal in more soberly terms a tour to Kilkenny that included a visit to Smithwicks Brewery in 1985."

"Like all professional organisations, patronage of the S. I. Section's activities in various forms over the years by various individuals and organisations was critical to the maintenance of a comprehensive programme. With a succession of Section honorary secretaries being employed by the semi-state dairy research organisation, Agricultural Institute (An Foras Taluntais), later to be known as TEAGASC, it was mutually beneficial from a dissemination perspective for both SDT and AFT to operate side-by-side. The latter organisation, along with the Dairy and Food Science Faculty, University College Cork, were always important resources when it came to the selection of speakers from many disciplines (dairy husbandry, economics, technology, chemistry, microbiology) for various SDT programme events. Likewise, the Dairy and Food Science Faculty, University College Cork played an important part in this organisational infrastructure, especially Prof. John Foley as Head of its Dairy & Food Technology Dept. With the future of SDT in mind, Prof. Foley kept an eye out for potential 'officer' material, and was instrumental in encouraging his college colleague, Ms. Maura Conway, to act as Section Secretary for a period during the mid-90s, and the participation on the committee of his newly-appointed College Lecturer, Dr. Alan Kelly. Equipment and other service suppliers kindly underwrote the travel costs of technical experts who were invited to speak at the Section's seminar, and availed of exhibition space at major conference events. Mr. Basil Booth, Chief Executive, APV Desco Ltd (later APV Ireland Ltd) and a founding committee member, took a personal interest in SDT and encouraged his executives to participate actively in its affairs. His manager of APV's Limerick office, Mr. Tom Dwane, was later to play an important role as a member of the organising committee in securing local sponsorship for Spring Conference 1991 and also served as Section Chairman in 1988. Dairy companies supported the Society's work in various ways from time to time. In earlier years, a few companies paid the annual membership subscription fee for its management executives. Financial sponsorship was made available on critical occasions such as the Spring Conference and Millennium events, and numerous industry representatives were elected to the office of Chairman from time to time while enjoying the support of their organisations. Major events organised by the Section were frequently honoured by the presence of Mr. Joe Walsh T. D., Minister for Agriculture, Food & Rural Development, who as a former dairy researcher employed at the National Dairy Research Centre, Moorepark took a keen personal interest in the affairs of the Section and always responded generously to invitations from its officers to address Society seminars."

ANNUAL 'FARMERS' MEETING

As explained earlier, the so-called 'Farmers Meeting' figured annually on the Section's programme, and came to life even more during the 90s largely thanks to the input by former (1972) Section Chairman and well-known Member Relations Manager with Golden Vale plc., Mr. Jerry Lyons. Skilful negotiation was always required to engage willing dairy companies to act as 'host' for this annual event. *'Dairying towards the year 2000'* was the topic addressed by the Minister for Agriculture & Food, Mr. Joe Walsh T. D. and Mr. N. Simms, Irish Dairy Board in November 1995 in an event hosted by Dairygold Co-op. An unique coincidence also occurred on that occasion with the reunion of two former Moorepark research colleagues (the then visiting SDT President, Dr. A. C. O'Sullivan, the first Republic of Ireland-born person to hold that position, and the Minister for Agriculture & Food) on a public platform after many years. Jerry's opposite number in Avonmore (now Glanbia), Mr. Michael O'Reilly was to be a central figure in the organisation of a major Farmers Meeting in Kilkenny that was addressed by the then Minister for Agriculture & Food, Mr. Ivan Yates T. D. along with Avonmore's Managing Director, Mr Pat O'Neill. Newmarket Co-op hosted the 1997 event when Teagasc economists, Dr. T. Donnellan and W. Fingleton outlined their research on the development of the FAPRI model of Irish milk production and marketing. The Junior Minister for Food, Mr. Ned O'Keeffe,

T. D. addressed the meeting also. The Clonakilty-based meeting (February, 1998) on *'Developments on Dairy Policy and Regulations'* referred to earlier was hosted by Carbery Milk Products and its constituent dairies. In addition to the annual economic updates and prevailing market situation, this meeting placed particular emphasis on the increasing necessity for regulatory compliance in matters of hygiene and safety. A 'Farmers Meeting' in Woodlands House Hotel (February 2000) in conjunction with Golden Vale plc dealt with EU milk quota and industry competitiveness issues. Mr. G. Kearns, ICOS explained the New Milk Quota Regime while Drs. Phil Kelly and Donal O'Callaghan presented talks on *'Competitiveness in commodity Dairy Product Manufacturing'* - Strategic Issues in Milk Powder and Cheese Production and Issues of Economy in the Manufacture of Milk Powder and Cheese, respectively."

ACADEMIC CONTRIBUTION

"The Section's dairy scientists and technologists made a significant contribution over the years to the Society's international profile and standing. Leading academic and prolific writer, Professor P. F. Fox, Head of the Dairy & Food Chemistry Dept., University College Cork, was much sought after as a speaker at SDT major events such as London-based symposia and Annual Spring Conferences (Northern Ireland). Both he and others (Profs. J. Foley, D. M. Mulvihill, M. I. Keane and Mr. M. F. Murphy) as well as researchers at the Agricultural Institute (later TEAGASC) such as economist, Eamonn Pitts, and those at its Moorepark-based Dairy Research Centre (Drs. A. C. O'Sullivan, T. C.A . McGann, W. K. Downey, R. A. M. Delaney, J. A. Phelan, P. M. Kelly, J. J. Touhy) contributed periodically to the Journal of the Society with original articles and reviews. Interestingly, the origins of both the Section and the National Dairy Research Centre at Moorepark tracked each other closely and newly-appointed research scientists at the latter very quickly began to present papers of their findings to SDT Conferences during the late 60s and early 70s. In those days, the presented paper usually carried a detailed account of the discussions that followed each oral presentation when published subsequently in the SDT's Journal. Some examples of these 'live' peer review exchanges took place between Dr. W. K. Downey and Mr. G. M. Smith, Dr. T. R. Ashton on 'Physical and Chemical Characteristics of UHT Creams' (UHT Processing of Dairy Products Symposium, Kilcoran Lodge Hotel, 1969) and also between Dr. A. C. O'Sullivan and Messrs. D. T. Shore, C. A. Hall, S. G. Coton on the subject of 'Reverse Osmosis' (Membrane Technology Symposium, London, 1974)."

OFFICERS OF THE SOUTHERN IRELAND SECTION

"Eamonn McCormick, Dairy Liaison Officer, AFT was Hon. Secretary of the S. I. Section at a time during the 1970s when the dairy industry in the Republic of Ireland experienced tremendous expansion following the country's accession to membership of the European Commission. Eamonn had worked his way up through various management positions in numerous organisations such Dublin & District Milk Board, and Irish Creamery Managers Association (later known as the Dairy Executive Association). Consequently, not alone was his powerful networking capability of value to his employer (AFT) in gaining the confidence and support of farming and dairy processing leaders throughout the industry for its embryonic dairy research operations at Moorepark, but it also garnered great interest in membership of SDT. It is appropriate also to mention at this point that Eamonn's brother, the late Gerry McCormick, a former senior executive in Town of Monaghan Co-op and SDT member, provided good regional representation and cross-border liaison while a committee member of the Section for periods during the 1980s. Gerry's boss, Mr. John O'Donnell, former Chief Executive of Town of Monaghan Co-op and Secretary of the company until the time of his death was also a committed SDT member from the earliest days and supported the organisation of many events at the 'northern' end of the Section."

"Dr. Joe Phelan, Head, Dairy Technology Dept at AFT, Moorepark, succeeded as Hon. Secretary ca. 1979-80 and his first task was coordination of Spring Conference 1981, with Eamonn close at hand as Chairman. Joe continued as Section Secretary until 1986 when he took up an offer to work for the Food and Agricultural Organisation (FAO), Rome. During that time, he oversaw the organisation of several significant technological seminars. Meanwhile, Eamonn took up again as Treasurer in 1983 for almost another decade."

"Dr. Phil Kelly succeeded Joe in a similar AFT capacity and was elected Section Secretary in 1986. Like his predecessors, he continued in this capacity for the next 6 years and was in situ for the 1991 Annual Spring Conference of the Society. Fortunately, having been a member of the 1981 Spring Conference committee, he was able to bring the experience of the previous event to bear on the second time around."

"The position of Section Secretary changed hands more frequently during the mid-90s with Ms. Maura Conway and Mr. W. Charteris carrying out the task for shorter periods. Following this, Mr. M. F. Hickey, a senior executive of Golden Vale, brought his considerable talents and experience to the position following his election to Secretary. Institutional support was once again close at hand, with Dr. J. J. Tuohy, General Manager, Moorepark Technology Ltd, taking on the role of Chairman for a number of terms (1998-2000) and also as Treasurer (2001-02). Regular participation by the Section's committee member based in the Cavan-Monaghan region in the usual business routines was always constrained by the geographical distance. However, the commitment of Mr. Tadg Murphy, while based at Lakeland Dairies, to various tasks including that of Treasurer is worthy of mention."

"Over the years, Section Secretaries relied extensively on AFT and TEAGASC secretarial assistants such as Ms. Terry Rogers at the Dublin-based Head Office, Ms. Mary Alice Hennessy and Ms. Claire O'Sullivan at Moorepark who helped with the typing of correspondence, mail-shot 'stuffing of envelopes' and attendance at conference registration desks. This was quite a chore in the days prior to the arrival of word processors, electronic communication and computerisation of office routines. It was a bonus to get the pre-addressed envelopes from SDT Head Office. In one letter regarding the provision of facilities at the 1981 Society Spring Conference in Killarney, the Section Secretary assured the General Secretary, Peter Lee that there was no need for him to bring his trusty typewriter as a 'Mary Alice would be bringing one' that she was 'more familiar' with."

"It is appropriate to wrap up this historical review by posing the question as to what impact did SDT participation have on the whole fabric of the dairy industry in the Republic of Ireland. In this respect, it is worth recollecting the words of Eamonn McCormick at the time of the 50th anniversary "...he feels that the Society can justly claim a significant influence on those major (technological) developments through all its activities, but above all, its Study Tours to countries with advanced dairy industries". Today, open debate of topical issues whether technological, regulatory or economic via seminars/symposia is likely to be of greater importance to members. Another point of significance is that the personal and professional friendships established throughout the dairy industry on the island of Ireland in the course of SDT involvement endured through some of the darker moments in the island's history during the last quarter of the 20th century, and today provide a platform for future growth and business development."

PAST OFFICERS OF THE SOUTHERN IRELAND SECTION OF SDT

| Year 1 | Chairman | Secretary | Treasurer |
| --- | --- | --- | --- |
| 1968 | Mr. M.J. Mullally | Mr. E. P. McCormick | Mr. E. P. McCormick |
| 1969 | Dr. J. Foley | Mr. E. P. McCormick | Mr. E. P. McCormick |
| 1970 | Mr. C. O'Leary | Mr. E. P. McCormick | Mr. E. P. McCormick |
| 1971 | Mr. D. Corkery | Mr. E. P. McCormick | Mr. E. P. McCormick |
| 1972 | Mr. J. Lyons | Mr. E. P. McCormick | Mr. E. P. McCormick |
| 1973 | Prof. E. C. Synnott | Mr. E. P. McCormick | Mr. E. P. McCormick |
| 1974 | Prof. E. C. Synnott | Mr. E. P. McCormick | Mr. E. P. McCormick |
| 1975 | Mr. P. Dowling | Mr. E. P. McCormick | Mr. E. P. McCormick |
| 1976 | Mr. M. Lane | Mr. E. P. McCormick | Mr. E. P. McCormick |
| 1977 | Mr. M. Lovett | Mr. E. P. McCormick | Mr. E. P. McCormick |
| 1978 | Mr. D. O'Connell | Mr. E. P. McCormick | Mr. E. P. McCormick |
| 1979 | Mr. C. Quigley | Mr. E. P. McCormick | Mr. E. P. McCormick |
| 1980 | Mr. B. Milton | Mr. E. P. McCormick | Mr. E. P. McCormick |
| 1981 | Mr. E. P. McCormick | Dr. L. A. Phelan | Dr. L. A. Phelan |

| 1982 | Mr. J. Murphy | Dr. L. A. Phelan | Dr. L. A. Phelan |
|------|---------------|------------------|------------------|
| 1983 | Mr. S. Murphy | Dr. L. A. Phelan | Mr. E. P. McCormick |
| 1984 | Mr. M. Lane | Dr. L. A. Phelan | Mr. E. P. McCormick |
| 1985 | Mr. P. Dowling | Dr. L. A. Phelan | Mr. E. P. McCormick |
| 1986 | Mr. M. Brosnan | Dr. L. A. Phelan | Mr. E. P. McCormick |
| 1987 | Mr. M. Leenane | Dr. P.M. Kelly | Mr. E. P. McCormick |
| 1988 | Mr. T. Dwane | Dr. P. M. Kelly | Mr. E. P. McCormick |
| 1989 | Mr. T. Dwane | Dr. P. M. Kelly | Mr. E. P. McCormick |
| 1990 | Mr. D. A. MacCarthy | Dr. P M. Kelly | Mr. E. P. McCormick |
| 1991 | Prof. John Foley | Dr. P. M. Kelly | Mr. E. P. McCormick |
| 1992 | Dr. P. M. Kelly | Mr. W. Charteris | Mr. E. P. McCormick |
| 1993 | Mr. M. F. Hickey | Mr. W. Charteris | Dr. M. K. Keogh |
| 1994 | Mr. M. F. Hickey | Ms. Maura Conway | Prof. John Foley |
| 1995 | Mr. M. F. Hickey | Ms. Maura Conway | Prof. John Foley |
| 1996 | Mr. T. Dwane | Ms. Maura Conway | Dr. M. K. Keogh |
| 1997 | Mr. T. Dwane | Mr. M.F. Hickey | Dr. J. J. Tuohy |
| 1998 | Dr. J. J. Tuohy | Mr. M.F. Hickey | Mr. Tadg Murphy |
| 1999 | Dr. J. J. Tuohy | Mr. M.F. Hickey | Mr. Tadg Murphy |
| 2000 | Dr. J. J. Tuohy | Mr. M.F. Hickey | Mr. Tadg Murphy |
| 2001 | Mr. M. F. Hickey | Mr. Tadg Murphy | Dr. J. J. Tuohy |
| 2002 | Mr. M. F. Hickey | Mr. Pat O'Connell | Dr. J. J. Tuohy |

Note:

CHANGE FROM SECTIONS TO REGIONS.

Until the AGM in November 2000 there were 13 Sections and several Branches (many were moribund) and therefore reference to Sections and Branches refer to the period before November 2000. After that date the 13 Sections were reformed as 7 Regions, being Scotland, Wales, North & Midlands, South & East, South & West, Northern Ireland and Southern Ireland.

*Spring Conference Limerick 1991*
*Eamon McCormicks After Dinner Speech*

# CHAPTER THREE

## EDUCATION, TRAINING AND RESEARCH

From the beginning, the Society took a keen interest in education, training and research. The Education and Research Committee reported as follows at the AGM on 27 Oct. 1947 (JSDT Vol. 1 No. 2, Jan 1948):

"The Chairman (Professor E. L. Crossley) in presenting the report stated that the Committee had devoted its main attention to the urgent problem of improved educational facilities in dairy technology, in regard to which he was pleased to report a most satisfactory development in the greatly increased interest which had been shown by the Central Milk Distributive Council (CMDC). As a result of this it had been possible to form a joint committee composed of representatives of the Central Milk Distributive Council and the Society of Dairy Technology. The C.M.D. Council representatives had proved most helpful, and with their valuable practical assistance really solid progress had been made. He felt that an expression of the thanks of the Society should be made to the Council representatives for the very strong support and assistance they had given".

"This joint committee had undertaken a very considerable amount of work in order to formulate a draft scheme for the creation of the National Dairy Technological Institute which was one of the Society's aims, and also to study the ways and means whereby it might be brought into being….."

"Firstly, they definitely recommend that the proposed Institute should be established at Reading, whereby, along with other advantages, those of the University and to the National Institute for Research in Dairying would be secured. The University of Reading had shown very favourable interest in the scheme, and he felt able to say without any breach of confidence that, in the event of the scheme materialising, provision could be made for the new Institute in a very extensive scheme of development which the University was undertaking on a completely new site".

"The Committee also recommend that this proposed Institute should be equipped with the necessary classrooms, teaching facilities, and laboratories for teaching laboratory control of dairy operations, and for research work. In addition they recommend that it should also include a working dairy fully equipped for the reception of milk, washing of churns, pasteurisation, bottling, and so forth….."

"The training courses should be suited to the needs of men and women who aspired to the highest positions in the industry, whether managerial, technical or scientific, and the qualifications obtainable at this Institute should reach the highest level, that is to say, a university degree".

"He stressed further another recommendation, *viz.* that the Institute should function solely as a training centre and should not be a commercial undertaking in any form. The milk supplies required would be provided by the trade, and any output of milk products would also be disposed of by the trade".

"The Committee had lengthy discussions on plans and estimates, and, as far as it was possible during the present period of very unstable prices, tentative estimates had been prepared of the probable capital cost. The joint committee had reached complete agreement in principle on this problem of finance, a very vexed one, particularly since a very strong hint had already been received that the trade itself would be expected to contribute. The Central Milk Distributive Council members felt very strongly that any voluntary scheme of securing contributions would fail, but that, on the other hand, in view of the very peculiar structure of the industry at the moment, any compulsory methods such as a levy would raise some very awkward problems. In the opinion of the committee, under present conditions a satisfactory solution to this particular problem could best be devised by the Ministry of Food".

"In view of the progress which had been made, the committee had approached the Ministry of Education to initiate discussion of the project, and subsequently the Ministry received a deputation which presented an agreed memorandum setting forth the considered views of both the Society and the Central Milk Distributive Council. The deputation also expressed the view at the time that the Ministries of Education and Food should accept the major financial responsibility. The Ministry of Education extended a very cordial reception and showed considerable interest in the proposals. It was agreed that, since the Ministry of Agriculture now confined its interest with regard to milk to training in dairy husbandry, the provision of new training facilities in dairy technology was urgently needed. It was also agreed that, when new building was eventually permitted, a Central Institute of the new type recommended should be provided as the most satisfactory solution of this problem".

"After a very full discussion on all the problems which would be likely to arise, the Ministry agreed to give consideration to the precise form of this proposed Institute, and thereafter to indicate the financial contribution which would be expected from the industry….."

"He felt that, from the point of view of the Society, this question of the financial contribution to be made by the industry was rather vital, and now that so much progress had been made it seemed not unreasonable to press for clarification of this aspect".

In 1961 Professor E.L. Crossley, Professor of Dairying, Reading University reviewed progress in the provision of education for the dairy industry. One of the conclusions he reached was that the industry needed a specific course in dairy science and technology (JSDT, Vol. 14, No. 3 1961):

"During the past 15 years dairy education has been the subject of so much discussion that there might be some excuse for the view that there can be little new to say. Yet there can have been few periods in the past which afforded so much scope for argument, prophecy, and sheer guesswork as the present situation presents. For there is a general feeling that something is wrong".

"Outwardly all is well. The United Kingdom has one of the major dairy industries of the world, more than twice the size of those in the so-called 'dairying countries' such as Denmark, Holland, or New Zealand, and indeed there is much cause for legitimate pride. By and large milk production is a profitable activity. Milk is our most important single agricultural product and milk production is the major enterprise on the majority of farms. In 20 years we have increased milk production by 45 per cent, as against 10 to 20 per cent elsewhere, the average yield per cow has risen by over 30 per cent, the labour input per gallon of milk has been halved, and by way of bonus increased beef production has become a major by-product of the dairying industry. The high proportion of 70 per cent of this milk reaches the profitable liquid market with the aid of a processing and distribution system which has become a pattern for the rest of the world".

"Yet this facade conceals a very disquieting situation. Notwithstanding the offer of stable employment in generally clean and attractive conditions with the status of an essential public service, few young people are prepared even to consider a career in the dairy industry…"

"The lack of a five-day week is certainly a difficulty at the floor worker level, much less so in the supervisory grades…"

"The industry faces unceasing political and economic pressure to maintain low prices for milk and dairy products, and as labour is a major element in costs the present situation is a natural result. Because milk is an essential foodstuff, Governments of every political colour in every country do their utmost to enforce the lowest possible price levels at which the industry can be made to operate. In Britain, financial margins at every stage are controlled and held down to limits which discourage attempts to improve career prospects. We also suffer too much from the economic outlook which holds in effect that the only real justification for the dairy industry is to secure the maximum financial return for milk producers. This raises at once the basic fact that by world standards the cost of milk production in this country is far too high; of all the elaborate processes involved in providing the consumer with a bottle of milk or home dairy produce the major element is the cost of producing milk at the farm. So that although our processing and manufacturing industries have cut their costs to a minimum and probably utilise labour and equipment more efficiently than any in the world, they are hampered by the cost of their raw material. Nor is this any reflection upon our farmers. It is true that some farm units are economically too small and present a real social problem; it is also true that the very success of the Milk Marketing Boards has encouraged milk production on some farms where the conditions are unsuitable. But here again our farmers probably do, by and large, use labour more efficiently and adopt mechanisation more extensively than any others in the world".

"We solve the problem in Britain by charging consumers an inflated price - on world standards - for liquid milk in a monopoly market in order to permit the manufacture of additional supplies and to maintain producers' returns at a profitable level. Many economists maintain that the industry should be reduced in size and that we are producing too much expensive milk for manufacture into products which could be imported more cheaply nothwithstanding the high efficiency of our dairy industry. In their eyes <u>dairy manufacture is almost a sinful occupation."</u>

"Our concentration on the liquid market has had, in my view, some unfortunate results which do not enter into the calculations of economists. In the 1930s, the U.S.A. and Britain led the world in dairy technology. Whilst nobody could maintain that the liquid milk trade has no scientific or technical problems, it is broadly true that <u>manufacturing processes are technologically more complicated and require more highly trained people</u>..."

"Liquid milk consumption per head is now almost static and unlikely to increase significantly and unless drastic steps are taken to reduce milk production we must maintain a large manufacturing industry. During the war years this industry was completely protected, but is now about to face the fiercest possible competition, particularly from the Common Market, to combine high quality with low costs of manufacture".

"A new financial outlook is needed between the State, the Marketing Boards, and the dairy trade which will make definite provision for more remunerative careers, and bearing in mind that ultimately milk producers and the dairy trade will sink or swim together. In my view, the most disturbing aspect of the whole situation is that many responsible quarters seem willing to accept the position".

"Personally I am more optimistic because I believe that the industry cannot continue indefinitely as it is. Either it will break loose from its agricultural connections and become a first class technological industry in every respect, or it will become moribund and eventually disappear".

"Many countries would gladly supply all our requirements of dairy products and the liquid milk market may not always be a monopoly. Let us assume, therefore, that we are concerned with scientific and technological education in which traditional semi-agricultural courses have no real place".

"The dairy industry still attaches itself to agriculture in educational matters and has always been provided with special facilities at little or no cost to itself, first by the agricultural societies and later by the State through various channels. Until the recent inauguration of the Reaseheath scheme, the trade has given no direct assistance to education, but there are now signs of a very welcome if belated change of heart. Although dairy education may remain with agricultural education for purely administrative purposes, it is becoming increasingly remote from it and more closely allied to chemical and microbiological engineering. New technological developments and international economics will almost certainly bring future changes affecting both labour and management, probably with a decline of <u>family businesses which are still the basis of the industry</u>. Unpopular as some aspects of recent mergers may be, they do result in more education-conscious organisations, better able to introduce new equipment and to undertake the more extensive research which will be essential in the future".

"Our dairy industry has grown in size and complexity, but increasing mechanisation and amalgamations of businesses have produced fewer but larger dairy plants…. More high-level training is essential, but the provision of teaching facilities becomes ever more costly for a declining number of students. Consequently these facilities must be concentrated at a very few centres, and even then it is doubtful whether it will be possible to continue this special provision for a relatively small industry".

UNIVERSITY EDUCATION

"A proportion of graduates will be essential to the future of the industry. The number of graduates in the dairy industry is very small, but they have contributed to its advancement out of all proportion to their numbers. The arguments as to the merits of pure science and applied degrees (such as in dairying) is rather a sterile one….."

"For research work a graduate training is essential, and usually so in teaching. Dairying graduates generally make the best teachers of dairying, and increasing numbers are being attracted to rural science teaching in grammar schools. The official advisory services also favour graduates and offer attractive careers, although I do not believe that graduates are necessary for this work. Finally, a nucleus of graduates will be essential to provide the skilled professional managers of the future, particularly in the manufacturing dairies".

"Except perhaps in Denmark, this failure to use all the facilities available is evident all over the world and our problems are very familiar elsewhere. Consequently most university courses in dairying are being changed into wider courses in food science or technology. Even in New Zealand - a dairying country if ever there was one - this change is being made. At Reading a food science degree course will start in October 1961, although by way of a last chance the dairying degree is also being retained for the present".

"A general food course should attract larger numbers of students by offering a wider choice of careers; it may well attract more students with a scientific rather than an agricultural outlook, and will better justify the provision of costly teaching facilities. All the basic dairying teaching will remain and in the end the dairy industry may find itself with more and better potential recruits. In any case dairying itself is increasingly becoming one sector of the food industry".

TRAINING FOR OPERATIVES
"However, not all people aspire to become scientists or senior administrators, and indeed the intelligent, skilled people who keep things going day by day on the dairy floor are just as essential. For them the industry itself has now done a great deal by supporting the establishment of a sandwich course at Reaseheath".

"Although this residential sandwich course has proved its worth, it can provide for limited numbers only and not all deserving individuals are in a position to take advantage of it. We must therefore not forget the well-established, part-time City and Guilds Courses held at local centres; these account for as many dairying students as all the other courses combined. The industry has been well served by them since 1930 and our own Society has made strenuous efforts to inaugurate more courses. Some employers support them strongly and in the past have provided transport for students in outlying rural areas. No educational standards of entry are required and there is a choice of courses and options which permits the student to study the particular branch of dairying in which he is engaged during the day, whilst no cost is involved apart from a purely nominal fee. The courses suffer from the disadvantages of all courses which combine spare-time study with earning a living, and the general introduction of day release would be a big step forward…."

"Unfortunately the general quality of students has deteriorated and it is clear that the industry is drawing an increasing proportion of new entrants from the lower levels in the schools".

NATIONAL DIPLOMA IN DAIRYING (N.D.D.)
"I have left the N.D.D. course until last because I feel very puzzled and uncertain about its future. It occupies an intermediate place between degree courses and vocational training and is the oldest professional dairying qualification; for many years it was the only one. It has a long and useful tradition and is now the only course for which applications greatly exceed the places available, partly because the number of teaching centres has been reduced. It is quite frankly an agricultural course with a major emphasis upon milk production, but contains enough of the elements of dairy science and technology to give it a general appeal and provide a fair choice of appointments. The only careers which are really closed are research appointments and the higher levels of laboratory work. Entry standards are imposed but these are much easier than for degree courses and well within the ability of the average grammar school pupil. Unfortunately this means that most present-day students are taking the course because they cannot obtain a university place owing to their deficiencies in basic science. Nevertheless they are intelligent, able people who should occupy a very useful place in the overall structure".

64

"The course has now reached a completely unsound position and those attempting to teach it with an utterly impossible task".

"In only two years the student is supposed to cover both dairy husbandry and dairy technology, and to acquire a knowledge of dairy science without a real understanding of elementary basic science. In practice no college except Auchincruive makes any serious attempt to teach the technology. Nevertheless, the courses are hopelessly overloaded and there is a strong tendency to memorise an impressive superstructure of useful facts without a real understanding of the foundations - the exact reverse of the university approach. These good, adaptable, useful people are left with no time to think for themselves and must rigidly adhere to a strictly defined syllabus".

"In the eyes of the dairy trade the N.D.D. has lost much of its former value, and incidentally it also includes much less practical work than in the past. Something will have to be done about it, but what?".....

"There is no doubt that the agricultural content of the N.D.D. should be reduced and the science content increased".

What the industry really needs is a specific course in dairy science and technology. But we have already tried this unsuccessfully with the N.D.D.T; which failed partly because it offered only a narrow choice of careers in a not too attractive industry, and partly because it was educationally unsound. Once again the mistake was made of imposing too advanced a standard in too short a time, on the basis of inadequate entry requirements. It is really grossly unfair to both student and teacher to cram science and technology into an individual who is really only interested in animals and wants an outdoor life, especially if at the same time he has done no physics and has no knowledge of mathematics beyond simple arithmetic".

"However, supposing that the industry becomes more attractive, as it will, and reaches a position of being able to require better standards of entry, there remains the difficulty that few agricultural colleges would be able to afford the necessary facilities or be interested in a form of teaching outside their scope. In addition a technological course might reduce student numbers to an uneconomic level (particularly by discouraging women), although this could perhaps be counterbalanced by an alternative course in laboratory control. There is a growing opinion that the real place for these courses may be in the colleges of advanced technology and not in the agricultural colleges. The National College of Food Technology springs to mind at once, although at present this College maintains entry standards which are little below those for degree courses. Ultimately these dairy courses will probably be absorbed into general, wider courses of food technology. In any case the colleges of technology are designed, staffed, and equipped for their special type of teaching. Such a change would free the agricultural colleges for their proper function and perhaps encourage them to develop first class milk production courses which would be invaluable to the advisory services and the rising generation of modern farmers. Agriculture is rapidly becoming an individual technology too".

"Our present generation of young people is equal in intellectual ability and intelligence to their predecessors and there is much fine material amongst them, although vast social changes have produced some striking changes of outlook. They are frankly more money conscious, and more interested in good financial prospects by say the age of 35 than in nebulous distant wealth at 55. In an age of full employment they are not particularly concerned with security, once the strong card of the dairy industry".

In 1965 Miss Mary Corbridge gave a paper to the North Eastern Section entitled 'The Dairy Industry and the Rising Generation' (JSDT Vol. 19 No. 3 1966). In it she described the educational work of the then National Milk Publicity Council.

"In 1954 the National Milk Publicity Council was re-established and began to develop campaigns to publicise milk and milk products. Initially these were directed through various forms of advertising and public relations activities, but in 1956 a programme was introduced whereby entry into the world of pure education was possible".

"As the programme developed, contact with schools, educational establishments and organisations widened and strengthened. This progress led to enquiries of many kinds outside the scope of the original project. Among the most important have been those concerned with careers. One of the early efforts, to make school children aware of the wide range of opportunities which the dairy industry can offer, was the production of a school project book entitled *'Don Dawson Reporting on the Dairy Industry'*. Thus, many thousands of children in secondary schools were indirectly introduced to many jobs which they had not associated with dairying".

"Another early publication was a leaflet distributed at exhibitions and youth employment offices as part of the June Dairy Festival activities. It was soon realised that such a leaflet had a wider use and it was revised and re-issued to become the careers leaflet - *'Careers in the Dairy Industry'* - which is widely used at the present time. The industry became increasingly aware of the need to recruit the right quality staff who would receive proper training and the Education and Research Committee of the Society of Dairy Technology, in co-operation with the Joint Committee of the Milk Marketing Board and the Milk Distributive Council, produced a draft for a careers booklet. This was eventually published by the National Milk Publicity Council on behalf of the Joint Committee under the title *'Careers in a £1,000,000 a-day Industry,'* and approximately 50,000 copies were distributed in four editions".

"It was subsequently decided that a completely new presentation was required to meet the need for increased knowledge of developments in the field of dairy education. Thus, a second careers booklet - *'Opportunities in the Dairy Industry'* - was produced. The design and presentation met with considerable praise and a second edition was published within a year. As the interest of schools grew and the organisation of careers' meetings and talks developed, so it was found impossible to meet all the requests with personal visits. At the same time, careers advisors in schools began to make considerable use of films in order to give their pupils as much information as possible".

For this reason, the film *'Careers in Dairying'* was produced and became one of the more popular films in the National Dairy Council's library..."

"It must be appreciated that however good a film is, and the National Dairy Council's careers film has received high praise, it is no true substitute for personal attendance. Every effort is being made to encourage and strengthen personal ties between schools and the dairy industry, and in this field members of the Society of Dairy Technology have always played a most important part. Normal channels of publicity have not been overlooked and articles dealing with careers have been arranged as editorial features in several magazines. Information has been provided to authors of books and producers of radio and television programmes while advertisements have been placed in a number of careers publications such as the annual *'Careers Guide for Young People'* issued by the National Union of Teachers and *'Opportunities for School Leavers'.*"

## EDUCATION

Education continued to attract much attention in the Journals. It was sadly also an area in which promising developments on technical education were stifled by zealous reformers with little knowledge of the needs of the industry. S. A. Hands (one of many to write about education) of Seale Hayne College reviewed the partnership between colleges and industry and examined the role of the Dairy Industry Training and Education Committee (DITEC) (JSDT Vol. 36. No. 4 October 1983).

"How does industry ensure it gets the trained people it needs? Colleges that specialise in dairy technology are fortunate in the partnership with industry that has been developed over the years. Certainly others are envious of our relationship with the Dairy Industry Training and Education Committee. (DITEC); no comparable organisation exists in the non-dairy food sector. DITEC, whose membership is drawn from all parts of the industry, seeks to establish the needs of industry. These requirements are then discussed with the colleges in the Colleges Sub-committee. Here the views of industry are put by industrialists. It serves, too, to keep the professional educationalist's feet on the ground".

"The Committee has had a great deal of work to do in recent years with the formation of the Technical Education Council (TEC). TEC is responsible for courses, now called programmes, for technicians; in practice, this means courses formerly known as ONC, OND, HNC and HND

(Lawton, 1979). TEC was established with the aim of redirecting education to serve the needs of industry. The sector and programme committees are made up largely of industrialists. Programmes are not to be approved unless they can be shown to produce technicians who are prepared to meet identified job functions. Neither are they approved if they do not clearly identify the objectives of the learning process".

"So now we have Diplomas and Higher Diplomas developed on these lines. Of course, forward-looking colleges did not await the advent of TEC before adopting this approach. In the case of the dairy colleges, this change has been a formalisation of existing practice".

BY DR J ROTHWELL, DEPARTMENT OF FOOD SCIENCE, UNIVERSITY OF READING, REVIEWED THE FACILITIES FOR EDUCATION IN THE MILK INDUSTRY.
(JSDT Vol. 32 No. 2 April 1977)

The teaching of dairying and, later, of food science and technology at Universities in the British Isles dates back nearly a century. To understand the present situation properly it is helpful to consider its development briefly.

The real beginning was probably in 1882 when the British Dairy Farmers Association set up an organization for dairy teaching, although even before this travelling dairy schools had been organized by, among others, the Bath and West of England Agricultural Society (Crossley, 1957). In 1888 the BDFA established the British Dairy Institute (BDI) in Aylesbury.

The story on the University side began somewhat earlier with the founding of classes in Reading for the teaching of art in 1860 and for science in 1870, the two uniting in 1882 as the Reading School of Science and Art. University Extension Lectures were arranged in Reading in 1885 by the Oxford Delegacy and Christ Church, Oxford, offered special assistance to further these. An amalgamation between the University Extension Centre and the Reading School of Science & Art in 1892 resulted in the formation of the Reading University Extension College.

At this new College agricultural teaching was one of the original activities. In this context, dairy courses were provided, including one for a dairy teacher's certificate, and travelling cheese and butter schools were set up.

For some time, similar classes by the BDI and the College ran in parallel, but after discussions the BDI was transferred from Aylesbury to Reading, where new buildings erected by the College were leased to the Institute.

In 1902 Reading University Extension College became University College, Reading, with arts and science students reading for external degrees of London University. Agriculture and dairying students were awarded joint Oxford and Reading diplomas, but not degrees at first, although, eventually, external degrees in agriculture were introduced by London.

From this time, and until 1926, the main dairying qualification was the National Diploma in Dairying, taken by students at County Farm Institutes and Agricultural Colleges in addition to the BDI. The NDD continued with a short entry into technology with the NDDT in the early 1950s

until the OND and HNDs in food technology superseded it.

However, in 1926 Reading University College was awarded its University Charter, and became able to award its own degrees, among which for the first time in any University in the British Isles was a BSc in dairying. The course extended over three years and became the first degree in any food subject in the United Kingdom. The first Professor (E. Capstick) was appointed in 1938.

Despite all the problems of war time, the BDI kept going, but in the late 1950s, and even more the early 1960s the number of students wishing to read for a degree in dairying declined seriously. A similar drop in numbers occurred at Nottingham University where degrees in dairying had been offered as an option of the agriculture degree from 1955. A course for the NDD had been available from about 1945, consisting of one year following a degree in agriculture, based on the then Midland Agricultural College at Sutton Bonnington. The University College of Wales at Aberystwyth also had a department of dairying but here students were not awarded degrees, taking the College Diploma in Dairying and the NDD.

All these departments suffered similar problems of falling numbers. This was due to the inability of the dairy industry to project a sufficiently attractive image; lower financial rewards to newcomers were coupled with the impression (not really justified) that a dairying course limited scope and opportunities through over-specialization. At Aberystwyth and Sutton Bonnington the final and nearly fatal stroke was the refusal of the University Authorities and the University Grants Committee to allow teaching to be limited to diploma level only, and at Reading the last students were accepted for the BSc (dairying) course in 1965.

The dairying courses at Nottingham and Reading were merged into the new food science courses which have since then attracted increasing numbers of students.

However, these were not the first degree courses in food science. At Strathclyde University, Department of Food Science & Nutrition, food science degree courses were developed from the long established bakery and associateship courses of the former Royal Technical College. The Leeds University Procter Department of Leather Science changed to Food and Leather Science and offered courses in food science from 1962 onwards.

Another very interesting development has been the National College of Food Technology, now situated at Weybridge and part of the Faculty of Agriculture and Food of Reading University. The College has been in existence since 1951, when it evolved from the Smithfield College of Food Technology, which had been earlier the Smithfield Meat Trades Institute. The College now offers degrees not only in food technology, but also in biotechnology and biochemical engineering.

*Undergraduate courses* in food science or technology are offered by the University departments shown in Table 1.

#### TABLE 1
**University departments offering food science or technology degrees**

*Single Subject*
Belfast, Queens
   Faculty of Agriculture & Food Science
Cork, Irish Republic
   Faculty of Dairy & Food Science
Leeds
   Procter Department of Food Science
London, Queen Elizabeth College
   Department of Food Science and Nutrition
Nottingham
   Department of Agricultural Biochemistry and Nutrition
Reading
   Department of Food Science
Weybridge (Reading)
   National College of Food Technology
Surrey
   Nutrition/Food Science/Catering
Strathclyde
   Department of Food Science and Nutrition
In addition a CNAA degree in Food Science is offered by the Polytechnic of the South Bank, London.

*Two Subject*
Leeds
   Food Science combined with another subject; physiology biochemistry or microbiology.
London (Queen Elizabeth)
   Food Science and Management
Reading
   Food Science/Economics
   Chemistry/Food Science

Although the courses listed in Table 1 have many similarities, there are some major differences. Cork is alone in offering a BSc in dairying for example, and the methods of teaching vary from institution to institution. Nottingham provides a good proportion of meat science, while at Reading dairying and cereal science are taught as commodity courses.

The major differences are that NCFT, Polytechnic of the South Bank and Surrey courses are all 'sandwich' courses where students spend one year in industry, making a total of four years spent on the course.

The honours degree in Food Science at Leeds University takes four years, the others three, although Strathclyde has a basic four year scheme with exemption from the first year for suitable candidates.

However, the basic pattern of teaching is usually very similar, and the courses can be expected to include subjects which may be grouped as shown in Table 2.

#### TABLE 2
**Grouping of courses in a typical food science curriculum**

1. Basic courses in chemistry, biochemistry and microbiology.
2. Applied biochemistry and applied microbiology relating directly to the composition of foods and to changes during storage, processing and distribution.
3. Human nutrition and nutritional evaluation of foods.
4. Quality appraisal by physical, chemical and sensory methods; food hygiene; food standards and legislation; statistics.
5. Courses in the principles and practice of food storage, preservation and processing.

In addition to these the NCFT course includes mathematics and management.

Lectures and tutorials on all these subjects are supplemented by laboratory courses, including work with pilot plant. More than half of the course time is normally allocated to practical work, and experience in the food industry during vacations is usually required in addition. The dairy industry as a whole gives considerable help in the placing of students, both for sandwich courses and for the two to three months normally available in summer vacations.

### Relevance of food science degree courses to the dairy industry

Over the past eight years 22 Reading BSc graduates in food science, out of a total of 123, joined the dairy industry for their first posts, and this is still continuing.

NCFT students are also being recruited for the industry, and DITEC offers scholarships for suitable students on the degree course in food technology.

The industry itself is the major food industry in the United Kingdom, and has been and still is diversifying into further branches of the food and catering industries. A student who combines a good grounding in basic sciences with the ability to apply this to food processing and quality control is therefore more likely to be a useful member of a large organization from the moment he arrives than someone with a pure science degree. One of the major benefits claimed for a food science or technology degree is that it is an integration of many sciences and has the advantage of the incentive of a final product based on the application of all these sciences. To complete the picture it is necessary for the industry to provide the motivation in the form of job satisfaction and adequate financial reward for what can be a seven-day-a-week job for those in higher management. It is to these posts that the degree students – at first in dairying, and now in food science – need to be attracted and they in their turn should in due course be able to rise to the challenge the posts provide.

## MSc courses

Several of the departments which provide BSc courses in food science and technology also organize MSc courses. Most of these are relevant to the dairy industry, and they include the following:

1. MSc Food Engineering – Leeds.
2. MSc Food Technology, Process Engineering, Food Microbiology and Food Quality Control – NCFT.
3. MSc Food Science (Dairy Science); MSc Food Science (General) – Reading.
4. MSc Meat Science – Nottingham.

These courses are mainly conversion courses and entry to them depends on the successful completion of a first degree in, say, chemistry, agriculture, applied biology, or some other appropriate science. Some, but not all, are courses approved by the SRC for grants, and suitable candidates for the Reading MSc (Dairy Science) may be considered by the Dairy Industries Technical Education Committee for grants; one student is being assisted in this way for this academic year.

## Higher degrees by research

All of the above departments have active schools of research. Suitably qualified students are thus able to understand studies in depth leading to the degrees of MPhil or PhD, provided appropriate facilities, supervision and financial backing are available.

The topics of research range widely; some have more and some less relevance to the dairy industry. Several projects are sponsored by the industry itself. There are several ways by which this may be done:

1. Science Research Council CASE (cooperative awards in science and engineering) research studentships.

2. Other research studentships.
3. Post-doctoral fellowships.
4. Undergraduate and post-graduate taught course projects.
5. *Ad hoc* investigations.

1. *SRC CASE research studentships.* For these a minimum of three contributions, by the sponsoring organization, of £300 are made to match the SRC contribution. In addition the student is supported for three months in the industrial organization. This would probably cost about £400, making, at present, a total cost to the organization of about £1,300. For this the firm obtains three man years of work by a graduate, with the use of university facilities and a university supervisor also.

Regular reports on the work have to be made, and a thesis is finally produced. There are some disadvantages from the industrial point of view, of course, including the fact that the work is not under the complete control of the firm, and that preservation of confidentiality is difficult. Publication of the work is essential from the academic view point and a delay of two years after completion of the work is usually the maximum that is acceptable. Results may not be so speedily obtained as when the work is carried out within the firm's own organization.

2. *Other research studentships.* These are wholly sponsored by the firm and would probably cost between £4,000 and £5,000 per annum.

3. *Post-doctoral fellowships.* These are more expensive still, and could cost between £9,000 and £10,000 per annum. For this the firm can expect a more mature and a more direct approach without the commitments of working for a higher degree.

4. *Undergraduate and post-graduate taught course projects.* Final year students in Food Science departments are normally expected to carry out a literature survey from which proposals are made, after discussion with a supervisor, for a research project of 10 weeks duration (about 8 hr/week).

Suggestions for topics for these would be welcomed, and the cost would normally be limited to materials.

Similarly, students on taught MSc courses are expected to spend at least two months in an industrial situation and a full report on the work is submitted to the examiners. For this purpose, some students have carried out short research projects which proved to be of appreciable value to the organization concerned.

5. *Ad hoc investigations.* All the departments have laboratory and pilot plant equipment which can be used for *ad hoc* investigations and, although at times the availability of such equip-

ment may be limited, discussions of requirements are very welcome.

It will thus be seen that the departments are far from being 'ivory towers' of detached learning. Indeed they depend to a very great extent on liaison with industry for progress in their teaching and research. Contact with industry and discussion of its problems are essential parts of technological education.

## CONCLUSION

The students of food science and technology being trained at the existing University Departments fill many of the requirements of the food industry. Some of these – a small proportion – find their niche in the dairy industry, and at the moment the students do not appear to have undue difficulty in obtaining employment.

However, it must be queried whether the dairy industry is obtaining sufficient students to meet the needs it will have in 5 or 10 years time. Some authorities go so far as to say that graduates are not really required for production management or for quality appraisal and control. What is perhaps forgotten is that at present there still remain some people in the industry who have been in it all their working lives and have 'grown up' with the ever increasing size and complexity of creameries and company organization. In their way they came to understand it and to cope with it.

New entrants must at least have something of a similar exposure before they can be really useful. They must also have full motivation and interest in the job, and have to realize that in many cases they may be situated far from a large town.

To help in most of these aspects adequate liaison between department and industry is essential (as has already been pointed out for higher degree research work). In particular one major problem looms very large – where are the teachers, at an adequate level, of dairy science and technology to come from in the not too distant future? Here again industry can help. Some assistance can be given with teaching of special aspects much as cheesemaking, on short courses, but soon there will be a shortage of good cheesemakers too. In any case it is essential that there should be an adequate foundation of fundamental scientific knowledge and of knowledge of the dairy industry to build on and this can be obtained via a degree course in food science and technology, coupled with sandwich vacation experience in the industry – but further training in the details of operation of, say, any one particular creamery or dairy company will be required in the industry itself.

The author wishes to acknowledge with thanks considerable help from Professor H. Nursten in the preparation of this paper.

## REFERENCE

Crossley, E. L. (1957) *Journal of the British Dairy Farmers' Association*, **61**, 53.

In 1966 Dr. E. J. Mann, Present Chairman of the Technical & Editorial Board, reviewed international and national aspects of dairy research and development at a meeting of the Society (JSDT Vol. 19 No. 4 1966)

"Before I define the vast range of subjects covered by the term 'dairy research,' let me make it quite clear that, unlike some people, I find it very difficult to distinguish a clear dividing line between dairy science and dairy technology".

"Good recent examples of the interdependence of science and technology in dairy research are the development of two continuous cheesemaking processes, one in the Netherlands and one an Anglo-French effort, both of which are based on the finding by a British biochemist, Dr. Berridge at the NIRD, that rennet action proceeds in cold milk without coagulation of the milk, but that such milk coagulates instantly on heating. Similarly, work in progress at a number of Dairy Research Institutes on the elucidation of the nature of cheese flavour requires close collaborative work between technologists, engineers, bacteriologists and biochemists".

"Having, I hope, illustrated clearly that dairy research embraces both science and technology, I want to take you briefly through the large number of subjects which, to a greater or lesser degree, are covered by the field of dairy research: the breeding of cattle for milk production; feeding and metabolism of dairy cattle; physiology of milk secretion and techniques of milk production; technology of market milk processing and dairy products manufacture and packaging; engineering and equipment; economics of milk and milk products production and distribution; nutritional aspects of milk and milk products; bacteriology of milk and milk products manufacture; veterinary and public health aspects; and, last but not least, the increasingly important field of chemistry and physics in relation to milk and dairy products...."

"Because dairy research can encompass so many disciplines, it is not surprising to find that the location for such work is not by any means confined to dairy research institutes".

"However, in spite of what I have just said there is little doubt in my own mind that the dairy research institutes throughout the world, including the 40 North American universities with food or dairy departments, where research is carried out, provide the solid background of true dairy research".

"Here, I have been fortunate enough to come across an excellent survey on this subject, carried out by Professor Frank Kosikowski of Cornell University on behalf of FAO and published in the June, 1965 issue of the *Journal of Dairy Science*. Professor Kosikowski obtained information for his survey from 113 dairy science and technology research centres in 25 countries and found among other things, to his surprise, that nearly three quarters of the 108 FAO member countries were entirely without dairy science and technology research facilities, this group including all the Latin American countries. Of the 113 research centres, 76 used the English language for publication which, incidentally, ties up with our own findings in the production of *Dairy Science Abstracts* in which approximately 55-60 per cent of papers abstracted are in the English language...."

"Returning to Professor Kosikowski's survey, he found that in 67 per cent of the centres research activities were deliberately oriented to strike an equal balance between practical and basic research. The dairy products receiving the greatest attention were liquid milk and cheese, dried milks and sterilised milk; concentrated milks appeared to receive the least attention. While the majority of research centres in Europe and Oceania were active in butter research, this subject did not command much interest in North America where the consumption of butter has decreased markedly during the past decade....".

"Since the NIRD at Shinfield, Reading employs about 130 professional staff, a simple calculation reveals the astonishing fact that <u>one out of every 10 professional dairy research workers throughout the world is employed by the NIRD</u>".

"The prior formal academic training of dairy research staff showed a relatively even balance between the 544 scientists receiving their training in the basic disciplines such as chemistry, physics, bacteriology and nutrition, and the 532 scientists receiving their prior training in the more applied areas of dairy technology, dairy chemistry, dairy microbiology, and dairy engineering. The largest number of staff members received their formal academic training as dairy technologists, followed in descending order by chemists, biochemists and dairy chemists...."

"Now, so far I have talked about general and international aspects of dairy research but this is a British Society of Dairy Technology and I am sure that you would like me to say something about past, present and future trends in dairy research in this country..."

"It would appear that the foundation stones to modern dairy research were laid early in the eighteenth century when the first agricultural societies were established, and much of the credit for these early efforts must go to these societies, their staffs, and to enterprising individual cattle improvers...... During the nineteenth century, there was still very little concerted Government action to initiate and co-ordinate dairy research in this country and what activity there was in this field was stimulated by the Royal Agricultural Society of England, the Bath and West Society and the British Dairy Farmers' Association. Voelker, the consulting chemist to both the RASE and the Bath and West Society, figured prominently in dairy research in the middle of the nineteenth century.... Many of his papers in the early agricultural society journals still make fascinating reading today. Even his death in 1884 was not as tragic a loss to the agricultural societies as it might have been because he was promptly succeeded by his son, who carried on the good work right into the twentieth century".

"It is probably true that organised dairy research on a national scale was born with the passing of the Development and Road Improvement Fund Act in 1909. This was followed, three years later, by the establishment of the <u>National Institute for Research in Dairying (NIRD)</u>. This institute which commenced its activities with a staff of three - a chemist, a bacteriologist and a laboratory assistant - in 1912, has grown rapidly during the last 50 years and is today, with its staff of 450, the largest, best known and most widely respected dairy research institute in the world. A little later, in 1931, the Hannah Dairy Research Institute was established in Ayr, Scotland".

"The early work at the NIRD was orientated towards clean milk production, the most pressing problem of the day. Although the interests of agriculture and the fundamental sciences, rather than the more applied problems of the dairy industry, have tended to dominate the work of the

Institute as a whole, the last 10-15 years have seen a gradual change in emphasis towards more applied dairy research. Batch processing has been largely replaced by continuous processing, and the traditional concept of an industry whose main purpose has been to supply liquid milk for consumption to a rapidly growing consumer market, leaving only relatively small quantities of 'surplus' milk for manufacture into dairy products, has undergone a gradual change..... It is, nevertheless, an established fact that during some of these years considerable quantities of dairy products were manufactured (last year, the figure was nearly 800 million gal) and Great Britain has emerged as a major dairy products manufacturing country. Furthermore, during the last few years, a number of products new to the British market, have been introduced successfully, notably cottage cheese, fruit-flavoured and other special yoghurts, cultured cream and instant milks. Will this trend continue?"

"With such a large quantity of milk available for manufacturing, it is to be expected that, apart from the traditional dairy products, more attention will be given to the newer types of dairy products as far as this country is concerned. However, the newer methods of selling such products in attractive packages from the shelves of self-service stores are opening up new possibilities in this field and I believe that independent research on improving the quality of such dairy products, as well as on the development of new types of dairy products, is highly desirable".

"However, be that as it may, diversification in manufacture and adaptability to rapid milk supply are likely to be the keynotes to the successful future development of the dairy industry and these will, undoubtedly, also be reflected in British dairy research".

"Already, the National Institute for Research in Dairying has made some very notable contributions of direct interest to the dairy industry in the fields of mastitis eradication, milk metering and bulk tank design, automatic milk analysis, aseptic bottling of market milk, continuous Cheddaring, and the scientific basis of cheesemaking, while the Hannah Dairy Research Institute has made valuable contributions in the fields of milk analysis and dried milk research".

"I have no doubt whatever that the British dairy research institutes will play an ever increasing part in providing the scientific and technological data necessary for healthy progress and development in dairy industry...."

POSTER COMPETITIONS

The Society has been always aware of the importance of interesting young people in the dairy industry. To this end poster competitions were held for many years, usually as part of a spring conference. At the Spring Conference in Harrogate in 1990 "there was a very good response to the Young Persons Poster Competition and all the participants are to be congratulated on their very high standard of work. The judges had a difficult task to find an out and out winner, and because of this it was decided to award two first prizes of £100 and two second prizes of £50".

With Dr John Gordon, President of the Society of Dairy Technology, in the photograph are the prizewinners. From left to right: Susanne Cherry, Brackenhurst College (Joint First Prize); Dr Gordon; Collette Casey, University College, Cork (Joint First Prize); Alison Widgery, Seale Hayne College (Joint Second Prize); and Alison Findlay, University of Bradford (Joint Second Prize). (JSDT Vol. 43, No. 3 1990).

*Spring Conference 1990-Crown Hotel, Harrogate. Young Persons Poster Competition.*

*Dr. A. C. O'Sullivan with the winners of the Student Poster Competition at the Autumn Conference, Bistol in 1994.*

*Mr E MacMillan-Scott, MEP addressing the Society's Spring Conference at Harrogate in March.*

*Mr Ian Hague, Chairman of the North Eastern Section of the Society, addressing the Spring Conference at Harrogate in March.*

The results of the Student Poster Competition 1994, held in conjunction with the Autumn Conference and AGM in Bristol, were reported in the Journal (JSDT Vol. 47 No. 4 November 1994).

The judging this year was carried out by members of the Society's Technical Committee, and although the response from the Colleges was disappointing, the standard was, as usual, extremely high, and two first prizes were awarded.

| 1ST PRIZE | 1ST PRIZE | 2ND PRIZE |
|---|---|---|
| Jean Burke | Julie Daniel | Ian Ross Carmichael |
| Seale Hayne Faculty | Seale Hayne Faculty | Seale Hayne Faculty |
| University of Plymouth: | University of Plymouth: | University of Plymouth: |
| *'Comparison of mastitis on first lactation heifers and third to fourth lactation cows'* | *'Fortification of a fermented milk (yoghurt) with deodorised fish oil'* | *'The use of vegetable oils in 'Camembert Type' cheese analogues'* |

3rd PRIZE
Kastanas/Ravanis/Lewis/Gradison
Gradison
University of Reading:
*'Effects of raw milk quality on fouling at UHT conditions'*

RUNNER UP
Jerapapon Ruangwijit
Cannington College:'
*'To advertise the importance of Dairy Products'*

All participants were awarded free membership of the Society for the duration of their studies.

The winners of the 7th Annual Poster Competition 1995 were reported in the Journal (JSDT Vol. 49 No. 1 February 1996).

Presentation, objectives, methodology, results and commercial viability formed the basis of the judging, which was carried out at Huntingdon by Mr Brian Hatt, Mr Garry Royall, and Mr Alan Stack, members of the Society's Technical and Editorial Board...

| 1ST PRIZE | 2ND PRIZE | 3RD PRIZE |
|---|---|---|
| Harriet Phipps | Pascalis Kastanas | Robert Allinson |
| Seale Hayne Faculty | The University of Reading: | Seale Hayne Faculty |
| University of Plymouth: | *'The effect of storage time on fouling of heat exchangers by goat's milk during UHT treatment'* | University of Plymouth: |
| *The effects of microbial growth on the quality of whipping cream with particular reference to Pseudomonas species'* | | *'The role and survival of bio cultures in a health associated frozen dairy product'* |

The last poster competition was held in Kinsale on the occasion of the Millennium Spring Conference in May 2000.

EDUCATION

Further encouragement for young people was given by the UK Council for Food Science & Technology.

"The United Kingdom Council for Food Science and Technology hopes to make a number of awards from the John Coppock Memorial Fund to young food scientists, technologists and food engineers (under age 35) to assist in travel to the 8th World Congress of Food Science and Technology to be held in Toronto from 29 September to 4 October 1991. The awards will not suffice for more than about 50% of the total cost of attendance but, depending on available funds, are anticipated to range up to £400. (JSDT Vol. 43. No. 3 August 1990).

A wide range of research educational and training facilities was available at several excellent centres some attached to universities and some free standing.

The work of some of these organisations is outlined below in extracts from the Journals of the Society.

THE HANNAH DAIRY RESEARCH INSTITUTE

J.A.B. SMITH (JSDT VOL. 5 NO.1 OCTOBER 1951)

"Before I refer to the Hannah Institute I should like to say a word about the general programme for the Summer Conference because, when it was first circulated, one criticism I heard about it was that it was all very interesting, but it didn't seem to have much to do with Dairy Technology! That criticism rather worried me, and I wondered whether instead of the 80 people we expected to attend the Conference, we might have only 20. I think, however, it is fairly clear from the present attendance of about 170 members and guests that the programme is in some way fulfilling what was expected of it!"

"I myself feel that we spend all the winter both in the London area and in the various Sections dealing with matters which are strictly dairy technology, and it seems to me therefore that in the Summer Conference we should have something different. In my view one of the main objects should be to see and hear as much as possible of what is going on in the dairy industry in the area where the Conference is being held…."

"It will be expected from what I have just said that the object of my paper on the Hannah Institute will be, first of all, to let you now something of what is going on in the Dairy Research Institute which is situated in this particular area of Britain…."

THE FOUNDATION OF THE INSTITUTE

"Round about 1927 the Development Commissioners of that day were examining the whole question of dairy research in this country, and they decided that, in addition to what is now the world-famous National Institute for Research in Dairying at Shinfield, Reading, there should be one other institute of a similar type in Britain, and since many of the problems were of a local nature it was agreed that the new Institute should be in Scotland…."

Just about that time, the late Mr. J. M. Hannah, of Girvan Mains, whose farm we shall pass tomorrow on our route to Stranraer, presented part of the estate of Auchincruive to the West of Scotland Agricultural College for educational purposes, and the remaining 135 acres or so were set aside for the foundation of a dairy research institute. When the Institute was constituted it very rightly adopted Mr. Hannah's name in recognition of his generous gift. The Institute buildings were opened in 1931, and they have been almost doubled during the past two years, the new extensions being opened in April, 1951 by the present Secretary of State for Scotland...".

"Possibly Britain is the only country in which an Institute of this type could be successfully controlled by three separate bodies, but at any rate in this case the system works extremely well!"

"The first of the three bodies is the Council of the Institute. The Institute has Articles of Association and this Council forms its Board. It consists of three members nominated by the University of Glasgow, three by the West of Scotland Agricultural College and two by the Secretary of State. The Principal of the University is the Chairman, and the University's other nominees are always professors of some branch of medicine or science. The Agricultural College usually appoints three of its Governors, and the Secretary of State some prominent farmers...."

"The second of the three bodies is the Agricultural Research Council, to whom we submit our programme of work every year and to whom we also send annually a report of the work which has been done in the previous year...."

"The third of the three bodies is the Department of Agriculture for Scotland. It is through this Department that almost all the money required for running the Institute is provided, and naturally therefore the Department concerns itself in some detail with how the money should be spent. When the Institute started, part of the necessary funds were raised locally and the Development Commission added to the sum raised in that way, but nowadays the cost of research work has increased so much that it has to be borne almost entirely by the Treasury, and at the present time about 98 per cent of our funds come from this source through the Department of Agriculture for Scotland".

"What is the object of an Institute like this?... Sir John Russell writing about Rothamsted some time ago said that the object of that great research station was "to broaden the scientific foundations of agriculture." Well, there is much of that same idea in the objects of the Hannah Institute; and while endeavouring to broaden in a general way the scientific foundations on which much of dairying is based, we also tackle problems which are of immediate practical importance. We have therefore two types of work in hand: (1) the fundamental problems which may take a very long time to solve and a few of which may inevitably turn out to have relatively little application in practice - we must face that possibility in research or else it is not research - and (2) we have various investigations in progress which are of immediate practical importance".

"For administrative purposes, the Institute is divided into seven sections and I should like to deal very rapidly and briefly this morning with each of these sections, and first of all, with the work of the farm. About 12 years ago Dr. Norman Wright, who was Director of the Institute at that time, realised that, unless some avoiding action was taken, it would be only a matter of time before this country experienced a very great shortage of feeding stuffs. War, if it came, would be one reason for the scarcity. But there was another and more lasting reason; for a long period we had depended greatly on imported concentrates, but these imports were coming from countries in which the standard of living was rising and the populations increasing, with the result that these countries were becoming less and less able and inclined to export feeding stuffs. At that time, therefore, Dr. Wright initiated a long-term experiment to determine to what extent a dairy farm such as ours at the Hannah Institute could be self-sufficient in feeding stuffs. That experiment was continued for 10 years. At the outset some people did not appear to take it very seriously; they took the view that Britain would never be short of feeding stuffs. But as the war went on the problem became increasingly important. Even then many inclined to the view that after the war the supply would be adequate again, but as things have turned out Norman Wright was correct and the attainment of a high degree of self-sufficiency in feeding stuffs has become more important than ever before....."

"At the Institute farm we have about 75 head of cattle and about 30 cows in milk at any one time, and during the last few years of the ten-year period we had an average yield of about 20,000 gallons, and, although some of the cows reached 1,000 gallons, most of them averaged round about 650 to 700. At an early stage we depended greatly on beans, kale and crops of that sort, but it soon became apparent that the outstanding crop which would be of most value nationally in maintaining a high degree of self-sufficiency was grass. We have therefore had several long-term experiments in hand for a number of years on methods for the intensive management of grassland in this area and have applied these methods on our farm".

"When that ten-year experiment ended in March 1949, we passed to a year in which the entire stock was fed on nothing but the grass and grass products grown on an area of just over 100 acres. The grass and grass products included pasture, silage, dried grass and hay. During that year the milk yield approached 25,000 gallons, half of it being produced in the winter. The yield of milk per acre of land amounted to a little under 250 gallons, which compared favourably with a figure of 243 gallons per acre for ten typically well managed farms in Ayrshire where imported concentrates were used".

"It is interesting and important to stress that on this all-grass programme the main shortage was one of carbohydrate. So often in discussing the management of a dairy herd stress is laid chiefly on the necessity for protein, but although we were able to produce sufficient protein, we should undoubtedly have had higher milk yields if we had had more carbohydrate in reasonably concentrated form".

"The average yield per cow of about 700 gallons on the all-grass programme was certainly lower than it would have been had imported concentrates been fed. During the present year, therefore, and probably for the next five years or so, we are dividing the herd into two parts; one is being managed on a self-sufficiency basis, while the other will have the self-sufficiency regime supplemented with what we think or find out to be, the minimum amount of concentrates required to give maximum yields. In that way we hope to find to what extent the use of imported concentrates can be expected to increase the milk yield of cattle maintained mainly on home-grown feeds".

"All this involves much experimental work on the intensive management of grass…."

"Experiments are also proceeding on the utilization of grass, for often in practice much pasture is wasted. A very well controlled comparison has been made between two methods of grazing, one in which the cows are close-folded, and one which might be termed "rotational" grazing…..."

BIOCHEMISTRY (Dr. E. C. Owen).
"I have dealt at some length with the farm because you are not likely to see much of it in the brief time available for your visit to the Institute this afternoon".

"Three main topics are being studied in the Biochemistry Department. The first is a continuation of the work which has been in progress for some time on certain aspects of digestion within the rumen of the cow. It is now well established that the amino-acid composition of at least a portion of the dietary protein ingested by the ruminant can be changed by the rumen micro-organisms, and that Lysine is one of the many amino acids which these organisms can synthesise. The nutritive value of the dietary protein can therefore be considerably modified in the rumen before it is actually 'digested' and its products absorbed into the bloodstream of the cow. This is one of the rumen problems that are being studied".

"The second topic consists of a very intensive study of the metabolism of carotene and vitamin A in the lactating cow and goat, and of some of the factors which affect the conversion of carotene to vitamin A and the secretion of these substances in the milk".

"The third subject which has been studied recently in this Department and which is closely allied to the second one, is the effect of substances such as thyroxine and thiouracil on the composition of the colostrum and milk secreted by cows and goats, with particular reference to some of the more important minerals and vitamins".

NUTRITION (Dr. K. L. Blaxter)
"In addition to the extensions which have now been made to the main building at the Institute, extensions are also being made to the buildings at the farm, and one of the main additions is a new metabolism house for cows in which it will be possible to study the digestibility and nutritive value of feeding stuffs. This building will be ready for use in the near future. In the

meantime a detailed study is being made of the nutrition of the very young calf, a subject of the utmost importance in a country like Scotland where the calf-mortality rate is so high. Investigations on protein metabolism of the young calf have included work on the nitrogen-sparing effect of carbohydrate, on nitrogen metabolism during the initial stages of vitamin E deficiency, and on the effect of the type of dietary fat fed to the calf on its nitrogen metabolism. The staff concerned have constructed excellent equipment for the detailed study of energy metabolism in the calf. This equipment has already been used to determine the effect of the plane of nutrition on the mode of energy utilisation. A series of experiments on acute vitamin E deficiency, and the muscular dystrophy which accompanies it, has been carried out. Cardiac abnormalities and abnormalities of energy exchange, of posture, gait and behaviour have been recorded. Post-mortem studies of the calves and analyses of the affected muscles have been made".

PHYSIOLOGY (Dr. J. D. Findlay)
"The main object of the work of the Physiology Department is to study the effect of climate on the physiology of the bovine. At present little fundamental knowledge is available on the subject, although it is obviously of the utmost importance in connection with milk production in tropical and semi-tropical countries, areas which include many of the British Colonies. During the period under review the construction of a psychometric room was completed...."

"Detailed histological studies have been made of the sweat glands, the pigmentation in the skin and the blood capillary supply in the skin of cows, all of which may affect the power of the cow to adapt itself to its climatic environment. Histological methods have been used to compare the skin of Zebu cattle with that of Ayrshire's".

PATHOLOGY (Mr. P. S. Blackburn, M.R.C.V.S.)
"In the early years of the Institute's history, and in more recent times it has collaborated in research work on various aspects of diseases such as mastitis, abortion and sterility".

BACTERIOLOGY (Dr. Constance Higginbottom)
"In the bacteriology section the main underlying object is clean milk production. In addition to long-term studies a number of *ad hoc* experiments are carried out, sometimes in collaboration with our Chairman of this morning, Mr. G. H. Chalmers of the Department of Health. We are always willing to take part in collaborative work in that way. Recently, for instance, we made a comparison of immersion-cooling and surface-cooling in this district. On another occasion, in collaboration with Professor Smillie at the Agricultural College, we tested the use of hypochlorites on 17 farms and compared the results with those obtained on comparable farms in which steam was used. The hypochlorites appeared to be effective provided they were used really efficiently but it appeared that under typical farm conditions steam was more fool-proof".

"To give another example of the work of this section, I might refer to the question of temperature changes which occur during the transport of milk from farm to creamery".

TECHNICAL CHEMISTRY (Dr. R. Waite)
"The Technical Chemistry Section has three main lines of research. One is to collaborate with the farm in its extensive grassland experiments, particularly with reference to the chemical analysis which is required. The second is to study certain problems associated with the composition of milk, and the third is to do research work on the condensing and drying of milk and on the products of these processes...."

"Regarding the third line of investigation in this Department, there is in our new extensions a large processing room equipped with a small milk-condensing plant and other apparatus connected with milk-processing. I should like to say here that we feel we have a very close link with milk producers in the animal husbandry side of our work, but that our link with those engaged in the processing of milk and the manufacture of milk products is not so strong, and personally I am keen to see this connection strengthened. The facilities and staff are there, and we are very willing indeed and very keen to have a closer link with what I might call the dairy technology side of the dairy industry".

# THE WEST OF SCOTLAND AGRICULTURAL COLLEGE
PRINCIPAL J. KIRKWOOD, O.B.E., B.SC.
(JSDT VOL. 5 No. 1 OCTOBER 1951)

"As you may know, Scotland for the purpose of agricultural education, like old Gaul, is divided into three parts. In the North of Scotland there is a group of counties, a very extensive area, which is served by the North of Scotland College of Agriculture with Aberdeen as the centre. Again in the East of Scotland a group of counties there is served by the Edinburgh and East of Scotland College of Agriculture with the city of Edinburgh as the centre. In the south west of Scotland a group of some 14 counties is served by the West of Scotland College with Headquarters in Glasgow and an extensive sub-centre at Auchincruive, Ayr".

"In each case the Agricultural College is closely associated with the University in the same city, and all the three Universities grant the degree of B.Sc. Agriculture. Also each of these three Colleges provides courses for the Scottish Diplomas in Agriculture and Dairying. I will refer to the latter. Two of them provide courses for the Diploma in Horticulture. We are the only one that provides a course for the Diploma in Dairy Technology".

"The Agricultural Advisory Services in Scotland are based on the three Colleges, and the Advisory Staffs who do the work in Scotland are members of one or other of the College Staffs. That is different in England. In England the new Agricultural Advisory Service instituted at the end of the war is quite distinct from the Agricultural Departments of Universities, Agricultural Colleges or Farm Institutes. This is not the place to discuss the relative merits of these two systems, but I do think our system in Scotland is the better. Of course Scotland is a small country. It was already divided into three areas and it was the natural thing to develop and extend these when the Government was putting extra emphasis on providing a complete Advisory Service. There are just two comments I would like to make on that. I always think that teaching and lecturing and that sort of thing will be more alive and more practical if the people who are responsible for same have also the advantage of close contact with the farming community engaging in advisory work. The other comment is this: when young men and women, who have studied at an Agricultural College or Farm Institute, set out on their life's work of farming and encounter problems, surely it is the natural thing to turn for advice to the same source. I think that these two points are in favour of our system…"

"Our College provides three chief services:

(1) Education or teaching.

(2) Advisory.

(3) Experimental and research.

As the name implies the greatest of these is teaching".

*"Teaching:* Courses are provided for the B.Sc. in Agriculture, the Scottish Diplomas in Agriculture, Horticulture, Dairying (both Husbandry and Technology) and Poultry Husbandry".

"Our College is the chief training centre in Scotland as far as Dairying is concerned and holds a position which might be regarded as analogous to Reading in England. Students who wish to study for the Diploma in Dairy Technology come to us for the whole course. Incidentally, we are just completing the modernisation and re-equipment of the Dairy School and already, both in February 1950 and 1951, short courses of a week's duration in ice cream making have been held; these seem to be meeting a need…."

"Dairy education in the south west of Scotland is not a new thing. Away back about 90 years ago, some of the landowners and the progressive farmers in the south-west of Scotland wanted to improve methods of dairying and cheese making and they brought up from Somerset a Mr. and Mrs. Harding to give instruction in Dairying. That was in 1860. Then in 1885, about 25 years later, Mr. R. J. Drummond was brought across from Canada for the same purpose, to try to teach and improve the methods of dairying. For a few years an arrangement was made whereby he went round farms and held classes here and there, and the neighbours were invited in. This met with a lot of success. Then something more in the way of permanent premises were obtained at Holmes Farm near Kilmarnock. That was prior to the College taking over that farm. Classes were held there and were well attended and it was decided at the beginning of this century that a Dairy School should be built and in 1904 a Dairy School for Scotland was opened at

Kilmarnock with the late Professor Drummond as head. He continued until 1925, so that from 1885 to 1925 he laboured and worked to develop and improve Dairying methods in the south-west of Scotland, and there is no doubt that dairying not only in the south west of Scotland but all over Britain benefited greatly from his methods, because students came from all over the country to the Dairy School at Kilmarnock".

"Similarly, we happen to be the chief centre in Poultry Husbandry. Again students may take the first part of their course at Aberdeen or Edinburgh, but the last year is spent in the Poultry Department at Auchincruive".

"Some years ago the decision was taken by our Governing Board and approved by the University and the Department of Agriculture that, as soon as may be, the Central College in Glasgow and all the activities carried on there will be transferred to Auchincruive. The broad plans for this have been made and the first step towards this goal has commenced; the site for a hostel for men students is being prepared and we hope that the building programme will commence almost any time. At the present time the chief teaching departments are located in the Central College, Glasgow - the teaching of Agriculture (Crop and Animal Husbandry), the Applied Sciences, Botany, Chemistry, Bacteriology and Zoology. As already indicated the more practical aspects of the teaching for the Diplomas in Dairying, Poultry Husbandry and Horticulture are carried out at Auchincruive".

"In addition to the Degree and Diploma courses mentioned, our College provides a number of shorter courses and perhaps the most important of these is a six months' course for farmers...."

"Incidentally, we do not take students on the College Farm for the one year's practical training which is insisted on for students going forward to the B.Sc. Degree in Agriculture or the Diplomas in Agriculture or Dairy Husbandry, the reason being that with 20, 40, 60 or 80 students on one farm you simply could not give them the individual practice at the various farm operations. We do, however, take two or three for specialised training; for instance, in field experimental work, pig husbandry, dairy husbandry and so forth".

*"Advisory:* In the area served by the College we have Advisory Officers in Agriculture, Horticulture, Poultry keeping and Bee-keeping. We have specialist Advisory Officers in Animal Husbandry, Crop Husbandry, Grassland, Plant Pathology, Milk Production and Utilisation, Farm Economics, Soils, Veterinary Problems and Farm Buildings. The County Advisory staff can call on these specialists and they have also the whole resources of the Central Staff behind them".

*"Examples of Specialist Advisory Services:* Our Bacteriology Department, of which Dr. James Malcolm is the Head, provides a fairly complete service in the examination of milk, water, ice cream. Last year 40,000 specimens in all were examined.... A further 17,000 samples were tested for mastitis and some 10,000 samples for research purposes.... I think we can modestly say that our College has made a substantial contribution to the great advance which has been made in the purity of the milk supply".

"Similarly, in Professor Smillie's Milk Utilisation Department, problems which arise in the manufacture of milk are investigated. Considerable success has attended an attempt at shortening the process of cheese making. Another function is the provision of starters for cheese and butter making and an increasing number of requests for starters is now being received even from England and Wales, which would seem to indicate that the lactic cultures prepared and issued from Auchincruive are proving satisfactory in creameries".

"The Veterinary Advisory and Investigation service is a comparatively young one, being instituted about the beginning of the war. Veterinary practitioners throughout the College area and stockowners are making full use of the service and our Advisory Officers are kept very busy. Great progress, as you know, is being made with many of the sheep diseases; work is being done in connection with acetonaemia in dairy cattle, fluorosis in cattle, worms in sheep, bracken poisoning, etc."

"In the Soils Analysis Department at Auchincruive, over 8,000 samples were submitted last year for ordinary tests and report for phosphate, lime content, etc. and about 1,000 for the special report on trace elements. We have a laboratory fully equipped with modern spectrographic equipment...."

*"Plant Pathology:* The Plant Pathology Department provides a service to the area on disease problems in connection with crops and many diseases have been investigated and useful control measures suggested. The commonly accepted practice of treating seed grain with mercurial compounds such as Coresan and which in many cases makes a tremendous difference in the health of the crop was introduced by the late Dr. O'Brien and has undoubtedly led to a substantial increase in the financial value of the oat crop which is the staple cereal grain in Scotland. A problem which has been and still is occupying much attention is that of potato root eelworm. It is a serious problem but progress is being made".

"In the Clyde Valley in our College area, the strawberry growing industry was a very important one for many years, but after the first World War in the early 1920s trouble came in the form of strawberry disease or red core disease. This spread and gave almost a knockout blow to this profitable line. Our horticultural staff under the late Mr. D.V. Howells did a lot of investigational work. There is no known method of controlling the disease as far as the treatment of soil or plants is concerned; work on these lines proved abortive. Then in the early 1930s a new line of approach was made. The Department of Agriculture appointed a specialist, Mr Robert Reid, who set about trying to breed immune varieties. This work has been very successful. Mr. Reid never claims any variety is immune, but already quite a number of varieties have been put on the market which are highly resistant. The latest and perhaps the most successful is the one known as Auchincruive Climax which is now being grown all over Britain and abroad. This is an excellent cropper of good quality and holds out promise of restoring to a considerable extent this special industry".

"Then we have quite a big Economics Department of which Mr. Gilchrist is the head. Numerically it is the largest department in the College. Financial data are collected from farmers and reports issued on the profitability or otherwise of types of farming. For some commodities cost of production figures are worked out for the Annual Price Review of agricultural products, e.g., cost of milk production...."

"A more recently introduced advisory service is in Farm Buildings. We now have two advisory officers - architects who have made a special study of farm buildings..."

"Lastly at Auchincruive we have the College Farm. This now consists of some 392 acres. We have over 200 head of dairy cattle, about 80 cows mostly Ayrshires but with some British Friesians, and a large herd of pigs which we breed and fatten for the bacon market. The uses of the College Farm are manifold. It provides our teaching and advisory staff with opportunities for getting first hand information and knowledge about stock and crops; we try to grow almost every crop which may be grown in the South-west of Scotland and even a few acres of crops which are not ordinarily grown in the South-west such as sugar beet. We make arable and grass silage and when any new varieties of farm crops are introduced, or new systems of manuring suggested, we try these out against some of the best known existing varieties and existing manurial practices".

"The Farm is being used to an increasing extent in connection with our Dairy Husbandry Diploma classes".

"The two chief sources of revenue on the farm are the dairy herd and the pigs. The first purpose of a College Farm is for education and practical experimentation. These objects should come first and are much more important than trying to make profits..."

"I would just say finally that our staff, particularly our County staff, participates in all the rural social activities of the countryside. If I were asked to suggest a motto for our College it would be: *"Service to the Farming Community."* Actually we have a College motto (I do not know when it was designed) and a very good one. It is in Latin but translated means: *"To make the Earth fruitful by Science."*

## THE CHESHIRE SCHOOL OF AGRICULTURE, (JOURNAL OF THE SOCIETY OF DAIRY TECHNOLOGY, VOL. 12, NO. 1, 1959)

"The Cheshire School of Agriculture, Reaseheath, is situated about one and a half miles from Nantwich and about six miles from Crewe. It forms the central part of a beautiful estate extending to about 400 acres".

"Formerly a country mansion, Reaseheath Hall now comprises the Men's Hostel and the Administrative headquarters of the School. The adjacent buildings were adapted as lecture rooms, laboratories and workshops. The Gardens were extended to form a horticulture department and a poultry department was founded. More recently a hostel for women, a Dairy Department and a range of farm buildings were erected.

"The Agriculture department consists of two farms, namely the Hall Farm and the Experimental Farm:

(a) HALL FARM. This is a dairy holding extending to about 250 acres worked on a system of arable crops and temporary leys. A pedigree, attested herd of sixty Ayrshires, maintained by home breeding, is housed in modern buildings. There is also a pedigree herd of 200 Tuberculin Tested Large White pigs and a flying flock of sixty grey-faced ewes.

(b) EXPERIMENTAL FARM. A seventy-acre holding of which forty-five acres is under the plough. Livestock comprises an 'Attested' herd of twenty-eight pedigree Friesian cattle. Milking is carried out by an electrically operated self-delivery 'Auto-recorder' type machine. Part of the farm is devoted to small plots to investigate intensive management".

"THE POULTRY DEPARTMENT forms a separate holding within the estate and extends to some fifteen acres. The main work of the department consists in the breeding, rearing and maintenance of Light Sussex and Buff Rocks for ultimate crossing - the resultant progeny being housed in Deep Litter and hen batteries for table egg production".

"THE HORTICULTURE DEPARTMENT. The land under cultivation is two acres. There is a block of glasshouses and a comprehensive range of Dutch light frames and cloches, as well as a small apiary. The school grounds on some twelve acres also form part of this department".

"THE DAIRY DEPARTMENT. This is a carefully planned building designed for the training of students in dairy work. It has recently been completely modernised and is furnished with all the modern equipment required for cheese-making, butter-making, pasteurising and separating on a creamery scale".

"COURSES OF STUDY. In common with other Farm Institutes an one-year course in Agriculture for men is provided. There is also a course in Agriculture and Rural Domestic Subjects for girls who wish to take up positions on farms. A Farmhouse Cheese-makers Course is offered and there is a two-year sandwich course for employees in milk processing and distributive dairies who aim at occupying posts of responsibility in the dairy industry".

VISIT OF HRH THE DUKE OF EDINBURGH TO N.I.R.D.

A significant event took place on 6 July 1954 when HRH The Duke of Edinburgh paid an informal visit, at his own request through the Agricultural Research Council, to the N.I.R.D. J.A. Irvine reported on the event in the Journal (JSDT Vol. 7 No. 4 October 1954):

"We had been informed that he wished to talk to 'the man at the bench' and would probably not be able to visit every department; a tentative programme was drawn up, based on the general theme 'from cow to consumer', in order to display a combination of the scientific and applied aspects of the Institute's activities. His Royal Highness completed practically the whole programme and from his numerous and apposite questions there can be no doubt whatever of his very real interest in all he saw......"

"The Duke, accompanied by Sqn. Ldr. Beresford Horsely, Equerry in Waiting, was welcomed by Professor H. D. Kay, introduced to the heads of all the departments and sections, and then went into the Director's office for a brief talk about the scope of the Institute's work and discussion of the programme".

"During the morning, His Royal Highness was concerned mainly with milk production studies".

"Next came some of the work of the Feeding and Metabolism department, where the influence of the quality of the diet and the consequent changes in rumen digestion on milk quality are being studied in detail. As a final item from the Dairy Husbandry department's programme, some advances in machine milking technique were shown and discussed in the main cowshed".

"Then, leaving the farm, a visit was paid to the goathouse and laboratories of the Physiology department and the Isotope section. Here the Duke talked with those who have thrown new light recently on the synthesis of the main milk constituents and also on the manner in which the milk is released from the udder.... This was followed by a discussion of radio-active tracers as tools in lactational studies".

"At the end of the morning came the Dairy Engineering department's test laboratories and workshops. Here the Duke saw some of the dairy equipment that had recently been tested or was actually undergoing performance tests, and a selection of recent reports. He also heard about and saw the apparatus being used in the fundamental engineering research into high-temperature short-time pasteurisation".

"After a short interval for a buffet lunch in the Senior Common Room at which he and Professor Kay were joined by Mr. J. F. Wolfenden, Vice Chancellor of Reading University, Lord Iveagh, Chairman of the Institute's Board, the Heads of Departments and Mr. H. F. Burgess, the Secretary, His Royal Highness planted a young chestnut tree to commemorate the occasion. This followed the precedent set by King George V when he visited the Institute in 1925 shortly after its establishment at Shinfield".

"Immediately after this brief ceremony the Duke went to the Stenhouse Williams Memorial Library, where he signed the Visitors' Book, and to the Commonwealth Bureau of Dairy Science. He then set off on the afternoon tour of those laboratories concerned mainly with milk handling and manufacturing problems and with the place of milk in nutrition......"

"This was followed by short visits to laboratories of the Nutrition, Chemistry and Physics departments to see research in progress on the nutritional quality of milk - for example how vitamins get into the milk and the effect of processing and of light on the nutritional value of milk....."

"Lastly, some of the problems connected with the improvement of cheese were considered - the connection between physical properties and quality, the identification of off-flavours by means of the techniques of absorption spectroscopy, and fundamental studies of flavour in cheese".

"Before leaving, His Royal Highness, after a brief interval for afternoon tea, paid a short visit to the Reading Cattle Breeding Centre of the Ministry of Agriculture and Fisheries, which is closely associated with the Institute....."

Professor B.G.F. Weitz OBE, DSc. MRCVS, FI Biol. who retired in 1981 as Chief Scientist to the Ministry of Agriculture, Fisheries & Food was a distinguished Director of the National Institute for Research in Dairying (NIRD). His retirement and career was reported in the Journal (JSDT Vol. 34 No. 3 July 1981).

"Professor Bernard Weitz has retired as Chief Scientist Ministry of Agriculture, Fisheries, and Food".

"After qualifying at the Royal Veterinary College Professor Weitz joined the Central Veterinary Laboratory, Weybridge, where he worked on the immunological aspects of the control of brucellosis and mastitis. In 1947 he moved to the Lister Institute of Preventive Medicine, becoming head of the serum department in 1952".

"In 1967 he became Director of the National Institute Research in Dairying (NIRD) and Professor in the University of Reading. The many years of research into mastitis at Shinfield were coming to fruition, and it was in no small measure due to his enthusiasm that the NIRD recommendations for the control of mastitis were adopted by the Milk Marketing Board and veterinary profession".

"Following the recommendations in the Rothschild Report for a customer/contractor relationship between the Ministry of Agriculture, Fisheries and Food (MAFF) and the agricultural research institutes the Joint Consultative Organization (JCO) was set up, and Professor Weitz became a member of the Animals Board. This new machinery for establishing priorities in research necessitated the introduction of procedures for the management of the NIRD research programmes. Professor Weitz instituted an annual review of the research projects, which enabled a detailed examination of the short and long term objectives, progress, resources, and costs of each project. It was he who invited the Director of the Hannah Research

Institute to join the Institute's Academic Committee and who began to co-opt representatives from the dairy industry, who could take an objective view of the Institute's longer-term strategy for research".

"His involvement in the JCO led him in 1977 to accept appointment as Chief Scientist to the Ministry. His understanding of the objectives of the Ministry, the Agricultural Research Council (ARC), and the body of research scientists enabled him to advocate, and to see effected, a simplification of the JCO structure, and he played his full part in bringing about closer co-operation between MAFF and the ARC".

"Professor Weitz has made notable contributions to the activities of the International Dairy Federation and of its national committee in the United Kingdom (UKDA). He is currently President of IDF Commission A, Production, Hygiene and Quality of Milk, and a member of the United Kingdom Dairy Association's Technical Committee, having also served on the Council of the Association while he was at Shinfield. He has maintained his contact with Reading University and is Visiting Professor in the Department of Agriculture and Horticulture".

"Honours and awards have come his way. He gained his DSc in 1961 and was awarded the OBE in 1965. He is a Fellow of the Institute of Biology and an Honorary Fellow of the Royal Agricultural Society of England and of the National Institute for Research in Dairying....."

RESEARCH & DEVELOPMENT IN THE DAIRY INDUSTRY

In 1964 Mr. W. R. Trehane (later Sir Richard), Chairman, Milk Marketing Board for England & Wales delivered a paper on *'Research & Development in the Dairy Industry'* (JSDT Vol. 17 No. 2 1964):

"It is always a temptation for me to accept an invitation to speak to the members of a learned society, and on this occasion, the prospect of a visit to Northern Ireland added its own stimulus..."

"Ours is a large industry, by any measurement, whether we count the number of people involved, the capital employed or the part which milk and dairy products play in the diet of the people..."

"To do this it is quite certain that the resources used for research behind and in support of all our activities must be adequate to meet our needs as a vital part of the food industry. I felt that you might be interested to consider this aspect of our activity with me here today".

"In the year 1962, the value of the gross domestic product in the United Kingdom amounted to £24,000 million and total research expenditure was probably around £600 million or about $2^{1}/_{2}$ per cent of the gross domestic product. The funds for research are provided mainly by Government departments (about 60 per cent), partly by private industry (about 35 per cent) and the remainder come through the universities, public corporations and others..."

"In so far as agriculture is concerned, it is always difficult to obtain accurate figures of research expenditure but rough estimates can be made. The net total expenditure of the Agricultural Research Council in 1961/62 was just over £6 million and the Ministry of Agriculture also spent an estimated £3 million on research in the same year. Together, these amount to about $^{1}/_{2}$ per cent of the value of agricultural output for 1961/62. The research contributions of the universities, commercial firms, etc, have still to be included but even if we double the estimate to take account of these, the total still amounts to only just of 1 per cent of the gross value of the farm products of the United Kingdom".

"For the dairying industry alone, figures are even more difficult to obtain, but in 1961/62 the expenditure at the National Institute for Research in Dairying and the Hannah Dairy Research Institute, the two A.R.C. institutes concerned primarily with dairying, was £680,000 or 0.2 per cent of the value of dairy output. These are, I know, only two of the many research units interested in dairying problems but the N.I.R.D. is a particularly large and important one. Consequently I think it is unlikely that total research expenditure in dairying will exceed 1 per cent of the annual output...."

"I can think of no industry of comparable size, or indeed of any size, that rivals our own in complexity. There are in the United Kingdom between 130,000 and 140,000 producing units widely scattered throughout the country sending their milk to a large number of creameries and handling plants. There are some 130,000 points of retail distribution making a daily delivery to the 17 million households in the country... Since the cow must provide the milk before it can be processed, I shall deal with research problems in that order".

*Research on dairy husbandry and milk production*
"The scope and standard of research on dairy husbandry and milk production in the United Kingdom can probably stand comparison with most other countries. The resources devoted to work in this particular field appear commensurate with the size of the industry".

"At the N.I.R.D. and the Hannah, work is now being done on energy metabolism, the biochemistry of the mammary gland, the microflora of the rumen, disease, milk composition and various aspects of dairy cattle management. The three Scottish Colleges of Agriculture are engaged in research on the housing, feeding and management of dairy cattle. Work is in progress at Wye College in Kent on milk composition and at Nottingham University on rumen capacity and milk yield. Here in Northern Ireland the Institute at Hillsborough is carrying out work on the conservation of feeding stuffs and the effect which such products have upon milk production. Some extremely interesting twin studies and crossbreeding experiments are being conducted by the Animal Breeding Research Organisation in Edinburgh and the contributions made by Dr. Alan Robertson of the Institute of Animal Genetics and his co-workers are well known. Finally I should mention the work of the Board's own breeding and production organisation in England and Wales, which is naturally aimed mainly at the solution of questions of fairly immediate practical importance...."

"Let me give an example of the kind of cooperation I have in mind. During the past few years there has been increasing general interest in the solids-not-fat fraction of milk and as a result much research has been stimulated. The results showed that S.N.F. percentage is less affected by environmental differences between herds than is milk yield, but that it is affected more by these differences than is the percentage..."

"Our Board are supporting these experiments and we have along the years done our best to help research workers to find answers to problems of practical importance to the industry. If our own resources are not sufficient to tackle a particular question then we must promote and support research elsewhere in an attempt to obtain an answer".

MARKETING, PROCESSING AND MANUFACTURING RESEARCH PROBLEMS
"Beyond the farm gate, the dairy industry continues to expand on the marketing, processing and manufacturing sides. We need much more work done on the manufacture, storage, packaging and distribution of milk and milk products, on the development of new products or processes. Let me give you just two examples. The extension of the market for prepacked cheese is hindered by the necessity for refrigeration to prevent leakage of fat during distribution: we want to eliminate this problem and experimental work needs to be done on methods of manufacture and packing. Then there is the poor spreadability of butter after storage at low temperatures compared with many good quality margarines: we would like to overcome this difficulty".

"In the sales field there is also much to be done...."

TECHNOLOGICAL RESEARCH INSTITUTE
"When we survey this vast field for scientific investigation, there seems to be ample facilities for fundamental research in our research institutes, but the missing link is a national institute which can take the results of this fundamental research with the object of adapting and developing them to commercial practice. Work is being carried out by many of the dairy organisations, but in our opinion, on an entirely inadequate basis for such a large and important industry. The staff and the funds are not sufficient and all too often the information obtained is not readily available to the industry as a whole. The most satisfactory solution is for the dairy industry to accept responsibility for setting up a national technological research institute, which could at any rate deal with development as far as the stage where a project can be handed over to the industry at a point from which individual organisation can develop further".

"It is not envisaged that such a unit would carry out fundamental research work in pure science as applied to dairying, but it could be expected to review the literature and follow up any fundamental investigations which appeared to be of practical significance".

"The industry should, I suggest, have advice on the latest improvements in the marketing of milk and milk products, supported by investigational results. The necessity for this has been shown by the work of the National Milk Publicity Council. For example, there should be full and up-to-date products advice and a technological institute should be in a position to give this....".

"While it is not envisaged that the proposed institute would compete with the existing research institutes, there would have to be close contact between staff of the technological unit with research workers carrying out investigations of a fundamental nature. In the same way, it is hoped that the facilities of the technological unit would be available to the research institutes to assist them in carrying out fundamental work. This co-operation could lead to a new era in dairy research and a full partnership between fundamental dairy science and dairy technology".

"If a technological unit could be developed, it would provide a centre for educational facilities. The cost to the industry of providing a technological institute will be considerable. The provision of the basic building and equipment might be as much as £250,000 with annual running costs of the order of £100,000 and these items might well be increased as more and more use was made of the unit. However, there is no doubt in my mind that it is essential to go ahead - even if we start modestly - to fill the gap in our facilities for full development of our industry…."

ORGANISATION OF RESEARCH & DEVELOPMENT IN THE DAIRY INDUSTRY

In 1970, Mr. S. G. Coton, a Past President of the Society, reviewed the organisation of research and development in the dairy industry (JSDT, Vol. 23, No. 4, 1970).

"My intention is to examine what place the research and development scientist should occupy in the dairy industry and how his function should be organised…."

"For my purpose it well also be useful to consider the dairy industry, that is milk production and utilisation, as two industries; the milk production industry which is concerned with milk up to the point where it enters a creamery, and the dairy processing industry which is concerned with the manipulation of milk or a milk by-product continuing up to that point where the milk becomes a relatively non-significant item in the total raw material to be processed. This definition of a dairy processing industry is consequently a good deal wider than would be thought of in the context of traditional dairy products, such as butter, cheese and cream".

"Any industry which fails to innovate suffers a decline and eventual extinction. …Research and development expenditure on food is very low at 0.3 per cent of net output (incidentally this figure has scarcely increased since 1958 the data year for this figure), compared with aircraft at 36 per cent. Growth rate for the food industry is correspondingly low…."

"A basic problem of management in any industry is, therefore, to apportion the available capital to these three activities; from the point of view of one engaged in industrial research and development the problem is to ensure that the innovative activity receives its fair share of available funds. But what is its fair share? This is difficult to answer…"

"I should like now to consider how these general principles relate to the dairy industry. The 1969 UK domestic expenditure on food was approximately £5,580 million and on dairy products approximately £960 million".

"I have assumed that the turnover of the UK dairy industry is about equal to the domestic expenditure on UK produced dairy products that is about £739 million. The figure is overstated in that the dairy industry as such does not have a monopoly of the retail sale of its products, but the figure is understated to the extent that certain products, for example cocoa crumb and the majority of skim milk powder, is not sold retail and does not show in these particular statistics…"

TABLE 3

Estimated expenditure on research and development in UK dairy industry, 1969

£ Million

| | |
|---|---|
| National Institute for Research in Dairying | 1.1 |
| Hannah | 0.2 |
| Industry (very approx) | 1.0 to 1.4 |
| | Total 2.3 to 2.7 |

"The amount spent on dairy research and development in 1969 is not known with any accuracy, but my estimate, which I admit has a large element of guess work, is £2.5 million made up as shown in table 3 above".

"Now £2.5 million is 0.3 per cent of the turnover of the UK dairy industry. No one can say with certainty that this is not adequate, but in relation to the size of the industry it is small… The average spent on R & D in the UK for all industries is just over 3 per cent, that is 10 times greater than for the dairy industry".

"The work of the NIRD and the Hannah Research Institute is directed very largely to basic research and I think it would not be unfair to regard this as taking, say, £1 million of the £1.3 million spent by these two bodies. This is 40 per cent of the total available research and development fund. The corresponding figure for the food, drink and tobacco industry is 10 per cent (Statistics of Science and Technology, 1968, HMSO). Whether we believe that the total sum spent on research and development in the dairy industry is appropriate or not, I think that there is no doubt that there is far too little emphasis on applied research and development, particularly in the dairy processing industry, aimed at producing new products and, to a lesser extent, new processing techniques".

"Whatever sum is deemed appropriate to the research and development activity, it must be spent as effectively as possible… Fundamentally this is achieved by a proper initial choice of project and a review of chosen projects at fairly frequent intervals, all in monetary terms".

"There is an increasing amount of literature on how this may be attempted, and many of the more sophisticated procedures suggest the use of discounted cash flow techniques".

"In the case of a new product, the estimation of net return requires an estimate of potential sales and this leads to an appreciation of the vital importance of a marketing orientated approach to the business of product development. I emphasise very strongly the importance of the marketing function in product development, but deplore that so often the whim of the marketing manager is regarded as an appropriate substitute for marketing policy derived and applied with market research techniques. Whilst the majority of new products should be initiated via the marketing function, new processes which often, of course, lead to new products, are likely suggested by the research and development function. However, whatever the source of the idea it should be judged in financial terms only….."

"In any form of product or process development, it is extremely difficult to fix a time scale to each project. Despite the difficulties it is very desirable to attempt this; the use of a simple form of critical path analysis is probably well worthwhile, and there is an excellent monograph on the subject by L.J. Rawle, published by Unilever…."

"An invasion of the food industry by the dairy processing industry would seem in fact to be one of the most promising ways of expanding sales of milk and dairy products in combination with other food materials".

"If the dairy industry is to be much more closely related to the food industry, it is well to establish if there are any general trends in the food industry of which account should be taken when developing hybrid food and dairy products. There are, I think, five discernible trends which can be listed as follows:

1.  First of all, there is a general move towards increasing shelf life.
2.  The second trend, and perhaps the most widely publicised, is a move towards convenience foods.
3.  The third trend is towards the increasing sale of 'brand imaged' product compared with anonymous products.
4.  Fourthly, there is a move towards much more complex packaging which to a considerable extent is necessitated by the first three trends which I have described of extended shelf life, convenience and brand image.
5.  The last trend, which I am sure will gather momentum, is an increasing diet consciousness".

"Factually, milk has a high nutritive vale at relatively low cost, has a highly valuable protein content, and is fairly easy to manipulate and incorporate in a variety of products".

"One liability, however, which the dairy industry has nurtured is a wealth of restrictive legislation…. More permissive dairy legislation is essential in order to give a greater freedom for the creation of new products although legislation requiring accurate labelling of products must be regarded as a necessity".

"My paper has been concerned entirely with abstracts and perhaps I could finish by briefly commenting on three specific possible areas for development in order to make some of the generalities concrete".

*"Reverse osmosis.* My first example is reverse osmosis, which I choose because we have in progress a considerable amount of work on the subject and believe that it may well be a process innovation of very great consequence… By suitably modifying the membrane some solutes, in addition to the solvent, migrate through the membrane. Consequently, in addition to being a concentration technique, reverse osmosis may also be considered as a means of fractionating or separating material…."

*"Cheese manufacture.* My second example concerns cheese manufacture. A considerable amount of effort is employed in attempting to make cheesemaking a continuous process with the object of reducing costs, principally by labour saving. The effort is concentrated more towards the initial phases of cheese manufacture, particularly those up to and including the Cheddaring operation rather than the latter end of cheesemaking. Without wishing to decry the excellent effort which has been put into this, I would suggest that attention to other aspects of the cheesemaking process could be more profitable….."

"Looking at cheese in terms of the consumer, that is with a market orientated approach to product development, one will arrive at a conclusion that the public, or a large sector of the public, wishes to purchase a cheese product with particular flavour, texture and other eating properties".

FRAMEWORK FOR GOVERNMENT RESEARCH AND DEVELOPMENT
- CMD 4814 (JSDT Vol. 25 No. 2 April 1972)
"There can be few Green Papers which have given rise to so much controversy in scientific circles as the Rothchild and Dainton Reports dealing with Government research and development".

"The reorganization of Central Government involves a review by Ministers of the function of Government Departments to ensure that responsibility and accountability are clearly defined and allocated. Accordingly, Lord Rothchild (Head of the Central Policy Review Staff) was commissioned to report on Government research and development. Government also sought advice through the Council for Scientific Policy on the most effective arrangements for organizing and supporting pure and applied scientific research and post-graduate training. This task was undertaken by a Committee, under the Chairmanship of Sir Frederick Dainton, which was appointed because a proposal to transfer the Agricultural Research Council to the Ministry of Agriculture, Fisheries and Food."

"The Rothschild Report distinguishes between basic research - the end product of which is an increase in knowledge; and applied research and development - the end product or objective of which is either a product, a process, or a method of operation. On the other hand, the Dainton Committee refers to basic science-having no specific application in view but necessary for the advance of scientific knowledge; tactical science - which is application of science to immediate executive and commercial functions; and to strategic science which is general scientific effort necessary to give a foundation to tactical science".

"Basic research is almost exclusively financed by the Department of Education and Science through the Council for Scientific Policy and its five Research Councils which are subordinate to the Council for Scientific Policy in respect of the allocation of funds but are autonomous in respect of their research programmes. Rothschild disapproves of this autonomy in the case of applied research, which he states forms the major part of the work of the Agricultural Research Council."

"The principal recommendation of Rothschild, and one which has raised considerable comment, is that Government applied research and development should be done on a customer/contractor basis because, he claims, the scientist cannot be so well qualified to decide the needs of the nation as those responsible for ensuring that those needs are met. The customers would be Government Departments and the contractors would usually be Research Councils which would not be able to refuse commissions unless they managed to satisfy the department concerned that they had good reason."

"Government Departments would decide on programmes, expenditure and priorities and would have their own Chief Scientist organisations which would advise and be closely associated with the departmental decisions. The Chief Scientist's opposite number would be a Controller R & D who would be the Chief Executive of the contractor responsible for providing the R & D service. It is essential that there should be continuous liaison between the customer, the Chief Scientist, the Controller R & D, and the research scientists."

"Lord Rothschild recognizes that virtually all applied R & D Laboratories will engage in research which is not directly concerned with the programmes commissioned by customers. He terms this General Research because it is not necessarily applied or basic and may comprise: (a) basic research which is relevant to the applied R & D and not being done elsewhere - this appears to correspond to the 'strategic science' of the Dainton Committee, (b) testing of new unprogrammed ideas of the scientists, (c) maintenance of expertise, and (d) work facilitating the transition from academic life to that in an applied R & D organisation. Rothschild approves of such General Research, but considers it important that it should be formally quantified, and that the average expenditure on it should not amount to more than a 10 per cent surcharge on that sanctioned by the customers. Rothschild recommends that if the change in financing of the Research Councils is implemented the amount paid to them by Government Departments in the first year under the new regime should not be less than would have been spent if no change had occurred, and that after the first year the Departments should not reduce their payments to Research Councils by more than 10 per cent per year for three years; the payment could, however, be increased. There would thus be a four-year transitional period."

"For the year 1971-72 the five Research Councils are being financed by the Department of Education and Science to the extent of £110 million. Rothschild proposes that under his system the DES vote should be reduced on average by 25 per cent, which reduction would correspond to expenditure on applied R & D to be commissioned by the appropriate Government Departments. It is noteworthy, however, that the Science Research Council would continue to derive its expenditure of £51 million from the Department of Education and Science, whereas the Agricultural Research Council now receiving £18.7 million from the Department of Education and Science would suffer a reduction of 77 per cent equivalent to £14.5 million to be funded for commissioned research by the Ministry of Agriculture, Fisheries and Food. This is to be compared with the present direct expenditure of £6.2 million by MAFF on Research and Development. It should be noted that the arrangements for applied Agricultural Research and Development in England and Wales differ from those in Scotland where the Department of Agriculture and Fisheries already finances the bulk of agricultural research with advice from the Agricultural Research Council."

"The Dainton Committee accepts that the Research Council structure has the disadvantage of limited ability of adjustment to changing circumstances and that Government Departments should play their part in the formulation of scientific policy, but considers it essential that the Research Council structure with its principles of scientific responsibility and judgement of scientific merit should be preserved. To retain the advantages of the Research Council system while providing for stronger links with the needs and policies of Government Departments and to give greater flexibility, the Committee proposes the replacement of the Council for Scientific Policy by a new chartered body referred to as a Board of Research Councils. This Board, on which Government organisations would be represented, would be responsible for financial allocations to the individual Councils and would have the overall responsibility of ensuring that the requirements of executive departments for scientific support from Research Councils were being properly met on a service or contract basis where appropriate, and by the provision of independent and authoritative information and advice in other cases."

"In a memorandum preceding the two reports Government endorses the customer/contractor principle for Government applied research and development. Subject to this principle Government believes that it would be right to preserve the Research Councils under the sponsorship of the Department of Education and Science. It would appear that the main debate centres not so much around the customer/contractor principle, but rather to the extent of the field of scientific effort to which it should be applied."

In 1977 Dr. L.A. Mabbitt (Head, Bacteriology Department, N.I.R.D.) critically viewed the research programme at the NIRD. He drew attention specifically to the problem of spreadability of butter and forecast the introduction of "spreads" (for example Clover) to the UK consumer. His comments on the new (at the time) customer contract system to research is worth quoting:-

"It is difficult as yet to observe the effect of the Joint Consultative Organization on research programmes for two main reasons. Firstly, for any particular investigation it is essential to have the relevant research expertise. Since this cannot be changed quickly, research projects tend to be changed slowly and the changes are consequently not easy to identify or attribute to a particular cause. Secondly, and probably more important, directors of institutes and heads of research departments are members of the Boards and Committees of the JCO. They naturally ensure by gentle persuasion that their research ideas and suggested programmes are well to the fore of the Committees' thinking and get the Committees' approval. After all they are in a strong position with their up to date scientific information and research background. It comes, therefore, as no surprise to the author to find that no change in the research programme of his Department can be directly attributed to the new customer-contract system. (JSDT, Vol. 30, No. 4, Oct 1977.)"

BRITISH STANDARDS

Another important area of activities to which the SDT contributed was the British Standards Institution. The late H. S. Hall contributed the following note on Standards for the dairying industry in July 1981 (JSDT Vol. 34 No. 3 July 1981 p.126)

STANDARDS FOR THE DAIRYING INDUSTRY

"The British Standards Institution (BSI) is an independent, non-profit-making body set up some 80 years ago and incorporated under Royal Charter in 1929, originally with the purpose of producing engineering standards. During the past half century the scope of its work has widened to include almost every industry, resulting in specifications for raw materials, techniques, and finished products appropriate to the needs of the United Kingdom. Through its close links with similar bodies in more than 70 other countries, BSI is responsible for co-ordinating and, where applicable, integrating British standardisation with international needs for world trade".

"The Institution has a committee structure, serviced by a permanent technical and administrative staff, which aims at securing representation of all interests affected by its work in the production of standards by voluntary consensus. British Standards are public documents, agreed through processes of public consultation in order to secure public acceptance. The scale of the operation is large: 30,000 committee members, 4,500 committees, 8,150 published standards, and 7,000 standards projects in hand. The annual standards-making budget of BSI is currently £7.5 million, half of which is derived from the sale of publications and half from subscriptions supplemented by Government funds on a pound-for-pound basis. These funds are augmented by the cost, unknown but considerable, incurred by committee members in attending meetings and in providing technical data, often by ad hoc research, to ensure that a draft standard is the best it is possible to produce with the knowledge available at that time".

"Planning is undertaken on an industry-by-industry basis by 72 Standards Committees. These committees appoint Technical Committees to prepare particular standards and, if appropriate, to be responsible for the UK input to international committees of such organisations as International Organisation for Standardisation (ISO) and International Electrotechnical Commission (IEC), of which BSI is the UK Member Body".

"There was little demand for standards in the dairying industry until the 1930s. Imperial conferences in 1930 and 1932 had laid great stress on the relationship between members and had recommended that wherever possible there should be common standards throughout the

British Commonwealth of Nations. As a direct result, BSI constituted a committee which set to work in four areas: (i) standard apparatus and method of test for determination of fat in milk; (ii) freezing point depression of milk; (iii) methods of chemical analysis; and (iv) methods of bacteriological analysis. The UK work was co-ordinated with similar committees set up at that time in Australia, Canada, New Zealand, and South Africa. In 1944, however, it was decided that a Dairying Industry Standards Committee (DAC/-) should be set up, to be responsible for work on milk and milk products, including methods of analysis together with certain items of equipment used in dairying. In recent years, some of this latter work, for example, on containers and packaging, conveyors, and milk piping, has been taken over by other BSI Standards Committees. From the beginning DAC/- has set up 32 Technical Committees, of which 16 are still in being. Of these, the Technical Committees with the heaviest workload at present are DAC/3 - Chemical analysis of milk, butter, cheese, and other dairy products, and DAC/4 - Microbiology for dairy purposes, each having a long list of subjects under study, mostly concerned with analytical procedures. Both committees are closely linked with the Joint Groups of Experts nominated by IDF, ISO, and AOAC".

"Because DAC/- is a planning and organising committee its membership comprises experienced representatives from over 20 organisations concerned with the dairying industry, including milk producers, processors and product manufacturers, equipment manufacturers, Government departments, research institutes, and technological societies. Its main tasks are to initiate new work, often at the request of a member organisation, by extending the terms of reference of an existing Technical Committee or by creating a new one; to allocate priorities and review these when necessary; and to arrange resources in consultation with BSI to ensure steady progress over the whole field of work. In past years, the Society of Dairy Technology, through its Equipment and Standardisation Committee, and the Milking Machine Manufacturers Association have provided initial working drafts of several proposed standards for dairy equipment. These are examples of the way in which member bodies of DAC/- can best help the industry it serves".

"With the growing importance of international trade and co-operation, DAC/- must be aware of relevant work in progress and in prospect overseas. In many instances, this is done by participation in meetings abroad either in an administrative role or as a technical representative. DAC/- is responsible for the selection of appropriate delegates to carry out these tasks who on some occasions are chosen from the staff of BSI and on others from the industry. In some cases BSI has undertaken the onerous work of acting as Secretariat to an international committee, the most recent example under the aegis of DAC/- being ISO/TC 23/SC 11/WG 2 - *Milking machine installations*. Much duplication of effort is avoided when standards generated abroad can be adopted verbatim for use in the UK, and DAC/- is constantly looking for these opportunities. Conversely, an established British Standard can he a strong negotiating factor in an international forum seeking to prepare an international standard".

"It is not only the experts serving on committees who influence the work. During the preparation of each standard, a Draft for Public Comment is issued, and all comments received are considered by the responsible committee. Details of the availability of Drafts for Public Comment are given each month in *BSI News*. But BSI is not only dependent on technical expertise from industry and users; cost-effectiveness of standards work depends greatly on correctly identifying problems that could be alleviated or eliminated by the preparation and application of a standard. Increasingly, Standards Committees need management and marketing advice from industry authorising new projects and allocating priorities. Those in responsible positions in the dairy industry are best able to identify problems where new standards, or revision of existing standards, could be of benefit. Almost all will be members of the constituent bodies of DAC/and so have a simple way in which to make their views known through their representatives".

"DAC/- is currently responsible for 51 British Standards, including the following published since 1975:

BS 1743:     Methods for the analysis of dried milk and dried milk products
Part 1:      1980 General introduction, including preparation of laboratory sample

| Part 2: | 1980 Determination of the solubility of dried milk, dried whey and dried buttermilk (reference method) |
| Part 4: | 1980 Determination of the dispersibility and the wetting time of instant dried milk |

| BS 3095: | Methods for the determination of the freezing point depression of milk |
| Part 1: | 1979 (HORTVET) method and thermistor cryoscope method |
| Part 2: | 1981 Recommendations for the interpretation of the freezing-point depression of herd milk |
| Part 3: | 1980 Storage of samples |

| BS 5545: | Milking machine installations |
| Part 1: | 1978 Vocabulary |
| Part 2: | 1980 Specification for construction and performance |
| Part 3: | 1980 Methods for mechanical testing |

| BS 770: | Methods for chemical analysis of cheese |
| Part 1: | 1976 General introduction (including the method for preparation of the samples) |
| Part 2: | Determination of water content (reference and routine methods) |
| Part 3: | 1976 Determination of fat content (reference method) |
| Part 4: | 1976 Determination of salt content (reference method) |
| Part 5: | 1976 Determination of pH value |
| Part 6: | Determination of phosphorus content (reference method) |
| Part 7: | 1976 Determination of citric acid content (reference method) |
| Part 9: | 1980 Determination of nitrate and nitrite contents. Method by cadmium reduction and photometry |

| BS 5522: | 1977 Milk and milk products - Determination of fat content - Mojonnier-type fat extraction flasks |

| BS 5475: | Methods for detection of foreign fats in dairy products |
| Part 1: | 1977 Detection of vegetable fat by the phytosteryl acetate test |
| Part 2: | 1977 Detection of vegetable fat by gas-liquid chromatography of sterols |

| BS 5305: | 1977 Recommendations for sterilisation of plant and equipment used in the dairying industry |

| BS 3441: | 1977 Specification for tanks for the transport of milk and milk products |

| BS 5539: | 1978 Specification for safety requirements in rotary milking parlours." |

"However, there are also many standards, produced by other Standards Committees, which have a direct interest for the dairying industry. The titles and numbers of these and of all the DAC/-standards are given in Sectional List SL31 newly revised. Single copies of SL 31 are free."

DECLINE OF UK INPUT TO RESEARCH

From time to time and currently Governments introduce new frameworks for research which often have the unintended result of diminishing research activity both in value and quantity. One well-known and controversial initiative was the investigation by Rothchild & Dainton reported in 1972 as *"A framework for Government Research & Development"* - CMD 4814 (JSDT Vol. 25, No. 2 April 1972). *(See P.83).*

For a Society whose objective has been, since its foundation, to foster research and development in the UK dairy industry, the most depressing aspect to emerge from the history of the last 60 years is the collapse in relevant research activity in the UK. It is no accident that the collapse in research activity has mirrored the decline in the industry right through from the milk producer to the processor and manufacturer. Dr Chambers has summed up this sorry story in his own inimitable way under the heading *"The UK surrenders its leading role in dairy science and technology".* In view of the critical importance of this subject to the future of the SDT and the dairy industry, I make no apology for putting on permanent record Dr Chambers' analysis.

"In September 2003 the International Dairy Federation (IDF) celebrated its centenary with a 'world dairy summit' in Bruges. The programme for that great event included a series of six conferences, involving twenty-one separate sessions and over 120 speakers on topics that are considered to be of prime importance to the international dairy industry today. If one excludes representation in a conference on machine-milking, only two of those speakers were from the UK, as compared with sixteen from Germany, nine from France, and five from the Republic of Ireland. Neither of the UK speakers is from the dairy industry in the UK or from an institute or other organisation that is traditionally associated with the dairy industry - they were from Southampton University and the Institute for Animal Health. The nationalities of the respective twenty-one chairmen of sessions is not given but the only one that is recognisable as being from the UK is Barry Wilson, editor of the *Dairy Industry Newsletter*. He is chairing a session on 'The Future'. The UK does not feature: even Tesco's view of the future is to be given by a speaker from the organisation's Republic of Ireland division! Worst of all, from the viewpoint of the SDT, the twenty-nine speakers and six chairmen in a six-session conference on dairy science and technology do not include anyone from the UK. Of the twenty-nine speakers, nine are from Germany, six from France, three from Finland, three from the Netherlands, and one each from Australia, Ireland, Russia, Japan, Canada, Switzerland, Belgium, and Denmark. To be fair, New Zealand - notably - is also missing from the list. The general theme of this conference is highly relevant to any progressive dairy industry: it is *'Advances in Fractionation and Separation: Processes for Novel Dairy Applications;'* and the sub-sections relate to Lactose and Whey Products, Proteins, Sterile Filtration (to remove mastitic cells from milk), Physiologically Active Ingredients, Milk Fat, and Waste Water and Environment. Likewise in another conference, on Energy Consumption and Life Cycle Assessment towards sustainability, the UK does not feature either amongst the six chairmen of sessions or the fourteen speakers. Finally, in another conference on leading-edge biotechnology - specifically the Effects of Probiotics and Prebiotics on Health Maintenance - only one of the nineteen speakers comes from the UK: he is from the University of Southampton. The UK involvement is absolutely minimal (See IDF publication - *IDF World Dairy Summit: Final Announcement and Registration Form (2003)*. What a contrast with the leading role in IDF exercised by the UK industry and its support organisations over some ninety of the hundred years being celebrated in Bruges in September!"

"Of course this embarrassing example of the UK industry's fall from grace in terms of contribution to scientific and technological development, of appreciation of the role of scientists and technologists in the industry, and of meaningful involvement in the work of IDF is but the latest in a long line of such debacles. The slide into relative backwardness in these matters began with the closure of NIRD in 1985 and was accelerated by the demise of the Milk Marketing Boards in 1994/5. And amongst the consequences is the struggle to keep our Society alive, in the interests of the professional development of its remaining members, of providing fora for technology transfer, and of restoring some cohesion in a fragmented industry. An excellent example of the way in which things have changed in the UK industry in relation to IDF was provided by the UK's level of involvement in the annual sessions ('world dairy summit' to use the modern term) in New Zealand in 2001, as compared with the same event in the same country thirty-eight years earlier - in 1963. The 2001 event attracted an attendance of around 800, of whom only about sixteen were from the UK. John Houliston, chief executive of Dairy Crest, was the only senior executive of any UK dairy company to be present; not a single one of the GB milk-brokering co-ops was represented; Barry Wilson of *Dairy Industry Newsletter* was present: so too were three representatives from Northern Ireland - leaving about eleven representatives from the rest of the UK industry (See an article by Barry Wilson in *'United News'*, the journal of United Dairy Farmers in Northern Ireland, December 2001). In 1964, on the other hand, the total international attendance in New Zealand was only about a quarter of that in 2001, BUT THE OFFICIAL UKDA DELEGATION ALONE WAS ALMOST TWICE THE SIZE OF THE TOTAL UK ATTENDANCE IN 2001. The UKDA delegation, led by Richard Trehane of the MMB and Frank Procter of Birmingham Co-op, was the largest after those of Australia and New Zealand; and members of our delegation provided office-bearers for virtually every IDF commission; Dr T J Drakeley was vice-president of Dried & Condensed Milk; Dr A L Provan (President SDT 1951/52) was president of Production of Milk; Mr J Matthews (President SDT 1947/48) was president of

Industrial Dairy Techniques; Dr J A B Smith (President SDT 1950/51 and Director of the Hannah Dairy Research Institute) was secretary of Chemical Analysis; Frank Procter (President SDT 1944/45) was president of Dairy Economics; and Professor R G Baskett (Director of the NIRD) was president of Education, with F C White (President SDT 1958/59) as secretary (See JSDT, Vol. 17 (1964), pp 117-9). We were playing in the premier division then!"

"The low level of participation of UK delegates in the lecture programme of the INTERNATIONAL DAIRY CONGRESS IN PARIS IN SEPTEMBER 2002 provided another example of the regressive state of dairy science and technology in the country. In preparation for a breakfast talk that I gave at the end of the Society's Millennium Spring Conference at Kinsale in May 2000, I scanned the main international journals of dairy science and technology for the previous five years… and I found that the output of published papers from the UK (the usual criterion of research activity) was negligible, apart from a few contributions from the Hannah in Scotland and the DARD Science Service in Northern Ireland. In the course of the conference I also found that the subject areas being investigated at the National Dairy Products Research Centre at Moorepark, Fermoy, Co. Cork were those that my survey of journals had highlighted as leading edge".

"I have also looked in detail at some of the most recent issues of the Society's own journal, beginning with issue No 1 of 2001 and ending with issue No 2 of the 2003. Those issues of the Journal contain ninety scientific and technical articles. Only nine of those articles originated in the UK and related to topics relevant to the UK industry - and three of that nine came from the Science Service of DARD in Northern Ireland, leaving six coming from GB. Two came from the University of Reading - one a review article on the significance of genetic engineering, the other about an analytical technique. One came from the University of Cardiff - it was about anticariogenic activity of milk proteins. And the remaining three came from the dairy industry in England & Wales: from Dairy Crest on milk fats as ingredients; from Volac International about whey protein isolates; and from Express and the Dairy Industry Association about the climate change levy. All of this means that there was nothing from the government-aided agri-food research institutes and laboratories in GB or from the food research associations. Contrast that with the steady input from the Republic of Ireland: over those issues of the Journal, nine major articles on relevant aspects of dairy science and technology. In drawing this comparison I have as indicated earlier regarded as UK output only those papers that originated in the UK, and that were relevant to the interests of the UK industry: that means that I have not included as UK output four articles where one of the authors was from Reading University or Seale Hayne and the other(s) were students or post-graduates from overseas working on subjects (e.g. types of cheese, or ovine, or caprine milk) which were relevant to their respective countries of origin but not to the UK. Again it is a sorry tale".

"All of the foregoing mirrors what has been happening in our Society: when the Society was set up and for the next forty-five years the UK was at the forefront of dairy research and had at least twenty dairy scientists and technologists who were of world standing. Where are their successors? Looking at my analysis of the programme for last year's dairy congress in Paris, some may say that the French took the opportunity to dominate it because they were the organisers - but the thing to remember is that they were able to muster one hundred speakers of sufficient quality, experience, and standing to address the top people in today's world dairy industry. By comparison the UK is now in the third division. Indeed when one reads the first two articles in the *"Focus on Dairy Research"* series in the Society's Journal one is inclined to say that the UK dairy industry is now a 'non-league side'! It's sad: who is going to do something about it? Meanwhile in Northern Ireland, where the record on these matters has been rather better than in the rest of the UK, the Department of Agriculture & Rural Development - with peripheral involvement of a direct-rule minister - is preparing to fragment and begin the withering process of a structure of cohesive educational and technical support of the agri-food industry that has served the industry well for over eighty years - without paying the slightest attention to the well-argued and well-researched opposition mounted by virtually every stakeholder in the business".

Pauline Russell reporting on the SDT Symposium, on "Dairy Ingredients in Food" commented on the subject of fat-filled powders (JSDT. Vol. 48 No. 2, May 1995).

"Who would have thought only a few years ago that the Society of Dairy Technology would ever consider this as a serious subject, let alone one worthy of discussion by this learned Society, the President questioned. Now we are in the position where the price gap between butterfat and vegetable fat is narrowing fast. The milk industry has to look to the future - and if that is the product food manufacturers want to put in their formulations, then that is what the dairyman must provide. The speaker Gerard Hayman, UK Business Manager for Pritchitt Foods, started with a history of spray drying. First patented in the 1860s and used to preserve eggs, spray drying has come a long way to its present levels of sophistication. When Pritchitts came up with their fat-filled powders for cold ice cream mixes using vegetable oil instead of butterfat, the industry looked askance. Farmers went up in arms and tried to boycott their supplies. Now the wheel has turned. The ratio of butterfat to skim milk in value terms has swung from 70:30 to almost 40:60. The expectation is that the price of butter fat (less EC subsidy) will all too soon approach that of the better quality vegetable oils".

The entrepreneur who was first to realise that the addition of vegetable fat to a low value product, skim milk powder, would produce a versatile high-value product was the late Dick Lawes. The opening paragraph of his obituary encapsulated his contribution:

"Dick Lawes, inventor, entrepreneur, opportunist and businessman, who has died at the age of 80, was acknowledged worldwide as a pioneer in the production of new and novel food products involving the spray drying of fat-filled milk".

"His career, which spanned some 60 highly productive years, embraced the development and production of a whole host of ideas including fat-filled milk powders, ice cream mixes, animal feeds and protected high energy diets. Throughout his career his business success was characterised by identifying opportunities to create value out of undervalued food by-products. He was responsible for the establishment of a number of UK and International companies, and will be principally remembered as the driving force behind Pritchitt Foods - the UK's premier catering and retail dairy-based products manufacturer - and as the founder of Volac International, a leading UK-based animal feeds/human foods' manufacturer. The group of

Mr. B. Cahill presenting cheque to Mr. R. W. Lanes for sale of MacCormac Products to Express Dairies in 1985.

companies, which is still family owned, is one of the UK's largest independent businesses, employing over 500 people and with a turnover of £100 million. It operates in 80 countries throughout the world".

"It was in 1933 that Dick Lawes, fresh from Roan School, Blackheath, joined a small, three-man chemical distributors, L E Pritchitt & Company Limited. The company was run by Bill West who quickly recognised Dick Lawes' potential and appointed him as a partner. The business was based in London's Mincing Lane".

"During the Second World War, Dick Lawes enlisted in the RAF Coastal Command, where he became a navigator, and subsequently a navigational instructor, throughout the War years".

"It was after the War that he pioneered the production of fat-filled milk powder which recognised the value in skim milk. It has taken the intervening 50 years for the European dairy market to recognise that there is more value in skim milk than butter".

"The post-War demand for powdered milk and the resurgence of the ice cream trade in the UK saw a booming business period for Pritchitts. It was during this time of rationing and shortages that Dick Lawes led Pritchitt Foods down the avenue of milk and ice cream powder production. The brand names developed, Millac and Comelle, are both market-leading products to this day. At that time, Dick went on to revolutionise consumer cream and dessert topping products by developing the UK's first synthetic cream product-Roselle."

"His first major manufacturing facility was set up in 1952 for L.E. Pritchitt at Newtownards, Co. Down in Northern Ireland, where he designed and built his own equipment to manufacture the products he had invented".

"A new factory was designed and built at Killeshandra in Ireland in 1962 to handle the rapid growth of the business and this factory was soon processing over a quarter of a million gallons of skim milk per day. Profits were reinvested, resulting in factories being built in Ballaghaderreen, Co. Roscommon in the West of Ireland, Bailieborough, Co. Cavan in the North East of Ireland, Liverpool and Felinfach in West Wales. His part in the development of the agricultural economy in Ireland was recognised in the rare award of honorary life membership by the Irish Farmers Association".

"In Newtownards in 1974, the first, fully aseptic one litre Tetra Brick machine was installed in the UK, enabling the company's UHT products to be packed quickly and efficiently for onward distribution around the world".

"Dick Lawes' thoughts also turned to other areas where the technology he had developed so efficiently could be put to good use, and the solution was found with animal feeds. Using the same process for calf milk and lamb milk replacer, after close co-operation with various universities, the Volac business was born.

"The first product was Volac Easy-mix, followed by the production of Megalac, a protected fat product. Today, more than 80,000 tonnes are produced worldwide, including production under license in America and Japan and, more recently, in the Czech Republic. Under Dick Lawes' guidance, the Liverpool and Welsh factories have progressed the production of animal feed and Volac International is today a major world class player in the animal feeds/human foods' industry".

"Pritchitt Foods is the UK's largest single producer of mini-pot products, making no less than 40 million pots per month".

"Dick continued working and investing to the very end of his life. His most recent project was the application of biotechnology for the prevention of disease in both humans and animals using Probiotics. This work is being done in Port Talbot, South Wales".

Mr. Bill McAskie joined Pritchets on 7 October 1957 as Chemist and Assistant Manager. Thus began a very productive partnership. In his own words:-

"When I joined the company, it was then called Dairy Cream Products, although it reverted to L.E. Pritchitt some time later.

"Initially there was one small spray dryer, with a capacity of $1/4$ tonne an hour of powder, and one of my first jobs was to help in the building of the second spray dryer, spray dryer 2, or SD2, with a capacity of $1/2$ tonne an hour."

"The problem then came of getting enough skim milk to feed the 2 dryers, so in 1959, Mr Lawes (RWL) went across the border looking for more skim. He tried Lough Egish and got a rather cool reception, with vague promises of some skim supply. Then he went to Killeshandra and met the late John O'Neill, who immediately suggested that it would be a far better idea to build a factory there because there were lots of grants available in the Republic. I was despatched immediately to have a look at the creamery set-up and test the quality of the milk, as the collection system was then fairly primitive. I was able to prove that, provided the milk was cooled after reception, the quality would be acceptable. (Killeshandra had a central creamery and some dozen small satelite reception centres)."

"The decision was made to go ahead with a factory to make Millac and skim milk powder, at the rate of 20,000 gallons of skim a day, through a $1/2$ tonne dryer, of the unique Pritchitt design. (The factory was called MacCormac Products, after the name of our then Dublin based accountant) I did most of the design work and was able to do my own drawings. The factory started production in 1960 and I was the factory manager. We set up collection from the satellite creameries using 1,000-gallon

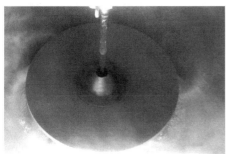

*Disk spraying Milk Powder. High speed flash.*

tankers, towed by farmers' tractors. We needed 2 more spray dryers in the sixties and by 1970 we were handling 200,000 gallons a day! We added evaporator centres at Monaghan, Longford and Bailieboro, all running 24 hours a day, seven days a week, same as the factory. A fourth dryer was added now with a capacity of 4 tonnes an hour and then a fifth, bringing peak capacity up to 400,000 gallons a day by the middle seventies and a work force of some 400 people."

"Demand was still outstripping our capacity so a whole new factory was built in Ballaghadereen, Co. Roscommon, with the same 4 tonne design dryer. From the first sod being dug, on a green field site, production of the dryer began in 3 days short of 6 months! (I was now General Manager)."

*Disk used in Lawes' style 'Upside-down' dryers.*

### NIRD AND THE AFRC INSTITUTE
### OF FOOD RESEARCH

A major blow to research in the dairy industry was delivered with the closure of NIRD in April 1986 and its absorption into the AFTC institute. This development is referred to in the Journal (JSDT Vol. 37 No. 2 April 1984):

"The National Institute for Research in Dairying at Shinfield, Reading, perhaps more generally known as NIRD, is renowned throughout the world for its research work. Increasingly in recent years the dairy industry has realised that the Institute is not an ivory tower and that scientists at Shinfield are pleased to apply their knowledge and experience to the solution of practical industrial problems. The Society of Dairy Technology was established through the initiative of a famous Director of the NIRD, Professor H D Kay, F.R.S., and since then the Institute has been closely involved in the Society's work at all levels: it has provided four Presidents and two Society God Medallists".

*Millac in Pritchitts own ship, Greenore, Co. Louth, Ireland.*

"It is therefore something of a shock to learn that after more than 70 years of existence the NIRD is likely to disappear in name and in its present form in a little over one year".

"The NIRD is a grant-aided institute of the Agricultural and Food Research Council (AFRC), a part of the University of Reading, but derives most of its income from the Government either directly through the AFRC or from commissions from the Ministry of Agriculture, Fisheries and Food placed through the AFRC. The Council has been facing great difficulties, stemming from a reduction in the true level of research funding from Government and a commitment to put a larger part of its effort into food research and other high priority work at the expense of agricultural production, which is now given lower priority".

"A policy decision has been made to concentrate food research at three existing AFRC institutes: the Food Research Institute at Norwich, the Meat Research Institute at Langford, and the NIRD. The changes at Shinfield will be far-reaching. That part of the institute's work which deals with milk production will be separated from milk processing and utilisation and will be combined with that of the Grassland Research Institute, Hurley, to form a new production research institute with its administrative centre at Hurley. The milk processing and utilisation parts of the institute will remain on the main NIRD laboratory site as the nucleus of a new food research institute. This will formally come into existence on 1 April 1985, but at the moment it has been given no name. The association with Reading University is expected to continue. The new institute will continue to specialise in milk and milk products but will also have wider food interests. What these will be has yet to be decided as a result of further policy discussions. The amount of effort to be devoted in the future to milk and milk products is also still uncertain; it will depend on the perceived needs for milk utilisation research within an overall strategic policy for food research".

"In recent years the amount of research at NIRD on problems related to milk utilisation has increased significantly, and formal relations with the dairy industry have been established through the Dairy Industry Research Policy Committee and its Contact Groups. The value of these links with the industry is well recognised by the AFRC. It is to be hoped that the changes to come will neither interfere with these links nor result in a reduction in milk utilisation research".

In his report to the 41st AGM on 23 October 1984, Dr D Muir, Chairman of the Education and Research Committee, commented as follows (JSDT Vol. 38 No. 1 January 1985):

"I am sad to say that the National Institute for Research in Dairying, to which this Society owes a considerable debt, will no longer exist. From 1 April next year NIRD will become the Food Research Institute (Reading). I personally think it is a dreadful name. I am pleased to say that we have some assurance that a very significant proportion of the new work at the Food Research Institute (Reading) will be based on dairy science and dairy research of the sort that we have expected from NIRD in the past. It remains to be seen whether the Institute will ever be the same again. I doubt it".

The Commemorative Booklet issued to celebrate 50 years of the SDT gives information about these developments, and also mentions the activities of Brackenhurst College, Cannington College, Seale Hayne, Auchincruive, Reaseheath and the Hannah Research Institute.

DAIRY RESEARCH IN THE FOOD SCIENCE DIVISION, BELFAST

Excellent work was carried out in this organisation which involved both the Dept of Agriculture for Northern Ireland and Queens University. A description of current activities was given in the Journal in 1997 by M T Rowe, Anna M Fearon, D E Johnston and John Early:

"This paper describes dairy-related research currently underway in the Food Science Division in Belfast. Topics include the heat resistance of *Mycobacterium paratuberculosis,* which has been implicated as the causative agent of Crohn's disease in humans, incidence of enterovirulent *Escherichia coli* in raw milk, shelf life of pasteurised milk, dietary manipulation of milk fat composition and the application of high pressure processing to dairy products. (JSDT Vol. 50 No. 2 May 1997)."

# THE AFRC INSTITUTE OF FOOD RESEARCH
## Lyndon Davies
Institute of Food Research, AFRC, Reading Laboratory. (JSDT Vol. 41 No. 3 August 1988)

"The last few years have seen a great deal of change within the Agricultural and Food Research Council (AFRC) and I am conscious that some of you are still unclear about the structure and precise role of the new institute. I will therefore explain the overall structure and organisation of the Institute of Food Research before going on to give you an overview of our research interests. I will also refer to the relationship between IFR's interests and those of the Hannah Research Institute and the research associations".

"Until fairly recently, AFRC's programme of in-house research was organised into some 20 separately run institutes and units mainly in England and Wales. Of these it is the National Institute for Research in Dairying (NIRD), which will be most familiar to many of you here".

"This large collection of research stations has now been drawn together administratively into eight new institutes (Fig. 2), each under a single Director of Research, with the intention of bringing a crisper focus and better coordination to AFRC's main research areas. The eight directors together with the Deputy Secretary of Council, The Director of AFRTC Central Office and a representative of DAFS (Department of Agriculture and Fisheries for Scotland) meet as a Management Board under the chairmanship of the Secretary (Chief Executive) of AFRC. Thus the Directors of institutes have been drawn much more directly into a central role with regard to the AFRC's policy formulation and implementation".

"The AFRC Institute of Food Research was formally constituted in November 1985, bringing together the former Meat Research Institute at Langford, Bristol, the Food Research Institute at Norwich and the Food Science Division of the National Institute for Research in Dairying at Shinfield, Reading. In addition, the Food and Beverages Division of the Long Ashton Research

Station, Bristol, was transferred to Reading. Some confusion in nomenclature occurs since each enjoyed a brief interim and independent identity as the 'Food Research Institutes' at Bristol, Norwich and Reading. However, since then, there has been the formal constitution of the IFR Bristol Laboratory, IFR Norwich Laboratory and IFR Reading Laboratory".

"As an aside and for those of you who are wondering what happened to the remainder of NIRD, the work on animal production was transferred to the AFRC Institute for Grassland and Animal Production (mainly at Hurley, near Maidenhead) except for research on mastitis, which has been transferred to the AFRC Institute for Animal Health at Compton, near Newbury".

## LORD WADE AND FARMHOUSE CHEESEMAKING

The Patron of the Society, Lord Wade of Chorlton, contributed an interesting article on future development in farmhouse cheesemaking in 1982. (JSDT Vol. 35 No. 4 October 1982).

"We extend our warmest congratulations to William Oulton Wade, who received the honour of knighthood in the Queen's Birthday Honours (1982)".

"Sir Oulton was educated at Queen's University, Belfast where he graduated BSc. Agriculture. He joined the family farming business of William Wild and Son (Mollington) in 1955 and during the next 20 years he gained practical experience in improving a large herd of pedigree British Friesians and establishing a large pig-breeding unit. He was fully involved in farm management and expanding a farmhouse Cheshire cheese-making industry. He became Managing Director in 1976 and the company now farms 500 acres near Chester, running 32 dairy cows and other cattle, 5,000 fattening pigs and 2 farmhouse dairies making Cheshire and Leicester cheese".

"He has been very active in the agricultural world, has held office in the Farm Buildings Association and the Cheshire Agricultural Society and has been Chairman of the Cheshire and Lancashire Farmhouse Cheese Federation. He is at present Chairman of the English Cheese Exporters Consortium. In spite of all these activities, he has found time to act as a Justice of the Peace and was a member of the Cheshire County Council for 4 years".

"Sir Oulton has been an active member of our Society for many years and has given authoritative papers on farmhouse cheesemaking and the development of the export cheese business. We wish him continued success in his many activities. (JSDT Vol. 36 No. 1 January 1983)."

## REORGANISATION OF ADAS

John Sumner reported on the Reorganisation of ADAS the essential point being the introduction of direct payment by farmers for the services of ADAS from the 30 March 1987 (JSDT. Vol. 40 No. 4 November 1987).

"As a result of a report by Professor Bell (ADAS Director General) in 1985, Government decided to seek savings in the cost of ADAS and endorsed the view that farmers and others who avail themselves of the services of ADAS should contribute to the costs".

"As part of the new look, commercial ADAS, the service has been structurally reorganised to proved a quicker, more flexible service to farmers. The changes are based on three principles:
1. A strong, simple management structure allowing the full range of ADAS services to be handled flexibly and efficiently.
2. Giving advisers in the field the best professional and scientific backup.
3. Building on the competence and integrity of the professional disciplines on which ADAS depends."

"The old ADAS Agriculture, Land and Water, Agricultural Science Services have been reorganised into two new, market-led services: the Farm and Countryside Service (FCS) and the Research and Development Service (RDS). The State Veterinary Service has not been affected by these changes".

"The FCS will bring together all the professional, technical, scientific, business and environmental advice previously provided by the Agriculture, LAWS and Science Services and it will also carry out the statutory duties covered by those services. The RDS will control all work carried out by Central Science Laboratories, experimental centres and specialist regional R & D units".

"From 30 March 1987, farmers have been asked to pay a fee for most of ADAS advice and other services. While every customer is given a quotation for the work required, it is difficult to give precise prices in general terms for such a diverse range of work. As an indication of level of prices, for a farm advisory unit the fee will be £28 + VAT for a visit lasting up to one hour and a further charge at that rate for additional time. However, a great deal of ADAS work will be done through contracts covering other services besides farm visits. Such prices are negotiated and based on the needs of the business, and there are many special schemes designed to meet special needs - e.g. an animal health scheme. In addition, advice and help is available by annual subscription".

"One important element in the reorganisation is that advice on conservation (including the prevention of pollution), farm business diversification and animal welfare will remain free".

"The RDS will continue to develop the basis of knowledge on which ADAS rests, and will provide R & D work on contract to agriculture, horticulture, food and associated industries…."

# CHAPTER FOUR

## THE TECHNICAL AND EDITORIAL BOARD, THE JOURNAL AND OTHER PUBLICATIONS

START OF JOURNAL

Dr. Mattick, the Society's President in 1946-47, in an Editorial in October 1947 (JSDT Vol. 1 No. 1 October 1947) heralded the progress of the Society and announced the decision to launch its own journal.

*DR. A. T. R. MATTICK*
*PRESIDENT 1946-47*

"When some four years ago a group of people, who were convinced of the desirability of forming a Society to facilitate contact and discussion between representatives of all sections of our many-sided industry, founded the Society of Dairy Technology, there were those who doubted the wisdom of such a move, especially in wartime. They have been manifestly confounded! The membership is now about, 1,200 and is increasing steadily. In addition to the London Headquarters, Sections flourish in Scotland, the North West, Wales, the Midlands and the West. The enthusiasm of the Section officers was encouraged by the indefatigability of the Presidents, Professor H. D. Kay, F.R.S., Mr. F. Proctor and Mr. Graham Enock to whom the Society owes a debt of gratitude".

"From the foundation of the Society, it has been obvious that material of such value to all branches of the industry should be recorded in a journal of good standing, available to all members".

"Thanks to the good offices of Dr. Cronshaw and the proprietors of *Dairy Industries* much has in fact already been published in that journal, but your Council has decided that the time is now ripe for the production by the Society of its own journal, of which this is the first number. Four numbers will appear annually and will contain all that is most valuable of the proceedings of the Society and its Sections, and will be available to all members. For the moment, the size is necessarily restricted by the rationing of paper supplies and other causes and, in any case, that judicious selection, which is the safeguard of all good journals, must be exercised".

"I am confident that all members will support our resolve to maintain the highest possible standard of material and production".

"Presidents have their privileges as well as duties and I am happy that it has fallen to me to introduce this first number of the Journal of the Society of Dairy Technology".

Further progress in the development of the Journal was reported at the AGM on 27 October 1947 (JSDT Vol. 1 No. 2 January, 1948) with the introduction of advertising and the limitation imposed by the control on paper use! Tribute is paid to the work of Miss Vera Watson.

EDITORIAL BOARD

The Chairman (Mr. Crowther) in presenting the Board's report expressed his regret at the impending severance of his active connection with the work of the Society.

"Referring to the Journal he recalled the announcement made by M. Ritchie in last year's report that it had been decided to recommend to the Council to accept advertisements for the Journal. He agreed with the President as to the wisdom of this step, which he felt was the only one possible. The alternative would entail quite a considerable and, in the view of the Board, undesirable increase in the membership subscription rates. The Board reserved the right to select advertisers and to check the copy which they submitted. He wished particularly to thank those advertisers who had prepared special copy for the journal, - copy, which, he felt, attained the standard of dignity which the Board would try to preserve all the way through. It was quite a business to prepare special copy for one journal, and to those advertisers concerned the Editorial Board would say a special 'thank you!'

"The Board's problem when it came to plan the journal was, of course, that so much of what it wanted to do was subject to controls. The first concern was the amount of paper available. The Board estimated that the Society could issue quarterly an 80-page journal, which it was felt would adequately cover its activities. Then it was found that other restrictions existed as to the amount of advertising space which could be let, and permission was finally obtained to let a total of sixteen pages out of eighty."

"In the first issue now in their hands members would see that the Board was following the policy of endeavouring to print in full the papers read at General Meetings of the Society, and to give particular attention to the discussions which follow those papers. They would note that a full report of the discussion which took place on the 'metals' symposium had been included, thanks to a very creditable effort on the part of the reporter."

"The Board aimed also at giving as much space as possible to Sectional meetings."

"He expressed the thanks due to Dr. Charles Crowther for accepting the Editorship of the Journal, and to his various colleagues, past and present, on the Board."

"He warmly endorsed the President's tribute to Miss Watson, saying that, but for her, the Journal would not have been out before the meeting, which was the deadline aimed at."

"The Board hoped that in future issues it might be possible to include in the Journal notes on various personalities concerned with the Society, and to give indications of forthcoming events. The aim of the Editorial Board would be to ensure that the Journal would be in every respect not unworthy of the Society."

The Editorial of Mr J Matthews, President of the SDT 1947-48 on the occasion of the second issue of the Journal, reports progress and apologise for mistakes. (JSDT Vol. 1 No. 2 January 1948.)

Mr. J. Matthews
President 1947-48

"In sending to you the second issue of the Journal of the Society of Dairy Technology we have at least no apologies to make as far as the scope of the contributions is concerned. Papers will be found dealing with subjects as wide as the general future of the milk supply and as detailed a matter as the construction of milking machines. Other papers cover as widely differing subjects as the question of the payment of quality premiums for milk as produced, and the spread of infections by milk as consumed…"

"The Society is now entering its fifth year of being and in spite of the difficult times through which we are passing we believe it is increasing in vitality and usefulness, as well as in membership".

"Time passes quickly and we feel that, as a Society, we must now lay the foundations of future work and accomplishment. May we then address an especial message to the senior members of the industry? - to encourage and give facilities to their juniors to begin to play a full part in the work of the Society, to attend meetings, both central and local, to take part in discussion, and in due course to take up their share of responsibility as officers on whose work the success of the Society so much depends".

"We have to apologise for a number of mistakes in, and omissions from, the last issue, - which was of course also the first, - and would explain that they were mainly due to a hastily arranged organisation and to working against time for early publication. The Editorial Board is striving to improve this and other matters concerning the Journal, and we shall be grateful for comments, criticism and suggestions from members."

In September 1965 a questionnaire on the society's Journal was sent to all members (JSDT Vol. 19 No. 2 November 1966.) There were 170 replies only 7 per cent of the total membership. Mr. F. C. White, the Chairman of the Editorial Board commented as follows:

"The Editorial Board is grateful to those who replied, and particularly to the considerable majority who went to some trouble to offer constructive suggestions and helpful criticisms. Varying points of view were expressed on every question"…..

"The main object is to produce a house journal, recording the activities of the Society in its endeavour to provide a technological and social link between members, and a record of the more interesting papers on all aspects of the dairy industry, which have been read at meetings of the Society".

"It is evident that most readers consider the balance fairly kept, although many would like more articles on their own particular branch of dairying. The requests for more scientific or technological articles can only be met if members arrange regionally, or ask Council to arrange nationally, for appropriate speakers; the Editor will be happy to print such papers. Original articles, other than those which are the subject of meetings, are always welcome".

"The appearance of the Journal brought forth some criticism of the 'muddy brown' covers, while others have 'grown accustomed to her face.' The Board finds no justification for the considerably increased cost of glossy covers, but the colour and format of the covers will be brightened up".

"There is fairly general agreement that, while the discussion is often a most important part of any meeting, there is scope for more severe editing, 'to eliminate the pleasantries and padding' inevitable from all but the most accomplished speakers. It is a difficult task, for anyone except the speaker, to decide how much of his contribution is worth recording for posterity. It is proposed, therefore, to try experimentally, at general meetings in the first place, distributing paper to speakers in the discussion and inviting them to record their comments. It is believed that this practice, which is followed successfully in some other Societies, will lead to conciseness and accuracy, but the tape recordings will still be available as supporting evidence when required".

"Section notes are obviously valued by many members, particularly those in more remote areas. They will be continued".

"Many would welcome more information from overseas. It is hoped to report on matters of interest arising from meetings of the International Dairy Federation, the Food & Agriculture Organisation, United Nations' Children's Fund and other international bodies, in so far as they may have an important bearing on dairying in the United Kingdom, especially in such matters as international standards".

"The Board will also consider inviting guest articles on the dairy industries of overseas countries and the developments which are taking place".

"It is evident that periodical reports on dairy education, the changes that are taking place in universities, colleges and institutes and the future trends necessary to meet the demands of an ever-changing industry, will be of interest to a large number of readers".

"New legislation appears from time to time and also new BSI standards and techniques. Apart from those few, who during the period of gestation have become only too familiar with the implications of such documents, there are many in the dairy industry who would welcome an independent interpretation and criticism from the point of view of the dairymen, milk producer or laboratory worker, as the case may be".

"It is hoped to provide this where the subject has not already become of such importance as to justify meetings, which would be reported in the usual way".

"Requests for a correspondence column invite the comment that the Editor can only publish letters received. It would be a little difficult to maintain a lively interchange of ideas through letters in a Journal published only quarterly, but there may well be some matters unsuitable for an article, which a member wishes to bring to the notice of others in the industry. In such cases the Editor will always be glad to receive contributions and to consider publication".

"The Journal of the Society occupies a unique position in the industry. It does not attempt or indeed wish to compete with its sister journals published by the trade press, nor with those purely scientific publications recording current research. It is a more personal link between those very diverse interests, which combine to make up the dairy industry".

"It will be the endeavour of the Editorial Board to see that the Journal faithfully reflects the changing interests and aspirations of all members. To this end, any constructive criticism or suggestions for improvement will always receive the careful consideration of the Editorial Board".

EDITORIAL BOARD

In 1979, P.H.F.L. wrote a brief description of the functions of the Editorial Board (JSDT Vol. 32 No. 4, 1979)

"Although the Society was formed, with 243 members, in 1943, wartime restrictions on the use of paper were not lifted until mid-1947, and consequently the first journal was not printed until October 1947, when membership was recorded at 1,200".

"The first issue carried, among others, papers from a symposium on *'Metals and Milk'* held in May 1947, and the proceedings of that year's Annual General Meeting. The journal has been published at quarterly intervals without a break since that first issue".

"The Editorial Board first met in January 1947 to get the project off the ground and consider printers, etc. It is interesting to note that the first printers, F. J. Parsons Ltd., quoted £157 10s to print one issue. It now costs over £2,000 an issue to produce, with postage costs averaging £360 per issue in addition".

"The present constitution of the Board, which meets every three months immediately after each issue, is basically nine members, one of whom must be a Council member, with power to co-opt, but they are joined by the Managing Editor and Honorary Assistant Editor together with an Honorary Scientific Editor and the President in office".

"The Chairman of the Board must be a Past President of the Society and is elected for a term of six years. The present Chairman is Dr. J. Rothwell, of the Food Science Department, University of Reading"....

"At its meetings the Board discusses the content of the previous published issue and deals in depth with the content which will make up the next issue. It also regularly reviews the financial aspects of producing the journal as well as the placing of advertisements and its revenue. Other tasks include the compilation and editing of the Society's publications. At present, work is commencing on a new edition of the *'Pasteurizing Plant Manual'* to replace the previous reprinted editions, which have been out of stock since early this year. A booklet on the proceedings of the 1977 Residential Course on *'Milk and Whey Powders'* is now nearing completion".

"A review of the year's work of the Board appears in every January issue of the Journal, included with the minutes of each Annual General Meeting".

"This, very briefly, is the history and work of the Board. Although it has always been the Society's policy to include proceedings of symposia and conferences, as it is felt that this is one of its prime duties to its members, original contributions are always welcomed by the Board and, although printing is not always guaranteed, each paper is judged on merit. Items of interest within the industry, i.e. honours, retirements, etc., are also always particularly welcomed. It should again be pointed out that preparation for each journal takes 2-3 months and items should always be with the Board for the meeting *before* the next issue. The journal is sent to all <u>2,600 members</u> and to over 300 outside subscribers, which include universities, colleges, libraries, firms, etc., around the world."

"While on the subject of the Editorial Board, Council in 1978 gave approval for the present uneconomical-to-produce Imperial size to be replaced by the more generally used A4 size journal. This change will take place from January 1980 with Volume 33. No.1".

"You will notice this is a Board not a Committee. Unlike the Committees it is not made up of Council members. Only 1 member of the Board need be a Council member, although often there are in fact more than that on the Board. Its secretary is automatically the Society's Executive Secretary. It has another distinction in that, whilst Committees are limited to Council membership and therefore one automatically relinquishes committee membership on retirement from Council after the statutory period of 3 years, election to the Editorial Board is for 6 years".

"Its sole function is the editing and production of the Journal…. The Journal is acknowledged throughout the world for its high technical and literary standard. To maintain this standard, exceptional editorial effort is needed".

"Again, this is all voluntary work, with the exception of a fee paid to the permanent Managing Editor for the overall task of collating the material and producing the Journal, and much time is spent by the members of the Board on the task of ensuring a high standard of publication".

"Some lectures which are good to listen to look dreadful in print, and in a technical journal might be quite unusable…. and one of the jobs of the Board is to make a printable article from an 'unprintable' one".

"Able and knowledgeable speakers and authors are sometimes unable to set their contributions out to the standard required, and the Board vets all contributions for technical content, grammar and spelling, often regrouping words, sentences or whole paragraphs".

"Let me also emphasize that the Editorial Board does not expect contributions to be cut and dried ready for printing without editing. We need the material and are glad to welcome it from any source, so whilst it is nice to have a perfect contribution please do not hesitate on the grounds that you might feel your article is not quite right. It will be sorted out and the proof offered to you before printing".

"Editing is done in members' spare time between meetings. The formal meetings are used to allocate articles to those best suited to edit them, and for agreeing all the business side - the format, finance, printing arrangements, advertising, etc".

As a consequence of the contraction of the industry and the decline in research and technical activity, the latter became a lesser part of the functions of the subsequent Technical & Editorial Board. Dr. E. J. Mann, Chairman of the Board since 1996 takes up the story:

DR. E. J. MANN
CHAIRMAN OF THE BOARD

THE BEGINNINGS

While several of the individuals who made major contributions to the Journal during the last 56 years, are mentioned by name, no attempt is made to mention everyone who has contributed in one way or another to making our Journal the successful publication it has become over the years, especially as the result of recent innovations which are described later.

There is little doubt that the Journal has become the flagship of the Society especially during recent years when other Society activities have had to be curtailed as the result of fundamental changes occurring in the dairy industry, such as the closures of the National Institute for Research in Dairying (NIRD) and of the Milk Marketing Boards (MMBs) and the general contraction of dairy industry activities which are described elsewhere in this book.

When, in 1943, a few farsighted individuals from research, Universities and the dairy industry got together and decided to give birth to a Society of Dairy Technology, there were sceptics who doubted the wisdom of such a move in the middle of a world war. However, their doubts were soon confounded for, by 1947, the new Society was well and truly off the ground and had a membership of 1200, and was encouraged by the enthusiastic membership to create Regional Sections, various Committees and to hold numerous Seminars on subjects ranging from milk production through to the end products. These Seminars, usually attended by up to 200 persons, generated much useful information, some of which was published by *'Dairy Industries'*, now *'Dairy Industries International'*. By 1947, encouraged and assisted by Dr. Cronshaw of *'Dairy Industries'*, the Society decided that it was time to produce a regular Society Journal, and the quarterly *'Journal of the Society of Dairy Technology'* was born.

The Journal's first editor was Dr. Charles Crowther and the first issue of the Journal appeared in October 1947. This issue contained the full papers of a Society Symposium on *'Metals and Milk'*, including verbatim discussions after each paper, as well as papers by already distinguished members of the dairy industry: Dr. Mattick, Professor. Crossley and Professor. Capstick, representing dairy research, education and the dairy industry. Significantly it also contained a paper by Rogers on American milk plant practise, giving an early hint that the Journal, although primarily concerned with Proceedings of Symposia and of General Society Meetings, was going to look increasingly beyond these shores for information and inspiration. The legendary Dr. J. G. Davis, as Chairman of the Editorial Board from 1947-1951, opened a thought-provoking editorial on *'The function of the Society's Journal'* in the April 1948 issue of the Journal as follows:

Quite obviously the function of the Journal of the Society of Dairy Technology must be to serve members of the Society, who will be interested in two distinct things - the Society and Dairy Technology. Our Journal must therefore serve these two distinct functions. It must contain a sufficiently detailed record of the activities of the Society to satisfy all reasonable requirements and in addition it will contain descriptive and critical articles on the advances of dairy technology.

The Editorial Board has decided that four numbers of 64 pages each will be published each year. Most of this space will be available for technological articles. Priority for publication will be as follows:

1. Papers read at meetings of the Society, including local Branches.
2. Contributions by members.
3 Contributions by non-members.

JG' goes on to define technology as a complex or summation of applied sciences, depending essentially on close co-operation between the scientists and the practical man. Significantly, he goes on to highlight the importance of education, of the technologist as well as that of the practical or businessman to appreciate the role and limitations of technology.

So, for a good many years, the Journal satisfied the objectives outlined in 'JG's thought provoking editorial, and about 80% of the Journal contents were devoted to Society matters, i.e. papers presented at numerous symposiums, with full discussions, reports of General Meetings and reports of study tours to different parts of the world.

It was noteworthy, especially during the early years, that the Journal covered the whole spectrum of dairying, from milk production aspects through to the chemistry, physics, microbiology and technology of milk and dairy products.

However, right from the beginning, the Journal contained reports of an international nature at least one per issue. Examples of these are articles on the International Dairy Federation, world trends in milk production, dairy research in India and many others.

The report on the International Dairy Congress held in the Netherlands in 1953 immediately after the terrible floods was particularly memorable for members of the Society, because it decided to hold its Summer Conference in conjunction with the Congress. Some of us still have vivid memories of the SDT Meeting held on a very hot day in a dark, unventilated cinema in Scheveningen, when a Dutch speaker read a very long speech about the Netherlands dairy industry using text which we had been given beforehand. The result was that most delegates, including our distinguished Chairman Professor H. D. Kay, fell asleep and the speaker had difficulties making himself heard above the crescendo produced by the snoring!

THE MIDDLE YEARS TO THE PRESENT

There have been noticeable and significant changes in the journal's content during the last 10-20 years, as manifested by the following:

1. Almost total exclusion of papers relating to dairy farming
2. Reduction of papers from the UK but compensated by more papers from Ireland
3. Increased input from emerging countries
4. Increased emphasis on new products and dairy ingredients

The reasons for these changes are many and complex, but include the following: (a) the closure of the NIRD (especially in relation to 2. above); (b) reduction in the number of papers generated from SDT Symposia as the result of the increasing unwillingness of speakers to provide publishable texts; (c) increased commercial secrecy resulting in fewer papers on 'cutting edge' technology/products. Fortunately, there have been some welcome exceptions to these trends in the Journal content - the most excellent one being the publication of the entire proceedings of the outstanding SDT Millennium Spring Conference in May 2000 in Kinsale, Ireland under the general heading of 'Dairy Technology' 2000 as the November 2000 issue (IJDT Vol.53 No.4 2000)

The above trends, together with the increasingly international nature of the Journal, the need to turn the deficit in the Journal account into a surplus, and the importance of keeping up with developments in electronic information transfer, have persuaded the Technical & Editorial Board to introduce some major changes in Journal production which are described below:

CHANGE TO COMPUTER TYPESETTING

In the early 90s we changed from lead typesetting to computer typesetting resulting in significant cost savings in journal production and resulting in the income derived from the Journal account exceeding expenditure by up to £5000 for a number of years, at a time when Society income from other sources of revenue was declining.

## CHANGE TO INTERNATIONAL JOURNAL

In 1996, the Technical & Editorial Board took the momentous decision to change the title of the journal to the *'International Journal of Dairy Technology'* from February 1997 onwards. As has been explained above, the contents of the journal were becoming increasingly international in nature, with a significant decline in papers generated from Society symposia. It seemed a logical step to reflect this trend in the title of the Journal, giving it, and possibly the Society as a whole, a more international image in the coming years.

## APPOINTMENT OF REGIONAL EDITORS

In order to enhance this international image, the Board decided to appoint distinguished international dairy scientists as Regional Editors - Dr. J. Hill from the Fonterra Research Centre, New Zealand to cover Australasia, and Dr. J. Lucey of the University of Wisconsin, USA to cover the Americas.

## PUBLICATION OF SEPARATE QUARTERLY MEMBERS' NEWSLETTER

This was introduced to replace the Members Supplement (traditionally included in the Journal), which was regarded as too domestic for international readers, and to provide a vehicle for articles of more general interest as well as news of member activities. Mr. R. Hamilton, Past President and the first editor of the Newsletter, created an attractive and interesting publication and his good work has been continued by Dr. A. C. O'Sullivan, the Executive Director of the Society.

## CO-PUBLISHING AGREEMENT WITH BLACKWELL PUBLISHING

This is perhaps the most important step taken by the Society since its creation in 1943. Under the Agreement, the SDT through its Technical & Editorial Board retains full editorial control over the Journal contents, while Blackwell Publishing is responsible for the printing, publishing and distribution of the journal in its printed format, as well as for the creation and maintenance of the Online database of the Journal which now forms part of the much larger Blackwell Synergy database, enabling users to access information not only from our database but from related databases on the worldwide web. During the coming years, the Society should benefit increasingly from this association with a world-renowned scientific publisher with its worldwide publicity and marketing network. Under the Publishing Agreement, the Society is provided with a guaranteed annual royalty payment for its services, thus ensuring that the Society Journal account remains in surplus during the 5-year period of the Agreement and hopefully beyond. Already, the Journal has taken on a more professional appearance, especially from 2003 onwards.

## SOCIETY WEBSITE

Blackwell Publishing has also taken on the management of the Society's greatly improved website accessible on: www.sdt.org

## INTRODUCTION OF *'FOCUS ON DAIRY RESEARCH'* SERIES

The article on *'The Swiss Federal Dairy Research Station'*, which appeared in the February 2003 issue of the Journal, represents a new venture for the journal. During the coming years we are planning articles on dairy research from all parts of the world. These articles will appear in random order and their purpose is three-fold: to provide our readers with an interesting new regular feature and to provide them with useful contacts; to make the Institutions described aware of our journal and activities; and to persuade scientists/technologists in the institutions described that the SDT can provide a good publication medium for the results of their work.

## EXPANSION OF TECHNICAL PUBLICATIONS SERIES

With the appointment of Dr. Adnan Tamime as our Technical Series Editor, we are planning to expand our Technical Publication series considerably during the next few years. In addition to the 7 Technical Publications already available, the revision of one of which - *'CIP: Cleaning in Place'* - is well advanced, there are plans for Technical publications on: Fermented milks; Ice Cream; and Brined Cheeses, as well as for a monograph on Probiotic Dairy Products.

Using residual funds from the former Capstick Memorial Trust, the Society has contributed to the costs of producing an *'FAO/IDF Dairy Development Newsletter'* at FAO, which is edited by Dr. Ernest Mann. This has given the Society useful exposure and contacts in the countries with developing dairy industries. Through its membership of the UKDA the Society is in constant contact with the work of IDF and will hopefully become a publishing medium for some of its reports in the future.

Contacts have been established with the Danish Society of Dairy Technology and with the *'Australian Journal of Dairy Technology',* with a view to establishing collaborative agreements in the areas of information exchange and publishing.

Society membership of AEDIL- the European Dairy Technology Diploma Holders Association - should help the Society to focus its efforts on dairy education at all levels and to assist in the establishment of European guidelines on dairy education. AEDIL has already produced a professional Dairy Passport, which is facilitating the movement of dairy personnel between different countries. This is becoming increasingly important with the establishment of manufacturing bases by multinational companies in different countries.

THE PEOPLE BEHIND THE JOURNAL:

While it is not possible to mention by name the Society Executives, Chairmen and members of the Technical & Editorial board who have all played a pivotal role in making the Society Journal a success during the last 56 years, there are a few individuals who have made such outstanding contributions over the years that they deserve special mention here.

They include the following:

Miss Vera Watson who followed Dr. Crowther as Editor/Managing Editor for the first 25 years.

Dr. Harold Burton of the NIRD, who followed Vera Watson as Scientific/Managing Editor from 1963 to 1987.

Professor Donald Muir who succeeded Dr. Harold Burton as Scientific Editor in 1987 and who continued in this role for over 10 years until 1998, when our current Scientific Editor, Dr.

DR. RICHARD K. ROBINSON
SCIENTIFIC EDITOR

Richard Robinson, took on this daunting task.

Dr. Robinson's professional background has encompassed various aspects of applied microbiology, with the study of fermented dairy products as the principal component. He read Botany for a first degree (BA) at Wadham College, University of Oxford, and then obtained his post-graduate qualifications (MA and DPhil) from the same University. After an initial move to the Botany Department, University of Bristol to lecture on mycology, he was appointed to the Botany School, University of Oxford to teach the same subject. However, he was increasingly drawn to the more applied aspects of fungal behaviour, and transferred to the Forestry School, University of Oxford to investigate the nature of specific plant/fungal associations.

Contact with Professor Frances Aylward, who was then Head of the Department of Food Science at the University of Reading, persuaded Dr Robinson that food microbiology could be as fascinating as forest pathology. On moving to Reading, the study of fermented foods seemed a logical avenue for research and, given that Reading enjoys an excellent reputation for dairy science going back to the days of Professors Capstick and Crossley, fermented milk products like cheese and yoghurt became the new focus of Dr Robinson's research activities. Over the ensuing years, the results of collaborative projects with numerous research students have led to the publication of over 300 research papers and review articles - mainly about fermented dairy products, and provided the basis for a number of co-authored and/or edited books like *'Yoghurt - science and technology'* and *'Dairy Microbiology'* which are used by dairy scientists across the world. In 1987, Dr Robinson was elected a Fellow of the Institute of Food Science and Technology and, more recently, the value of his contribution to dairy science was recognised by his gaining a D.Sc. from the University of Stellenbosch, South Africa.

Although Dr Robinson has recently retired from the University of Reading (end-September 2003), he is continuing his role as the journal's Scientific Editor, and the Technical & Editorial Board, the Council and the Society are very indebted to him for this commitment.

Without the selfless efforts of these good people, assisted by several Managing and Production Editors, of whom the most recent was Sue Burkhart, our Journal would never have achieved the high standard which it has reached today, and which should be improved still further, if the various measures described above come to fruition.

The appointment of Sue Burkhart and details of her predecessors was reported in the Journal (JSDT Vol. 41 No. 1 Feb 1988.):

"Sue Burkhart has now taken over as managing editor of the Society's Journal. After taking the Bachelor of Science degree at the University of Michigan, Ann Arbor, Sue was a teacher for seven years in Long island, New York. She came to London in 1971 and has worked at the British Medical Association, first as a technical editor of two monthly journals and then as staff editor on the *British Medical Journal*, where she now works part time. Travel is her favourite past time and includes attending meetings on editing and publishing".

"Beryl Flitton resigned as managing editor of the Journal in June 1987 after serving continuously in that position since the January issue in 1981. She had taken over from Miss Anne Pinder, whose continuing ill health had resulted in her resignation in late 1980".

"Beryl had been with the British Medical Association as a technical editor of specialist journals for some 20 years and had previously been news editor of the *Journal of the American Medical Association* in the United States. On retiring from the BMA, Beryl accepted the post of managing editor of the society's Journal."

In her six and a half years of service with the Society, editing 26 issues of the Journal and two of the society's publications, Beryl was always meticulous and efficient in everything she carried out for the Society and her expertise in editorial matters was always greatly appreciated. She has a quiet and friendly manner and will be missed at future meetings of the Editorial Board".

"Beryl was unanimously recommended for the award of Honorary Membership of the Society in August 1987 and hopes to collect the certificate from the society's offices prior to her moving from Reading to the Cotswolds in the not too distant future".

"We extend our heartiest good wishes for a long and happy retirement in her new home".

JOURNAL COSTS

Throughout the years there was concern about the costs of printing the Journal and there were changes of printers for various reasons of both costs and competence.

Thus in the accounts for the year to 30 June 1982, it is reported that the net cost of printing the Journal has been reduced dramatically by almost £4,000 or 32% as a result of changing to Halstan's for printing the Journal (JSDT Vol. 36 No. 1 January 1983).

COMPANY AND EXHIBITION NEWS

Company News was first published in January 1985 as a service to members (JSDT Vol. 39, No. 1 January 1986). The news was obtained from press releases sent to the SDT office. Notice was also given of forthcoming exhibitions of interest to the dairy industry. As an illustration of the range of companies, products and services here is the list published in April 1986 (JSDT Vol. 39, No. 2 April 1986):

ACR LIFT TRUCK (SOUTHERN) LTD
ALPMA (GB) LTD
ANALYTICHEM INTERNATIONAL
E & A ASHWORTH LTD
BACTOMATIC LTD
BOOTS-CELLTECH DIAGNOSTICS LTD
BOWATER LIQUID PACKAGING LTD
BRAN AND LUBBE
CARRIER INDUSTRIAL REFRIGERATION
CARVER AND CO (ENGINEERS) LTD
CLIFFORDS DAIRY PRODUCTS of Bracknell launched, in September 1985, two new dairy

desserts in both orange and lemon flavours based on quarg produced from skimmed milk curd blended with fresh cream and fruit flavourings. These are being sold through British Home Stores under their own labels.

COUNTY DAIRIES OF Kidlington, Oxford, have become the first juice and drinks supplier to offer short-life orange juice in a 2-litre plastic bottle to the majority of leading multiples.

DELTECH EUROPE AND DELTECH ENGINEERING

DIONEX (UK) LTD

DRAEGAR SAFETY

ELECTRICARS LTD

GOODFELLOWS

HALE INSTRUMENTS

HANOVIA

HUSKY COMPUTERS LTD

INDUSTRIAL MONITORING EQUIPMENT

INTERLUBE SYSTEMS

KECOL PUMPS LTD

KELBER LTD

D D LAMSON PLC

LEVER INDUSTRIAL

MASS TRANSFER INTERNATIONAL

T R MCDONALD ELECTRONICS LTD

BOOK REVIEWS

An important feature of the journal was the book reviews over the years including "From Little Acorns", a history of APV and on the same page Food Control in Action (JSDT Vol. 35 No. 2 April 1982).

OBITUARIES

Another element of the journal has been the publication of obituaries of prominent members of the Society. Some of these are reproduced in Chapter 5.

THE JOURNAL AND THE SOCIETY

This review of the functions of the Journal in relation to the objectives of the Society is as relevant today as it was in 1986. (JSDT Vol. 39 No. 4 Oct 1986).

"The journal has two main purposes: firstly to act as a focus for the membership by providing information of various kinds and, secondly, to act as a display window for the Society to others".

"In the year ending July 1985, the last year for which figures are available to the Board, the net cost of producing and circulating the journal was £8,050, so that each issue cost a member about 88p, or about 57p when the cost of postage is deducted. The total cost of producing the journal is much more than this: £18,600 in the last year, of which 71% represented direct printing costs and 15% postage. The difference between the total cost and the net cost is accounted for by other income attracted by the journal, and in fact sales of the Journal to non-members last year brought in an income (£8,400) higher than the net cost to the Society".

"Non-technical information now reaches members through the Supplement, bound in with the Journal, which has been well received but which needs a constant supply of material from the Sections if it is to fulfil its purpose completely".

"The rest of the Journal provides technical information of various kinds, together with some personal information on members. Of the papers published, about two thirds originate from Society Conferences and Symposia, a few are versions of papers given to local Section meetings, and about a quarter are contributed papers, either reviews or reports of technical investigations".

"It is assumed that if a paper is given to a Society meeting its subject is of interest to members. However, such a paper may not be published if the Editorial Board does not think it suitable for publication".

"Contributed papers must meet two criteria. First they must be appropriate to the Journal:

Editorial Board policy is that the subject should obviously be of interest to members, or that it ought to be of interest to members who adopt a forward-looking attitude to their industry. Papers that are felt to be too scientific are not considered for publication and are returned to authors with the suggestion that they should seek publication elsewhere".

"The second criterion is that of merit. Every contributed paper is examined by a referee knowledgeable in the subject and, in some difficult cases, by more than one referee independently. In this way the Board tries to make sure that every contributed paper either presents good original information on a relevant subject or is a survey of a technical subject that is not influenced by commercial publicity".

"This strict approach to contributed papers has been responsible for the growing reputation of the Journal outside the Society, in this country and overseas. This in turn has led at least in part to the substantial income to the Society from the sales of the Journal to non-members".

"It has been suggested that the Journal should carry articles of a more down-to-earth character, descriptions of plant, personal experience of processes, etc. The Board would be very happy to publish such articles. The fact that they do not appear is simply a reflection of the fact that nobody writes them. The difference between our Journal and a commercial publication is important here. A commercial publication, and even the journals of some large societies and institutions, have full-time salaried staff who can seek information from others and turn it into an article if necessary. Alternatively, they can attract authors by offering a fee. Neither of these possibilities is open to our Society. Apart from our Managing Editor, who prepares the material finally for the printer and deals with him on production matters, all other duties connected with the Journal, such as preparing the Supplement, receiving papers, approving them, ensuring technical and factual accuracy, and so on, are carried out quite voluntarily by members of the Editorial Board and others in addition to their normal activities. If articles are to be published of a different character from those at present appearing, suitable authors are needed. If any members are prepared to write on topics of the kinds that have been suggested, or can persuade others to do so, the Editorial Board undertakes to receive them with enthusiasm and to consider them for publication".

"The Journal is, of course, not the only Society publication. The Editorial Board has been involved with others in the Society in publishing a wide range of technical material of a practical nature. For example, in recent years the 3rd Edition of the 'Pasteurizing Plant Manual' has appeared, and a monograph on 'Short-shelf-life Products' has been published in the last few weeks. Revisions of the 'Ice-cream' and the 'In-place Cleaning' Manuals are in train. These supplement the papers appearing in the Journal, and all are the result of voluntary work by Society members".

"The Editorial Board is proud of the Journal and of the Society's other publications. However, the Journal exists to meet the needs of members. To what extent do we succeed, and to what extent do we fail? The Board would like to receive your views".

PRINCIPLES, PRODUCTS AND PRACTICE (PPP)

This is an example of initiative taken by the Editorial Board to make the Journal as relevant as possible to the members.

"Readers of the Journal will all be aware that for two years or so the Journal has been featuring short articles in the 'Principles, Products and Practice' series. These are intended to be short, punchy papers which restate basic principles in a selected area, summarise a current situation, briefly review an area or detail a new approach, etc".

"The series has apparently met with general enthusiasm and the PPP subcommittee, consisting of Sue Hayes, Norah Shapton, Terry Tamplin and myself, have contributions on some 30 subjects so far".

"We invite you to suggest topics and authors for this series of short papers, and so keep this interesting series going". (Dr. D. Moran – Ed, Board (JSDT Vol. 43 No. 4 1990). An example of one of these papers was Mr F Harding's 'Milk Adulteration - Freezing Point Depression' (JSDT Vol. 43 No. 3 August 1990).

In a similar vein the Editorial Board invited *'Millennium Contributions'* on subjects ranging from biochemistry to marketing to celebrate the new Millennium.  For an example see Kanekanian, Gallagher and Evans *'Casein hydrolysis and peptide mapping'* (IJDT Vol. 53 No. 1 Feb. 2000).

NUMBERING OF JOURNALS

Those looking back at old issues of the Journal will discover an anomaly in numbering, which is explained in the Journal (JSDT Vol. 5 No. 2 January 1952).

"It is proposed to make Volume 5 extend from October 1951 to October 1952 inclusive in order that Volume 6 may start in January 1953 and from that time onwards each volume of the Journal will be correlated with the calendar year!"

INTRODUCTION OF INTERNATIONAL JOURNAL

Title change and ISSN number change for Volume 50, 1997:

"The Journal of the Society of Dairy Technology will change its name to the INTERNATIONAL JOURNAL OF DAIRY TECHNOLOGY in 1997".

New ISSN number: 1364-727X (JSDT Vol. 49 No. 3 Aug 1996).

# CHAPTER FIVE

## AWARDS, DISTINCTIONS AND MEMORIAL TRUSTS

The awards made by the SDT include:

GOLD MEDAL
PORTRAITS TO PRESIDENTS
HONONARY MEMBERSHIP
SECTION AWARDS

*The Gold Medal.*

Gold Medal

The SDT's Gold Medal was inaugurated in 1950 to recognise outstanding achievement in the dairy industry. The conditions for this award are briefly as follows:

1. To be awarded for an outstanding contribution to dairy science and/or technology throughout the world.
2. The award to be made by the Council when merited, but not more than three awards to be made per decade.
3. When practical the award to be made the occasion of a special meeting of the Society, or when the recipient is a person domiciled overseas to be made at a special meeting of the Society's members in the recipient's country or of a body with similar interests.

Note: With the present prohibitive price of gold, recent medals have been struck in silver-<u>gilt.</u>

*Dr. Richard Seligman, Chairman of the A.P.V. Company, who addressed the Society on the previous day, chats with Mr. J. White, President, and Mr. E. Capstick, Past President.*

The first recipient in 1951 was Dr Richard Seligman of A.P.V. Since then the Gold Medal has been awarded to:

| | |
|---|---|
| Professor W. Riddet | October 1953 |
| Professor H. D. Kay | October 1957 |
| Professor P. O. Kästli | October 1966 |
| Sir Richard Trehane | October 1969 |
| Mr G. Loftus Hills | May 1972 |
| Professor E. L. Crossley | October 1975 |
| Dr T. R. Ashton | February 1983 |
| Mr H. S. Hall | February 1983 |
| Dr J. G. Davis | October 1987 |
| Dr G. Chambers | April 1993 |
| Dr B K Mortensen | April 1993 |
| Professor D D Muir | November 1996 |

It has been decided to award Gold Medals to Dr E J Mann, Chairman of the Technical & Editorial Board, and Dr D L Armstrong Immediate Past President, on the occasion of the 60th Anniversary of the SDT in November 2003.

The presentation of the first Gold Medal to Dr. Seligman was an emotional and memorable occasion (JSDT Vol. 5 No. 2 Jan. 1952):

"The announcement was made a few months ago that our Society was about to institute what would be known as the Society of Dairy Technology Gold Medal. It was decided that this would be awarded in recognition of outstanding achievement in dairying or of some exceptionally meritorious service to the dairy industry".

"However, what I want now to stress is that the Council of this Society had the very same ideas in mind when they considered some months ago who should be the first recipient of the

Society of Dairy Technology Gold Medal. We were determined that the award of the medal should for all time be regarded as a very great honour and that it should be given only for the most distinguished and meritorious service. With that in mind the Council never had a moment's hesitation in deciding unanimously that the first <u>person to be honoured in this way should be Dr. Seligman</u>".

"In the early days of his progress, Dr. Seligman was not only an inventor, but he was able to say with amazing precision what he was going to invent and how long it would take him to invent it! It meant little what distractions went on around him. In fact one of the greatest advances he made, (for the development of a plate heat exchanger), was thought out and put on paper in the evenings in a crowded hotel ballroom in Switzerland to the accompaniment of a dance band and the games and singing of the other guests. This is how he spent the evenings of a few weeks' holiday with his family!"

"Dr. Seligman is a man of many accomplishments and of exceptionally wide experience. For what would now be called his 'secondary' education he went to Harrow, which has been responsible for the education of so many who later became famous in British history, and, unless I am greatly mistaken, he must have been there just about the same time as no less a figure than Sir Winston Churchill himself. From Harrow he went to the Central Technical College where he studied under that great chemist Professor H. E. Armstrong. From there he went to Heidelberg and Zurich and was awarded his doctor's degree at Heidelberg in 1902. In the following three years he appears to have defied both time and space. In 1903 we find that he was chemist to the British Uralite Company; the following year we find him first as research chemist to the Niagara Falls Research Laboratories in the United States and then as Chief Chemist to the United States Zinc Company in Colorado; late in 1905 he is back in Britain as Chief Chemist to the British Aluminium Company. After three years with that Company he left to undertake research work under Sir William Ramsay at University College, London, and then from 1910 onwards he was instrumental in founding and developing the <u>Aluminium Plant and Vessel Company</u> which we know to-day as the <u>A.P.V.</u> Company. The entrancing story, (one might almost say romance), of the development of that Company and of its entry into the dairy industry in 1924 has already been told elsewhere. It will suffice here to say that the main factors leading to its many achievements were the genius, determination and perseverance of Richard Seligman. It is on record that, when danger lay ahead and risks had to be taken, Dr. Seligman used to quote to his colleagues, who were more timorous than he, a passage from *'The Explorer'* by Kipling: *'Something hidden'. Go and find it. Go and look behind the Ranges. Something lost behind the Ranges. Lost and waiting for you. Go.'* It was in that spirit that so much was achieved".

"Time will not permit me to dwell in detail on the many inventions and developments for which Dr. Seligman is directly responsible, but perhaps it will suffice if I read the list which was before the Council when they so unhesitatingly decided to invite him to be our first medallist:

1. He was the pioneer of the autogenous welding of aluminium, and of applying it to vessels in the milk industry
2. He was the original inventor of the plate heat exchanger in 1923. In addition, he has been personally and directly involved in its developments up to the present day
3. He played a very large part in the commercial development of the circular positive holder invented by Mr. Tarbet, and in the development, manufacture and application of the principle of vacuum pressure to pasteurisation. These two inventions generally eliminated the possibility of forward leakage in holder plants
4. He was closely concerned with research on the welding and fabrication of stainless steel since the very early days, and was the pioneer in the application of welded stainless steel vessels to the dairy industry
5. He introduced high temperature milk pasteurisation in the United Kingdom
6. He has made many contributions to the sanitary design of all types of equipment now in common use in the milk industry
7. He is an acknowledged authority on the corrosion of metals used in the food industries".

"Ladies and Gentlemen, I think you will agree that we owe Dr. Seligman a great debt of gratitude for all he has done, and in my view while we are undoubtedly honouring Dr. Seligman in making this award, he is certainly honouring us in accepting it".

"Dr. Seligman, it is a great honour for me today as President to have the privilege of presenting to you on behalf of the Society of Dairy Technology the Society's Gold Medal in recognition of the meritorious service you have rendered to the dairy industry."
*(Applause).*

At a general meeting of the Society on 24 October 1966, Mr John Glover, President presided over the presentation of the Society's GOLD MEDAL to Professor Paul O Kästli, Director of the Federal Institute of Dairy Research, Liebefeld, Bern, Switzerland, (JSDT Vol. 20 No. 1 1967). Dr J.A.B. Smith, the Director for the Hannah Research Institute, gave the citation. This is worth quoting from as it explains the philosophy of the award of Gold Medals and gives information about earlier recipients.

DR. J. A. B. SMITH: "Mr. President, ladies and gentlemen, the Society of Dairy Technology Gold Medal was instituted just about 16 years ago and, as President of the Society in 1951, I had the honour of presenting the first medal to Dr. Richard Seligman, whom we are all so delighted to have here with us today".

"As I said in my address to the Society when the first award was made: 'We were determined that the award should for all time be regarded as a very great honour, given only for the most distinguished and meritorious service.' Indeed I pointed out that when we invited Dr. Seligman, who had made such outstanding contributions to the development of our modern dairy industry to be the first recipient, we intended that all who came after him would realise the greatness of the honour which was being conferred upon them".

"The second medal was awarded to Professor W. Riddet, who, over a period of 26 years, rendered great service to the dairy industry of New Zealand. The third medal was presented to Professor H. D. Kay, who did so much in the 1940s to form this Society, who was its Founder President and who, over a long period, had served so well the interests of dairy science and the dairy industry both nationally and internationally".

"When, a short time ago, the Council of the Society considered the possibility of awarding a fourth Gold Medal, the man whose name was uppermost in the minds of the committee concerned was Paul Oscar Kästli of Switzerland. As I shall show in a moment, Professor Kästli is receiving this award today because of the great contribution he has made to dairying over a long period of years, but it should be noted at this point that he happens to be of Swiss nationality. This fact is important because, when the medal was inaugurated, the Council at the time strongly believed that, in order to foster international fellowship, eligibility for the award should be world-wide. We therefore give Professor Kästli a specially warm welcome to London today".

"In the year 1926, just 40 years ago, Paul Kästli completed his studies as an undergraduate and became fully qualified as a Veterinarian. He then began what became a most fruitful career in research and was awarded the degree of Doctor of Veterinary Medicine in 1928".

"In 1943 he became Director of the Federal Dairy Research Institute of the University of Bern".

"During this period from 1929 onwards his research on the spread and methods of detecting and controlling udder disease, his discovery of non-infectious udder catarrh and his outstanding work on the survival of pathogenic bacteria in dairy products received the international recognition it all so richly deserved".

"In many countries of the world mastitis is generally regarded as the greatest single source of loss in the milk-producing industry, and after much work on this subject Professor Kästli described in 1949 the campaign which he has instituted in Switzerland against this disease - a campaign that met with very considerable success".

"In 1951, Professor Kästli's work on the bacteriology of milk and milk products and on improvements in milk hygiene was recognised in a very special way when in that year he was awarded the Werder Medal".

"Professor Kästli is a Professor of the University of Bern. Since 1946 he has been President of the Swiss Milk Commission which is the Swiss National Committee of the IDF, corresponding to our UKDA, and in 1958 he was appointed a member of the Federal Research Commission for Agricultural Education and Research."

"He is well known far beyond the boundaries of his own country. During the years 1956-59 he was President of the International Dairy Federation, and in that capacity he presided over the International Dairy Congress when that memorable event was held in London in 1959".

"Several countries have honoured him. In 1954 the Municipal Council of Paris announced that he was to be made an 'Ami de Paris.' In 1959 he received the honorary degree of Doctor of Science of the University of Reading, and was made an honorary member of the Society of Dairy Technology. In that same year the honorary degree of Doctor of Agriculture of the University at Vollebekk in Norway was conferred upon him".

*A highlight of the Golden Anniversary Celebrations was the award of Gold Medals to Dr. George Chambers and Dr. B. K. Mortensen.*

Citations on the presentation of the Society's Gold Medals at the 50th Anniversary Conference 20 April 1993

### Dr George Chambers
*Prepared and delivered by George Grier*

*The Presidnet, Mr. P. W. S. Fleming, presenting the Society's Gold Medal to Dr. G. Chambers*

"The presentation of this Society's Gold Medal award is a most significant occasion, not only for the recipient, but also for his or her peers who are assembled together to share in the pleasure that it conveys. The fact that the presentation is taking place at the Society's 50th Anniversary Conference must add a very special dimension to the proceedings, and there is no doubt that the person whom I am introducing has always had a very finely tuned sense of occasion."

"It has been my pleasurable experience to work for and with Dr George Chambers, CBE, for some 30 years during which period the dairy industry, not only in his native Northern Ireland, but throughout the United Kingdom, has experienced many revolutionary changes in technology, legislation, economies and marketing. George Chambers has been at the forefront or indeed the actual vanguard of many if not most of these major developments."

"Born of farming stock in a most attractive area of Co Down, George prepared the foundation for a notable career by obtaining a first class honours Bachelor of Science degree in Chemistry at the Queen's University of Belfast in 1949. Postgraduate research work and lecturing occupied the period 1949 to 1954 when a doctorate was awarded for a thesis entitled *'Reactivity of paraffin-carbons in relation to their structure and flexibility'."*

"After four years as a research chemist with ICI Alkali Division in Cheshire, he returned to Northern Ireland and the remainder of his career was entirely with the dairy industry and in particular as Chief Executive of the Milk Marketing Board for Northern Ireland, a post which he held for 25 years until he retired in 1988."

"Firstly, the Society has benefited from his membership over the past 35 years during which time Dr Chambers was for three years the Secretary of the Northern Ireland Section during the period 1960 to 1963, and Chairman for the year 1966-67. He was a member of Council between 1972 and 1975, becoming Vice President in 1974 and President during the year 1975-76."

*Peter Lee (centre), who retired as Secretary of the Society in July, was presented with gifts from the Society by George Chambers and the President, Eurwen Richards. (Photograph by Philip Abbott)*

"The debt which the Society owes to George was first recognised in 1987 when he was awarded Honorary Membership; and from 1988 to date he has been a Trustee of the Society."

"For his services to agriculture, industry and commerce over the years Dr Chambers was awarded the CBE in the 1985 New Years Honours."

"In the further progression of our Society and our industry one thing will never change and that is the need for people with the dedication, the energy, the initiative and the vision of which George Chambers is an exemplary model. It therefore gives me great pleasure to ask George to come forward and receive the Society's Gold Medal from the President."

DR BØRGE MORTENSEN
*Prepared and delivered by Dr R J M Crawford*

"When our Society, in 1950, inaugurated its Gold medal award for outstanding achievement in the dairy industry it did so recognising that to foster international fellowship, eligibility for the award would be worldwide."

"Today this Society has decided to honour a dairy technologist from one of the most famous dairy industries in the world - that of Denmark....."

*The Presidnet, Mr. P. W. S. Fleming, presenting the Society's Gold Medal to Dr B. Mortensen*

"Dr Børge Mortensen was a distinguished member of Hillerød's staff. He was born on 3 September 1940, a very difficult time in Europe and the Second World War in progress."

"By 1966 Børge Mortensen had graduated Master of Science (Dairy Technology) from Copenhagen and began his career as a dairy scientist at Hillerød, the State Dairy Research Institute. In 1972, he was awarded the Doctor of Philosophy degree of the same university; his field of study was *'Butter Consistency'* a topic of great importance, which was to be his interest during the whole of his career at Hillerød."

"In 1976 he was appointed Head of the Butter Department at Hillerød and in 1982 further promotion led to his appointment as Managing Director of the Institute. Changes in funding for dairy research in Denmark took place around the end of the 'eighties and his Directorship of the Institute ended in 1990 with the closure of Hillerød, a disappointing event for many."

"He also gained much satisfaction from his reorganization of the structure of the Hillerod Institute, but unfortunately the political decision to close the Institute deprived him of seeing the full results of his efforts."

"His career serving the dairy industry has now developed in a different direction. Since 1990 he has been head of the R & D Centre of MD Foods, Denmark, a major international dairy company now with interests in the United Kingdom."

"Other involvements of note to dairy technologists are that from 1982 until 1990 he was a member of the International Circle of Dairy Research Leaders, a body which arranged much important basic and applied dairy research. During his directorship of Hillerød, he was a member of the Ministry of Agriculture Committee on Agricultural Research; thereafter a member of the Board and eventually Vice President of the Committee. Since 1990 he has been a member of the Board of the Danish Dairy Research Foundation, the new body concerned with the funding of much of the dairy research in Denmark. He serves as an External Examiner in Dairy Technology of his old University."

"Dr Mortensen has been President of the Danish Society of Dairy Technology since 1980 and has helped promote fellowship between his society and others including our own."

"1987 he was admitted a member of the Danish Academy of Technical Science. In 1990 he received the award of the Gold Medal of the Academia Rolniczo Techniczna at the Agricultural University of Olsztyn, Poland."

"Today, this Society wishes to honour him with its own recognition of excellence. Mr President, Past Presidents, Ladies and Gentlemen, it is with very sincere pleasure that I present to you Dr Børge Mortensen."

"More recently the President, Mr Grahame Lee, presented the Gold Medal of The Society of Dairy Technology to Professor Donald Muir in a ceremony held at the 53rd Annual General Meeting on 13 November 1996 at The Farmers Club, London. (JSDT Vol. 50 No. 1 February 1997)."

*The following citation was made by Dr Tony O'Sullivan, Immediate Past President, on behalf of Council and the Society:*

"The Gold Medal is the most prestigious award that the Society can award... Today, the Society wishes to honour another distinguished scientist, Professor David Donald Muir, Head of the Food Quality Group at the Hannah Research Institute. The award is made on the basis of the following two main considerations".

"First, Professor Muir has made an outstanding contribution to the Society in the role of Honorary Scientific Editor of the Journal. Second, Professor Muir has made a substantial contribution to the field of Dairy Science and Technology in his own right. This has been recognised by a Visiting Chair at the University of Strathclyde and by an Honorary Lectureship at the University of Glasgow. Finally, Professor Muir enjoys transferring science into practice".

"In advancing Professor Muir as a 'role model' for aspiring dairy scientists and technologists, it gives me the greatest pleasure to invite him to receive the Gold Medal from the President".

PROFESSOR MUIR RESPONDED:

"I am delighted to receive the Society's highest accolade, especially for an activity that I enjoy. The Gold Medal elevates me to the ranks of a select few".

The award of the Gold Medal to the other distinguished recipients is also recorded in the Journals and in the 50-year Commemorative Booklet.

PRESENTATION OF PORTRAITS TO PRESIDENTS

The recognition of the services of Presidents by the presentation of portraits was inaugurated in 1948 (JSDT Vol. 1 No. 2 January 1948.):

*The President* Dr. Mattick: "I have now a particularly pleasant office to perform, namely, the presentation of portraits to Past Presidents of the Society".

"There were to have been three "victims," but, unfortunately, Mr. Procter is only now on his way back from America so the presentation on his portrait will have to be postponed. I think, however, that Professor Kay is in full view of the audience and if he will be kind enough to come up, I will present him with his portrait".

"The Society can congratulate itself on having secured - I nearly said procured - the services of so distinguished a scientist as Professor Kay as its first President. He played a leading part in laying the foundations of the Society, in formulating its objects and ideals and its rules of procedure. He has no official position in the Society now, but we know that he will retain a very active interest in its affairs, particularly as a negotiator in the matter of the new Technological Institute".

"It is my privilege on behalf of the Society to ask you, Professor Kay, to accept this portrait as a token of its gratitude and its regard for what you have done." *(The portrait was then handed to Professor Kay to the accompaniment of prolonged applause).*

*Professor H. D. Kay:* "I am not personally, I hope, unduly emotional, but I do feel very strongly indeed at this moment the kindness and the support which this Society has always extended to me from the very day upon which it was founded, and before. Some who are here today were very largely responsible for overcoming those very serious early difficulties of which the President has spoken, and indeed without a tremendous amount of help from a very large number of people this Society would never have been founded during those very awkward days. I still remember two occasions on which our discussions were carried out against a background of flying bombs; but the Society has, nevertheless, developed in an amazing way. When one considers that at the present time we have a membership of over a thousand, representing almost everybody in the whole industry who is concerned with the progress of dairy technology, it is truly a great achievement on the part of those who were instrumental in getting people together in those early days"...

"Thank you very much indeed for this portrait." *(Applause)*

*The President*: "In Mr. Graham Enock we have the benefit of a person of wide culture, technical ability and business acumen, characteristics which are extremely valuable in a Society of this kind. All those who sat under him during his term of Presidency know that his loyalty to the Society has been absolutely outstanding".

"It is particularly in financial matters, I think, that Mr. Graham Enock has served the Society so well…"

"May I ask you, Mr. Graham Enock, to accept this portrait on behalf of the Society as another token of our gratitude for your services." *(The portrait was then handed to Mr. Graham Enock to the accompaniment of prolonged applause).*

*Mr. Graham Enock:* "I cannot make a speech in regard to this matter and I think I should, in the main, like to reiterate what Dr. Kay has said. I do sincerely thank you all very much indeed for this portrait. I can assure you nothing like this has ever happened to me before, and I do very much appreciate it! Thank you very much indeed." *(Applause).*

"It is, therefore, right for us now to remember for a few moments what Dr. Mattick has done for us during his term of office".

"Dr. Mattick is, as you know, the Deputy Director of the National Institute for Research in Dairying, and has brought to our Society a mind of much learning and great ability"…

"Dr. Mattick, we are grateful to you for being our President, and we do thank you very much indeed for the time you have given and the trouble you have taken over our affairs. On behalf of all the members of the Society I should like to ask you to accept this framed portrait of yourself as a token of our sincere wishes and gratitude for your work as President this year." *(The portrait was then handed to Dr. Mattick to the accompaniment of prolonged applause).*

*The President:* "As you may imagine, I have given a certain amount of thought to what I should say to my victims but none at all to what I should say on being a victim myself."

"I have very much enjoyed the Presidency of this Society. I belong to many societies and hold office in some, but there is no society with which I am connected in the meetings of which I feel more at home and conscious of the friendly feeling which exists".

"I hope none of you will have to see this portrait of mine again; I think it is awful, but no man is in a position, so I am told, to judge his own portrait. They say the camera cannot lie, and if that is true, well, - here I am"

"Thank you very much indeed for your kind remarks, Mr. Graham Enock, and thank you, ladies and gentlemen, for doing me the great honour or presenting me with this lasting token of a happy year of Presidency." *(Applause).*

The practice of presenting portraits to Presidents has continued to the present day.

HONORARY MEMBERSHIP

The award of Honorary Membership is referred to in the journal in 1979. (JSDT. Vol. 32 No. 2 April 1979).

Until 1968, the subject of Honorary Membership and the award of the Society's Gold Medal, both of which were instituted in 1951, were dealt with when the occasion warranted it by Council. In 1968, an Ad Hoc Committee met and recommended to Council that an Awards Committee should be formed specifically to look into future awards and, in November 1968, the first Committee was elected. From then to the present date, the Committee has comprised the President, the Immediate Past President (now the Vice-President since the inception of that post in 1973), the Honorary Treasurer, the Honorary Secretary and two Council members (now two Past Presidents) who consider and make recommendations for the conferring of Honorary Membership, the award of the Society's Gold Medal and any other award that might be considered.

*Presentation of Honary Membership to two Past Presidents, Eurwen Richards and Mrs Suzanne Watson*

*Honorary Membership.* Briefly the following are the rules for awarding Honorary Membership:

1. Persons who have rendered signal service to the Society, either on the Council or its Committees can be considered, as can
2. Persons who have rendered distinguished service to the science or industry of dairying and
3. Persons abroad who have rendered services useful to the Society.

The number of Honorary Members was originally limited to TWENTY at any one time, but in 1983 the limit was raised to THIRTY. (JSDT Vol. 36 No. 2 April 1983). Present Honorary Members, shown with the dates that Council approved the Committee's recommendations, are: Professor P. O. Kästli (May '59); Miss V. Watson (August '64); Mr. S. Clifford, OBE (August '68); Professor E.L. Crossley (August '68); Mr. A. Graham Enock (August '68); Mr. F. Proctor, OBE (August '68); Dr. J. A. B. Smith, CBE (August '68); Mr J. Lewis (August '69); Mr. S. B. Thomas, OBE (August '69); Dr. T. R. Ashton (August '70); Mr. H. S. Hall, OBE (August '70); Mr. C. S. Miles (August '70); Mr. M. Sonn (August '73); Mr. S. W. Stilton (August '76); Dr. R. J. Macwalter, OBE (August '77); Dr. R. C. Wright (August '77); Dr. J. G. Davis, OBE (October '78); Professor J. Murray (October '78); Mr. J. R. Rowling, OBE (October '78); Dr. E. Green (August '80); Miss E. Pinder (August '80), Past Managing Editor of the Journal; Mr E. P. McCormick, (February '82).

Honorary Membership has been subsequently conferred on: Dr. J. Rothwell (February '84); Mr P. Hoare (February '84); Mrs. S. W. Barron (February '85); Dr. H Burton (February '85); Miss J Betts (February '85); Mr. W. J. Hipkins (February, '85); Mr. F. C. Weldon (February, '85); Dr. R. J. M. Crawford (February '86); Dr. G Chambers (February '87); Mr. H. E. L. Harz (February '87); Mr D Ralph (August '87); Miss B. Fitton (August '87), Past Managing Editor of the Journal; Mr. P. Lee (February '89), Past Executive Secretary; Mr. R. Lawton (February '89); Miss E. Richards (February '92); Mrs. S. Watson (February '92); Dr. P Young (February '93); Mr R Thompson (August '94); Professor D. D. Muir (May '98); Mrs. R . A. Gale (August '98), retiring SDT Executive Secretary; Mr. R. Early (May '02); and Mr. J. Lyons (September '03).

SECTION AWARDS

Reference is also made to the possibility of an award to recognise meritorious service of members to the Society at Section level. This award takes the form of a plaque, and was approved by Council in August 1979 (JSDT Vol. 36 No. 2 April 1983).

The guidelines for the award are as follows:

1. The Society shall not seek nominations, but each Section shall have the opportunity, at will, to nominate a member who has given outstanding service within the Section, to receive the award
2. The award shall be made not more than once in five years in any one Section, and there shall be a maximum of two awards made by the Society in any one year
3. The application, giving full details of the service on which the proposal is based together with full written reasons for the award, should be sent by the Section Chairman to the Society Secretary before the Annual General Meeting of the Society. Nominations will then be considered by the Awards Committee before any recommendation is made to Council…
4. The award shall take the form of a wooden plaque incorporating the Society motif, the cost of which shall be borne by Society funds
5. The plaque shall be engraved with the recipient's name, the year in which the award is made, and the inscription *'For outstanding service to the* (appropriate) *Section.'* Presentation should be made at a suitable occasion within the Section, and the award should ideally be presented by the President.

Among the Section award winners were:

1980 (2 submissions made)

| | |
|---|---|
| Mr. D. Ralph | North Eastern Section |
| Mrs. O. Stewart | Western Section |

1981 (1 submission made)

| | |
|---|---|
| Miss J. Betts | London, South and East Section |

1982 (2 submissions made)

| | |
|---|---|
| Mr. M. Boyle | Scottish Section |
| Mr. H. E. L. Harz | Midlands Section |

1983 (2 submissions made)

| | |
|---|---|
| Mrs. E. McKenszie | Welsh Section |
| Mr. J. Sproat | Northern Section |

1984 (Extra award)

| | |
|---|---|
| Mrs. R. W. S. Warren | Western Section |

1985 (Extra Award)

| | |
|---|---|
| Miss N. Pennie | Midlands Section |

1986 (2 awards)

| | |
|---|---|
| Mr. G. I. A. Lang | Northern Ireland Section |
| Mr. R. Lawton | North Western Section |

1987 (1 submission made)

| | |
|---|---|
| Mr. G. R. Thompson | North Eastern Section |

1993 (2 submissions made)

| | |
|---|---|
| Mr. J. Crombie | Northern Section |
| Mrs. H. Fowler | West Midlands Section |

The East Midlands Section also made a presentation to Bob Huck for his contribution over the years - not only to the East Midlands Section but also to the East Midlands Branch. (JSDT. Vol. 47, No. 1 February 1994).

MEMORIAL TRUSTS AND AWARDS

Over the years the SDT has been involved in setting up Memorial Trusts and Awards. Two of these were the Capstick Memorial Trust and the Stanley Clifford Milk Industry Award.

The setting up of the Capstick Memorial Trust is referred in the Journal in 1966 (JSTD Vol. 19 No. 4 November 1966) as follows: "In February 1964 an appeal was launched by the Society of Dairy Technology to friends of Professor Edward Capstick, MC, OBE, MSc, to support a fund to perpetuate his memory and to reflect his lifelong interest in dairy education, especially where it concerned the peoples of countries where malnutrition can be alleviated by the development of a dairy industry. This appeal was generously supported, not only by organisations in the dairy industry, but by many personal friends and business associates of Edward Capstick".

"The ad hoc Committee set up by the SDT was, therefore, encouraged to form a Trust Fund for the investment of the subscriptions, which with interest accrued have reached approximately £3,200 to date."

"A Trust Deed was accordingly drawn up, with legal assistance, and submitted to the Inland Revenue Department for approval of the Trust as a charity. The hope that this would speedily be concluded and enable the Trustees to take over administration of the Fund was short lived. Suffice it to say that negotiations with our legal advisers and with the Inland Revenue Department have taken over a year and considerable correspondence to complete."

"Not until July 1965, was a letter received from HM Inspector of Taxes concerning the Trust Deed of Covenant stating that: 'It is considered that the Deed establishes a charitable trust entitled to the exemptions afforded to charities by the Income Tax Acts. These include repayment of tax deducted from payments made under Deeds of Covenant which constitute effective dispositions of income for Income tax purposes'."

"The SDT ad hoc Committee was able, at this stage, to hand over the Trust Fund to the following Trustees:

Professor E. L. Crossley, nominated by the University of Reading; Mr. J. Ridley Rowling, OBE, nominated by the Society of Dairy Technology; Mr. F. Proctor, nominated by the United Kingdom Dairy Association; and Mr. F. C. White, OBE nominated by the United Kingdom Dairy Association."

"At the first meeting Mr. F. C. White was elected Chairman of the Trustees, Mr. J. Ridley Rowling Hon. Treasurer and Mr. M. Sonn Hon. Secretary. The objects of the Trust, as now approved, are 'the promotion of dairy education in countries developing a dairy industry and generally for the provision of any other charitable project related to dairy education which in the opinion of the Trustees would have merited the approval of Edward Capstick."

"The Trustees have decided that the terms of the Trust can be interpreted most effectively by using the first annual grant for the <u>purchase of technical and scientific books for the libraries</u> of Egerton College and of Naivasha Dairy School, both of which are in Kenya."

"With the assistance of the Dairy Branch of the Food and Agriculture Organisation and after consultation with the Institutes concerned, the Trustees have selected suitable books, which will be sent to Egerton and Naivasha from the Trust."

"A bookplate is being designed to bear the following words:

THE CAPSTICK MEMORIAL TRUST
17 Devonshire Street, London W1
This book is presented to..........................................................................
by the Trustees of the Capstick Memorial Trust, founded by the friends of Edward Capstick, MC, OBE, to promote dairy education.
Chairman of Trustees                                                   (Date:)"

"As it is understood that the provision of technical books on dairying in developing countries is a desperate need, the Trustees hope to make further grants for this purpose in future years. These will be reported in the Journal and the dairy press will also be informed."

STANLEY CLIFFORD AWARD

Stanley Clifford, the longest serving Chairman of the Central Milk Distributive Committee (CMDC), died on 28 October 1979 (he was 80 years of age). His brother Gordon took over from him (JSTD Vol. 33 No. 1 January 1980)

In recognition of his great service to the industry, the CMDC in 1962 decided to inaugurate a Stanley Clifford Award, affording anyone engaged in the UK dairy industry an opportunity to apply for selection as the award winner in any year, to enable them to undertake a study tour either at home or abroad of any aspect of dairying. Funded now by the Dairy Trade Federation [now the Dairy Industry Association Ltd (DIAL)] but continuing to be administered by the Society of Dairy Technology, it so happened that a panel representing DTF/SDT/DITED interviewed five applicants out of an initial 16, and selected Mr. M. B. Shaw, a cheese technologist based at Cardington, as the winner, on the day before the funeral.

*Presentation of FOSS Prize (from left) Mr Peter Watkins, FOSS; Miss Khawla Al-Haddad, winner; and Dr Tony O'Sullivan, SDT.*

**The FOSS prize was created to recognise the services of the late Mr Mike Griffiths.**

*The winner of the 1978 Stanley Clifford Award, Mr. Andrew Burgermeister receiving the cheque from Mr. J. Ridley Rowling, OBE (a Past President of the Society) with (right) Mr. J. R. Owens (Director General DTF) and Mr. L. J. Hall (Liquid Milk Director, DTF)*

As is clear from Professor Kay's account of the beginnings of the SDT, the intention was to provide a forum where people involved in the dairy industry could meet and discuss matters of common interest and at the same time facilitate progress in the industry of which they were a part.

At the start there were two types of membership: ordinary and associate. Ordinary members were defined as those 'holding responsible positions of a technological, scientific or educational-character, in or directly connected with the dairy industry'. Associate membership was available to those interested in those matters but not holding responsible positions. Essentially the difference between ordinary and associate members was one of position in the industry.

Later members were divided into groups and concept of associate membership disappeared. At present the Society has been giving consideration to the re-introduction of associate/affiliate members with the intention of facilitating joint membership of related organisations.

Scattered through the journals can be found interesting information about the members of the Society. It would appear that the list of members first appeared in the journal in 1949 (JSDT Vol. 2 No. 3 pp 127-47)

Almost all the companies in the industry in the UK were represented. Among them Dr. T. R. Ashton (a founder member), Express Dairy; Miss N. Bunion, Cheshire School of Agriculture; T. J. Bindley (Stilton cheese maker), Briscoe's; C. H. Brazened; Miss H. R. Chapman; Professor E. L. Crossly; Dr. C. Growth; Professor Dalgalarrando of Chile; C. L. Damoglou (Dobson Dairies, Belfast); Sir Ernest Debenham; Dr. T. J. Drakeley; A. Graham Enock; E. Falzon, Malta; F. N. Gascoigne; Lt Col V. Gates; R. G. Good; J. E. Heald; T. G. Henderson; E. P. W. Hill; C. Hillman; Wilfred Hipkins. P. A. Hoare; J. C. Hood; The Earl and Countess of Iveagh (Guinness); G. I. A. Lang; J. Lewis, J. McCartney (C.W.S. N. Ireland), S. Mackie, Dr. R. J. Macwalter, Miss K. D. Haddever, C. S. Miles; Dr. J. C. Murray; H. O. H O'Neill; K. W. Pocock; F. Procter; Dr. A. L. Provan; H. Rostern; Dr. S. J. Rowlands; R. A. Russell; C. V. C. Sekhar (India); P. W. Seligman; Dr. R. J. S. Seligman; F. S. Smith (Kirby & West); Dr. J. A. B. Smith; Felix Susaeta (Chile); W. H. Sutton; W. D. Thomson (Belfast); Dr. Bryon C. Veinoglou (Greece); N. Verlinsky (Israel), R. J. Warren (a founder member); E. White, J. C. D. White; and L. S. Yoxall.

Members were also drawn from the Commonwealth and Continental Europe, South America and the Middle East. There were representatives of animal feed manufacturers and equipment and laboratory suppliers. Members came from every corner of the UK and from all levels of activity in the milk industry. In May 1987 there were 75 overseas members.

It would appear that the last time a full list of members was published was in May 1977 as reported in the Journal in 1982 (JSDT Vol. 35 No. 3 July 1982). The report follows:

"From time to time the Society's office receives enquiries as to the level and progress of Society membership, and it is felt, in view of this, that the following may be of general interest to members".

"At its foundation in March 1943, the Society enrolled 233 Founder Members, and some 38 of these, 40 years on, are still included in the current membership".

"A complete check of membership records, carried out in May 1974, showed that in the 30 years since that modest beginning, membership of the Society had risen to 2540."

"A graph of the progress of membership kept since that time is reproduced below."

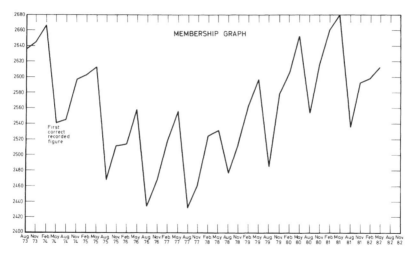

MEMBERSHIP GRAPH

First correct recorded figure

"It will be seen that peaks in membership occur four times yearly, when members are enrolled, and troughs annually in August when, in addition to normal losses by resignation, usually on retirement or through change of employment, the membership of those whose subscriptions have been unpaid for two complete years is terminated. Many such terminations arise from the fact that contact has been lost with members who fail to advise the Society's office of their change of address."

"Despite these regular, periodic losses, it is pleasing to note that, in general, there has been a steady increase in membership over the years. It is impracticable, for reasons of space, to publish a complete list of members, although such a list compiled in May 1977 is held in the Society's office and is available to members on request."

"Although the complete list cannot be published, it may be of interest to members to note the most recent applications, and these will be published regularly in future. In May 1982, 38 members were elected, bringing the current total membership to 2620."

"As the graph shows, in the 70s and 80s, the number of members fluctuated around 2200 to 2620 in May 1982, which was probably close to the maximum. During the Presidency of John Lewis a sustained effort was made to achieve 3000 members but that target was not met.

There was always anxiety about the number of members which was closely related to the ups and downs of prosperity in the industry and the consequences of rationalisation which is a more sophisticated way of describing plant and company closures."

Not for the first time the Society set up a Committee in 1994 to deal with the recruitment of members under the chairmanship of Dr. D. Moran. In 1994 (JSDT Vol. 47 No. 1 Feb 1994), Dr. Moran reported: "During the year the Committee has tackled the job of raising the profile of the Society and encouraging potential members to join in four ways."

"Direct Approach: A complete set of recruiting literature is available including explanatory leaflets, details of local Sections and scatter cards for distribution at open meetings".

"Indirect Approach: Following suggestions from section Chairmen, some 30-plus members in larger centres of employment and academia have been invited to help the Society in their local workplace. We hope these will act as 'recruiting agents' and that the numbers involved can be gradually expanded".

"We have also devised a one-to-one recruitment campaign whereby each member will be invited to recruit just one colleague, with a prize ballot for successful members. The Technical Committee have also been asked to co-operate in planning a 'project paper' competition for student/younger members with the intention of publishing the winning entry in the Society's journal".

"Recruitment Aids: The Committee has considered various options for promoting the awareness of the Society among non-members".

"Members' Services: One particularly attractive item the Committee is planning is a member's handbook (including a membership list). This will be distributed to each member and it is hoped it will be largely self-financing from advertising".

"During the year, several Sections completed a questionnaire among members on their attitude to the Society. Results will be collated in the office and published. This may well suggest some improved services to members. Some other options under consideration include discounts for 'quick joiners', a consultants list, a technical enquiry service, and a low-cost recruiting video tape for unmanned exhibition stands".

"Most Sections now have a delegated Committee member who is responsible for recruitment and good liaison exists between Sections and the centre".

"Because of the depressed situation in the industry, it is not surprising that there has been a modest fall in membership over the year; nevertheless, the committee has a vigorous programme in hand to help stabilise our membership numbers".

Dr Moran makes mention of a Members Handbook which I also pressed for in my time as President. The production of this has been agreed but has been held up for various reasons but should now appear in the coming year.

For many years and in the days when the industry had confidence in future growth, it was not unusual for companies to pay the subscription of employees on the grounds that membership of the Society would add to their knowledge and improve their performance. For example the Scottish MMB paid for 70-80 members but about 1986 this was reduced to 36 members (JSDT Vol. 40 No. 1 Feb 1987). The Scottish MMB was not alone in acting in this way.

About this time also a Residential Course and a Symposium had to be cancelled early for lack of support. In 1989 it was noted that attendance at the AGM was poor.

At the time the membership of the Society was 942 of which 886 were from the UJK and Ireland. (We now know this was an inflated figure as it included numbers not up to date with payment of subscriptions).

It is important to note that membership numbers are always subject to the qualification that they included members who were in subscription arrears (and possibly for some years). The culling of members from time to time for non-payment was a cause of fluctuation in numbers.

DETAILED MEMBERSHIP SURVEY: REPORT OF A SURVEY OF THE UK AND REPUBLIC OF IRELAND MEMBERSHIP OF THE SDT, MAY 3, 2000 BY MR. RALPH EARLY, SENIOR LECTURER IN FOOD SCIENCE, HARPER ADAMS UNIVERSITY COLLEGE, NEWPORT, SHROPSHIRE AND MEMBER OF COUNCIL OF THE SDT.

The Council decided to commission Mr Early to carry out a survey of members. A postal Survey of most of the membership in UK and Ireland was carried out in February and March 2000.

The questionnaire required an anonymous response and no provision was made for respondents to identify themselves. Some 770 questionnaires were sent out. By the end of March 2000 346 responses had been received and 341 could be analysed. The conclusions and recommendations of this detailed and very professional survey were as follows:

"The survey has shown that the respondents have a generally high belief that membership of the SDT is value for money. However, the Society faces a number of problems which bring into question its short-term viability. The membership of the Society is low, at under 1000. With nearly 20% of the membership population between 56 and 65 (assuming the responses are representative of the membership as a whole) and nearly 20% over the age of 65, the death of members during the next decade will bring about the demise of the organisation IF NEW MEMBERS ARE NOT RECRUITED QUICKLY. The result showing that only 0.6% of respondents are under 25 and 9.4% are in the 25 to 34 age group illustrates the deficiency in recruitment of young members. This is also reflected in the result showing that the majority of members are middle and senior managers - presumably people who have risen through the ranks, since joining, probably some time ago. That only 19.2% of members are female illustrates the deficiency in recruiting women into the organisation".

"When considering the recruitment of young members and more women, the following questions must be asked: what has the SDT to offer new members?: and what benefits will they obtain by being members? The journal is generally appreciated by members and is given as a major reason for being a member; however it might be accused of being too academic and of little use to those employed in practical dairy technology or in commercial areas of the dairy and food industry. The desire for a more 'newsy' publication is indicated by almost three quarters of respondents requesting news of members in the journal (or another publication)."

"The survey has shown that attendance at Section meetings is not strong, confidence in Section committees to organise an attractive programme is not high and the pressure of work and distance to travel prevent attendance".

"Sections depend on the generosity of some members to succeed and, with only a fifth of respondents saying they would like a greater involvement in the Society, it would seem that Sections are destined to struggle for viability. An optimistic point is the high response supporting the suggestion of creation of special interest groups. These may be an option to the traditional concept of Sections".

"The high proportion of respondents claiming Internet and e-mail access is an important finding".

"In conclusion it must be said that the Society of Dairy Technology has a major problem regarding its viability. If a more positive situation regarding membership numbers cannot be achieved in the next one to two years, the organisation's income and cash-flow status may force its demise. Areas which might be considered as important to securing the future of the Society are:

- Targeting employees under 30 years of age in dairy manufacturing businesses - possibly with the aid of company personnel managers
- Targeting further education and higher education institutions involved in teaching food and dairy science and technology
- Encouraging every member to recruit one person a year
- Creating national Special Interest Groups to reinvigorate attendance at meetings
- Producing a topical quarterly newsletter addressing UK dairy industry issues and promoting the activities of Sections and Special Interest Groups
- Presenting the newsletter on the SDT's website
- Improving the website by making it worth visiting because of the nature of its content, e.g., providing scientific, technical and commercial information about milk and dairy products and processing
- Producing policy statements along the lines of the Institute of Food Science and Technology's position statements on matters relating to dairy science and technology, e.g., a policy statement on *E. coli* 0157, and on *Mycobacterium avium paratuberculosis*
- Ensuring effective implementation of these recommendations and requirements, necessary to securing the future of the SDT, by appointing a full-time Officer/Executive of the Society
- Establishing a guiding principle that the Society shall be commercial in the way it operates, e.g., actively forging links with sponsoring companies and using the internet and the Society's website to promote both the Society and the discipline of dairy technology
- Seriously considering changing the name of the SDT to the Society for Food and Dairy Industry, in order to broaden the scope of the organisation and broaden the prospective membership base
- Targeting members outside the core target area of the dairy industry, i.e. seeking to encourage membership from the food industry broadly.
- Appointing to the board of directors an 'ideas' person to act in a strategic capacity and catalyse the actions of the full-time Officer/Executive of the Society."

RULES AND SUBSCRIPTIONS
At the AGM London 27 October 1947 the objects of the Society were restated and in essence they were similar to those in our current constitution. At the time of this meeting the annual subscription remained at £1.1s and there were 1188 members the vast majority being in

Groups 1 and 3.

At the AGM held on 11 May 1949 a "series of modifications" of the rules were approved. It was stated that most of them were merely improvement in drafting.

At this meeting it was decided to increase the subscription from one guinea to 30s p.a. and the fee for life membership from fifteen guineas to twenty guineas with effect from 1 July 1949.

In July 1978 subscriptions were increased from £4.50 to £10 and it was noted later that members were reluctant to pay the increased subscription. Those who did not pay promptly were liable to forfeit receipt of the Journal.

At the Council meeting in February 1990 (JSDT Vol. 43 No. 2 May 1990) it was decided to increase subscriptions to the following:

| | |
|---|---|
| Annual | £35 |
| Young Members | £22 |
| Students and retired | £12 |

Subscriptions in 1998 were as follows:-

| | |
|---|---|
| Members *(annually)* | £48 |
| Life Members *(one payment)* | £570 |
| Members aged under 26 *(annually)* | £30 |
| Students *(one payment for duration of studies)* | £15 |
| Retired Members *(annually)* | £24 |
| Life Membership Retired Members | £135 |
| *(Minmum age 60 - at least 10 years' membership)* | |

The current subscription for members is £58 and £29 for retired membes. This subscription includes the journal and is payable by variable direct debit.

MEMBERSHIP AND THE SOCIETY'S JOURNAL

The Journals of the Society are a dictionary of national biography of the dairy industry. There are reports of appointments, appreciations of outstanding service and obituaries the primary purpose of which is, of course, to commemorate the life of the deceased but which are also valuable sources of information. It is impossible even to summarise all the obituaries but extracts from those who made their mark in the SDT and the wider world are given below. These are:

Dr. James Macwalter (d16 Sept 1997 at the age of 87) (JSDT Vol. 50 No. 4 November 1997). (President, Trustee and Honorary Member).

Professor Edward Capstick, M.C. O.B.E. M.Sc. (d1963) (Founder member of SDT President 1948-49 (JSDT Vol. 16 No. 3 1963.)

Miss Barbara Fischer (1998) Hon Asst Editor of The Journal (JSDT Vol. 51 No. 1 February 1998.

Sir J. Hammond C.B.E. FRS

Dr. J. G. Davis O.B.E.

*Obituary Dr Macwalter*

"Dr Roy James Macwalter, who died on 16 September 1997 at the age of 87, was formerly the Chief Scientist of Unigate Ltd and will be remembered as one of the most distinguished and leading scientists in the dairy industry. He joined the Society of Dairy Technology in 1949 and, having served on various committees and the Council, became President in 1960. He also served as a Trustee of the Society and of the Capstick Memorial Trust. He was made an Honorary Member of the Society in 1997".

"Dr Macwalter (often misnamed MacWalter) graduated with First Class Honours in Chemistry at University College, London, in 1929 and went on to do research in chemical spectroscopy for which he was awarded a PhD in 1932. He was subsequently engaged in research on vitamins A, D and E with Professor J C Drummond in the Biochemistry Department of University College. He joined United Dairies Ltd in 1934 and at this stage took a Diploma in Chemical Engineering".

"In 1950 Dr Macwalter became Chief Research Chemist and Chief Chemist of United Dairies Ltd and was responsible for the administration of the Central Laboratory and some fifty quality control laboratories in the United Kingdom. Following the amalgamation of <u>United Dairies and Cow and Gate in</u> 1959, he became Chief Scientist of Unigate. Subsequently he undertook the transfer of the activities of the Central Laboratory in Wood Lane to a new and larger building at Park Royal which was opened in 1966 by the Minister of Agriculture. From 1950 onwards, in addition to his commitments with his employer, he became very much involved with general work on behalf of the dairy industry".

"Dr Macwalter served on various advisory committees to the Ministry of Agriculture, Fisheries and Food and was chairman of the Milk Hygiene Committee, the Dairy Effluent Committee, and the Milk and Milk Products Technical Advisory Committee, which was concerned with the research requirements of the industry; he was also an advisor to the Ministry's UK delegation to the FAO/WHO Committee of Government Experts on Milk and Milk Products at its meetings in Rome".

"Dr Macwalter served as chairman of the Scientific Consultative Panel, which was responsible for work related to the milk quality schemes operated jointly by the Dairy Trade Federation and the Milk Marketing Board. Also he was chairman of the Technical Committee of the Creamery Proprietors Association, and chairman of the Scientific Liaison Committee of the Dairy Trade Federation. He was active in the work of the British Standards Institution and in this capacity he was chairman of the Dairying Advisory Committee responsible for the work of various subsidiary committees and was involved in the work of the International Organisation for Standardisation. He made many friends overseas as a member of the UK delegation to the Annual Sessions of the International Dairy Federation (IDF) and was on the Council of the United Kingdom Dairy Association, which is the UK's National Committee dealing with IDF affairs. Dr Macwalter was appointed an OBE in 1971. In spite of all his achievements, Dr Macwalter was a modest man with a great sense of humour. He was well liked and respected by his contemporaries to whom he was known as 'Mac' and would have regarded misspelling of his surname as a matter of little consequence"…

*Obituary Professor Capstick*

"The death of Professor Edward Capstick, after a short illness at the comparatively early age of 64 years, has shocked his many friends both at home and overseas. His background was unique, for he had not only academic and commercial experience, but also service during the war as a temporary civil servant".

"The son of a Westmorland dairy farmer, Capstick was a countryman at heart and his practical understanding of the milk producers' problems often stood him in good stead. He joined the army from school in 1917 and was soon commissioned in the King's Regiment, with which he went to France where he was awarded the Military Cross and Bar. In 1919 he was awarded a scholarship to Durham University, where he read agriculture at Armstrong College obtaining a B.Sc. degree and the N.D.A. and, after subsequent work on bacteriology, an M.Sc. He spent a post-graduate year at University College, Reading, obtaining the N.D.D. and B.D.F.A. diploma, after which he was appointed as demonstrator in the British Dairy Institute at Reading".

"At this stage Capstick decided that there was greater scope in the field of commerce. He joined Aplin & Barrett Ltd. as a dairy manager and helped to create their new condensery and drying plant at Frome, Somerset".

"When <u>Reading University introduced the first Chair of Dairying in the country, Edward Capstick was invited to become the first Professor.</u> This was a job that appealed to him, providing infinite opportunity for his energy and enthusiasm in developing dairy education from a farmhouse course to an applied science, but the war intervened".

"In 1939, the Food (Defence Plans) Department was building up the nucleus of a wartime organisation to control the food industries. Once more Edward Capstick was an obvious choice to be responsible for butter and cheese, later being appointed <u>Deputy Director (Home Produce) in the Milk Products Division of the Ministry of Food</u>".

"The industry found him fair, but quite ruthless in dealing with any evasion of the regulations which operated for the common good".

"Although this was more than a full time job, Major Capstick commanded a company of the Ministry of Food battalion of the Home Guard at Colwyn Bay".

"Capstick was a Founder Member of the Society of Dairy Technology and in 1942 served on the original committee that formed the Society, subsequently becoming President in 1948-49. During his year of office he produced a notable Education Report and reorganised the Society's office accommodation. He was later chairman of the Education and Research Committee".

"In 1946, Capstick resigned from the chair at Reading University, which he had held in absentia while with the Ministry of Food, to join United Dairies Ltd. He became a Group Director in 1955 with responsibility for laboratory organisation and for the Inspection and Public Relations department. On the formation of Unigate Ltd. in 1959 he was appointed to the main Board".

"During these years Capstick was in great demand to serve on committees of both the Ministry, to whom he was a much-respected adviser, and of the industry. It was to the international field, however, that he devoted more and more of his time. As a member of U.K.D.A. Council he took a keen interest in the work of I.D.F., particularly from 1959, when he was elected to the chair of U.K.D.A. and a seat on the Executive Council of I.D.F. Here he worked ceaselessly to attain a better understanding between the dairy industries of the world and his work as chairman of the Cheese Commission of I.D.F. is now bearing fruit in the introduction of international standards for cheese through F.A.0. He was greatly interested in the work of UNICEF and served on the London Committee. He had undertaken missions to Kenya and the Eastern Mediterranean on behalf of F.A.0. and was delighted that, after four years negotiation, he had concluded arrangements for a course for overseas dairy technologists, to be held in this country at the end of this year".

"Capstick was a familiar figure at International Dairy Congresses where he was much in demand. At the London Congress in 1959 he was chairman of the Reception and Hospitality Committee and at Copenhagen a Section Chairman for liquid milk".

"In 1958 official recognition of his activities was given by the award of O.B.E.; seldom can such an honour have been so well merited or so widely acclaimed by the industry he served".

"The Ministry of Agriculture, Fisheries and Food invited him to become chairman of the Milk and Milk Products Technical Advisory Committee on the retirement of Professor Kay in 1960. He also served on the Board of the N.I.R.D., was a Council Member of the R.A.B.D.F. and found time to be on Committees of B.S.l., as well as to act as an examiner for universities and for the N.D.D. A strong advocate of quality control of dairy products, he was a member of the Management Committee and Chairman of the Butter Advisory Committee of the N.A.C.P. Grading Service. He was also elected an honorary member of the D.I.C.E. Association".

"Edward Capstick had now reached the peak of an outstanding career".

*Obituary Miss Fisher*

"The death of Miss Barbara Fischer brought to an end a very long connection with the dairy industry and with the Society. She and her sister were born in India, where their father was with the Forestry Service. She returned to England at the age of 5 to start school. Her connection with the dairy industry began more than 70 years ago, when she obtained a Diploma in Rural Economy from Oxford University. She moved to Reading University in 1923 where she obtained the National Diploma in Dairying, and in 1926 became a Student Demonstrator while studying Advanced Agricultural Bacteriology. This led to an invitation in 1928 to set up a bacteriological laboratory at Job's Dairy, then beginning a period of expansion, and she remained there for the rest of her working career".

"There were many technological changes in the industry during her career, and she was involved in the development stages of several of them in cooperation with the NIRD at Shinfield. Job's took part in one of the first pilot schemes for bulk collection of milk from farms, and Miss Fischer was concerned with the identification and solution of some of the problems that arose, in chemical and bacteriological quality determination of individual supplies, and in quantity measurement".

"She retired from Job's in 1967, leaving behind a group well able to continue to the meticulous standards she had set".

"She became a member of the Editorial Board of the Society in 1965, and an Honorary Assistant Editor in 1969. She remained in the latter post until shortly before her death, helping a succession of Editors and Scientific Editors with the preparation of papers for publication and with proofreading".

*Obituary Sir John Hammond, CBE, FRS*

"Sir John Hammond, CBE, FRS, a Founder member of the Society and an Honorary Member since 1955, died at Cambridge on 25th August, 1964."

"His deep interest in all matters concerned with the breeding and management of farm animals and his general zest for life made him a stimulating companion, and one whose wise counsel was sought by scientists and farmers the world over. In Hammond was combined the brain of a brilliant scientist and a knowledge of farming problems which stemmed from an upbringing on a Norfolk farm. It was this combination, together with a natural friendliness, which led to his opinion being sought by geneticists and physiologists on the one hand and farmers at a country meeting on the other. In the years between the wars and until 1954 when he retired, his laboratory and field station was a magnet which drew animal physiologists from many countries, for to have worked under John Hammond was a highly prized qualification."

"His work, which covered the wide field of animal physiology and animal husbandry, is mirrored in the books he wrote and the scientific papers of which he was author over a period of some 50 years."

"Hammond went up to Cambridge in 1906 and read natural sciences taking his Tripos in physiology, chemistry, botany and geology. Thereafter he spent a year taking the agricultural diploma course. He was awarded a research scholarship by the Ministry of Agriculture and joined Sir Thomas Middleton, T. B. Wood, and F. H. A. Marshall in the new Department of Agriculture at the University. After serving in the Norfolk Regiment in the first World War he returned to Cambridge to begin a life-long association with the School of Agriculture. Those of us who remember the years after the first World War will know of the magic which attached to the names of T. R. Wood, Marshall, H. A. D. Neville, Halnan, Hammond and Newman - pioneers in agricultural education and research and courageous in applying scientific knowledge and method to the solution of problems of the agricultural industry."

"In those early days facilities for research were meagre or almost non-existant, and little money was available for the purchase of experimental animals. Hammond used all existing records for the study of growth rate in farm animals and wastage in dairy herds. From these small beginnings developed Hammond's classic contributions to our knowledge of growth in farm animals, the effect of plane of nutrition on carcase conformation, and the influence of the dam on growth of the foetus. In the early twenties he and his colleagues began to experiment with artificial insemination. They devised techniques which could be adopted under practical conditions, and indeed it was Hammond's vision and enthusiasm which enabled the establishment of the first artificial insemination centre at Cambridge. Today over two-thirds of the calves born in the national dairy herd are the result of artificial inseminations."

"Hammond was a member of many scientific societies and it was characteristic of him to listen with close attention to the first paper of a nervous youngster and be one of the first to offer congratulations and constructive advise. Indeed it was an inbred modesty and consideration for others which endeared him to so many who will long remember the tall lean figure, the twinkle in the eye and the cigarette whose ash was eventually brushed off the front of the jacket."

"From 1936 until 1963, Sir John was a member of the Board of the National Institute for Research in Dairying and took a keen interest in its work. Members of staff will long remember the guidance and help which they received from him. It was only natural that Hammond's advice was sought by many countries overseas."

"Many universities both at home and abroad have recognised Hammond's great distinction as an animal physiologist and have been glad to include his name among their honorary graduates. Many agricultural academics have elected him a foreign member and Downing College will long remember him as one of its distriguished Fellows."

"When the history of agricultural science for the first 50 years of this century comes to be written, Hammond's name will be included among those of four of five agricultural scientists whose work has made a lasting impression on the agricultural industry of this period."

*Obituary* JOHN GILBERT DAVIS OBE
*A Tribute to his life and work (JSDT Vol. 47 No. 3 August 1994)*

"Dr J G Davis, who died in February two months before his 90th birthday, was a most remarkable man. He was a man of commanding stature, trusting of others and trustworthy himself. By nature and training an academic, he became one of the leading dairy scientists of his time."

*Mr George Grier, outgoing President of the Society, presenting the Society's Gold Medal to Dr J G Davis, OBE, before the AGM on 27 October 1987.*

"Dr Davis graduated in chemistry and physics at University College London in 1925. His postgraduate work in biochemistry leading to the degree of MSc was carried out under Professor Jack Drummond (later to become Sir Jack Drummond). He was awarded the PhD degree of London University in 1930, to be followed by the DSc (Bacteriology) in 1939. He received the OBE in 1975."

"In 1927, at the early age of 23, Dr Davis was successful in obtaining a position in the Bacteriological Department of the National Institute for Research in Dairying (NIRD), then under the headship of the much respected and knowledgeable Dr A T R Mattick, a man dedicated to clean milk production on the farm and during subsequent handling procedures from farm to consumer, his main aim being to limit or control the spread of milk-borne diseases - the fundamental concept of the first director of the NIRD, Dr Stenhouse Williams. As senior biochemist, later to become head of a newly created department of biochemistry, Davis' work led him to studies on bacteria found in milk and milk products, more specifically lactobacilli and in particular their role in cheese starters and the cheese ripening process. He also carried out work on the relatively newly discovered phenomenon of bacteriophage infection of cheese starters. In 1932 he was awarded a Rockerfeller Research Fellowship and for one year interrupted his work at Shinfield to study the implications of hydrogen peroxide production by lactobacilli."

"In the late 1920s to the early 1930s a National Advisory Body under Ministry of Agriculture authority was formed to advise farmers and others on methods of clean milk production; the body of men, advisory bacteriologists as they were called, were stationed at universities and colleges throughout England and Wales, often serving the dual function of farm advisory officers and university lecturers in dairy bacteriology for students preparing for the National Diploma and College Diploma examinations (NDA, NDD, CDD). At some time or other most of the advisory bacteriologists underwent more advanced training in the dairy bacteriological department of the NIRD and could be expected to come under the strong influence of Drs Mattick and Davis, particularly the latter. To promote the interchange of knowledge this body formed the Society of Agricultural Bacteriology (SAB), which later became the Society for Applied Bacteriology with a much wider scope of interest. Together with other members of the original SAB, Dr Davis published several papers on the various tests which were used to assess the hygienic quality of milk, e.g., studies on the correlation between the plate count, coliform content, methylene blue test, clot-on-boiling test, sediment test, the subjective tests of taste, smell and appearance, the alcohol test and the keeping quality test. Experience in this field of work and in particular the biochemical aspects led Dr Davis to develop the resazurin test. This was to prove invaluable during the forthcoming World

War II effort to produce more and more milk of an approved hygenic standard in order to augment the nutritional resources of the nation and especially to meet the needs of babies and younger children. Indeed, Professor H D Kay, the director of the NIRD at the time, was to say that Dr Davis had brought untold benefit to the nation in saving thousands if not millions of gallons of milk by perfecting the short platform test and the longer grading test."

"The need for milk as a staple food resulted in the appointment of Professor J C Drummond as Scientific Advisor to the Ministry of Food and Professor E Capstick (then Professor of Dairying at Reading University) as Head of the Ministry's Milk Division, and further the secondment of Dr Davis from NIRD to arrange and implement a Milk Testing and Advisory Scheme - at first a four counties trial scheme, later to become a national scheme (NMTAS) - to monitor the hygienic quality of farm milk supplies. The scheme was a great success and a great achievement on the part of Dr Davis, who by this time was popularly called JG in order to avoid confusion with his colleague Dr W L Davies, the head of the Chemistry Department of the NIRD, known as WL."

"Within a short time of the end of the war Professor Drummond had met his death while on holiday in France and Professor Capstick had taken up an invitation to join the United Dairies Ltd in the capacity of Technical Director, E L Crossley of Alpin and Barrett having became the new Professor of Dairying at Reading University."

"After some consideration and possibly vying with the appointment of Professor Capstick to United Dairies, Mr W A Nell of the Express Dairy Company Ltd decided to approach Dr Davis to become Technical Advisor to the company on the impending retirement of Mr A L Barton, the Head of the Central Laboratory, a laboratory run on the lines of that of the public analysts. Dr Davis accepted the position without a specific brief and given carte blanche to re-equip premises assigned to him."

"JG proceeded to build up a somewhat grandiose organisation on traditional lines, as a consultancy rather than an active unit involved more in the programme of expansion of the company - the need at the time amounting to the redesign of existing buildings to suit new plant, the design of new buildings and the choice of new equipment to meet the new demands of greater productive efficiency, etc. The lack of communication at the outset, combined with some poor and inappropriately qualified and experienced members of his senior staff, began after a time to make him the target of the cost-conscious accounts."

"JG became increasingly unhappy but uncomplaining as the situation developed until such time that he decided to set up his own consultancy in Gerrard Street, London under the name of Dr J C Davis and Partners. JG turned his experience of commerce to advantage in his consultancy."

"His association with Tom McClachlin over the years proved fruitful to both, Tom himself on the chemical side while JG was strong on biochemical and bacteriological aspects. (TMcC was a public analyst)."

"His verbal contributions at scientific and other meetings were to the point and gave little room for misunderstanding. JG was prolific in his writings. He wrote in excess of 300 research papers and articles on dairying topics for the press and scientific journals."

"His work was of considerable help to many research and other students in often providing short cuts to hours of laborious time consuming searches and perusal of reference. It should be admitted that he received immense help in this side of his work from Miss Doris Knight, the librarian of the NIRD. Perhaps JG's greatest contribution today was his *Dictionary of Dairying*, which first appeared in 1950 updating that of Cronshaw published circa 1930. This dictionary was virtually the Encyclopaedia Britannica of dairy science and technology."

"Dr Davis' move to consultancy practice quickly became extended to include work on foods. At first reluctant to support the formation of an institute of food science and technology, probably because of a clash of interests with the food group of the Society of Chemical Industry of which he was Chairman at the time, he was eventually persuaded to help and found himself elected to its council in 1964. From 1964-71 he was founder chairman of the publications committee and had responsibilities in connection with the Journal of Food Technology (the forerunner of the IJFST) and the Proceedings (forerunner FSTT). Dr Davis was made President of the Institute in 1967 and became honary fellow in 1981 in recognition of his outstanding work and distinguished services to food science and technology."

"In addition to the offices to which reference has been made above should be listed the following: Secretary, Society of Microbiology, 1948-54; President, Society for Applied Bacteriology, 1954-56; Chairman, Association of Consulting Scientists, 1959-61; Chairman, Society of Chemical Industry Food Group, 1961-63; President, Food and Nutrition Section Royal Society of Health congress 1962; Vice-Chairman, British National Committee for Food Science and Technology, 1963-66; President, Society of Dairy Technology, 1971-72; Chairman, various committees of the British Standards Institute."

"Dr Davis was awarded the Gold Medal of the Society of Dairy Technology in 1987 for out-standing achievements in the dairy industry."

"Dr John Gilbert Davis crowded much into his life, and the Society of Dairy Technology will honour his name for ever more. He gave much time and effort to establish the Society as a viable and virile organisation, helping to spread the knowledge and science of dairying throughout the UK and indeed the whole world."

### Prestigious Cheese Industry Award for Eurwen Richards

Eurwen Richards, a Food Consultant and an ex-President of the Society, is this year's winner of the cheese industry's lifetime achievement award, which recognises and honours individuals who have made an outstanding contribution to the UK cheese industry.

This cheese industry award is sponsored by the British Cheese Board; and the recipient is selected by a panel of indepenent judges seeking individuals who have had a positive impact on the cheese trade, through outstanding contributions in the areas of production, marketing, research, education or training.

Eurwen is the third winner of this prestigious cheese industry award, with the first two recipients being K D Maddever, OBE and Marc Adams. The Society congratulates one of its stalwart members in receiving this UK cheese industry-wide recognition.

# CHAPTER SIX

## RECOLLECTIONS OF RECENT PAST PRESIDENTS
## AND EX-EXECUTIVE SECRETARY

The following contributions have been submitted by recent Presidents of the SDT, and describe the highlights of their terms of office. These personal recollections are presented in chronological order, thereby giving a picture of the evolution of the Society over recent years as seen by the individual Presidents. This Chapter ends with a contribution from the Society's last Executive Secretary, Mrs Ros Gale, who continues to help the Society by providing information on previous practices.

EURWEN RICHARDS, PRESIDENT 1988-89

Many of us have cause to thank the enthusiast for kindling our own life interests. That is certainly my experience as far as the SDT is concerned. Among those enthusiasts were John Lewis, Director of the College Dairy at UCW Aberystwyth, Mena Jones who was Regional Dairy Husbandry Advisor, Wales - one of five women who headed the England and Wales NAAS regions - and S.B Thomas who was Regional Bacteriologist, National Agricultural Advisory Service, Wales. They encouraged participation, co-operation, learning, investigation and, most of all, promoted the objects and interests of the Society in the Principality.

*MISS. E. RICHARDS*
*PRESIDENT 1988-89*

In its heyday the SDT had an enviable mix of members, all being equal in status, whether they were captains of industry or the lowliest of employees. That I am sure was one of the reasons for its success. We, the younger members, were encouraged to participate in meetings, conferences and study tours. I recall a meeting at Seale Hayne and listening to Miss K.D Maddever talk about clotted cream. That was the beginning of many years of friendship. A group of 50 or more visited New Zealand in 1977. None of that group will forget Peter Lee selling the shoes on his feet for a crate of water - or was it something stronger? So many memories of happy times, learning from each other as well as from the people, places and countries visited.

It was accepted that there were confidential aspects and barriers not crossed within each organisation, but this did not inhibit the hospitality and willingness to share information and experiences. I was privileged to be involved in an exciting, expanding industry and an active Society of Dairy Technology. This culminated with my being elected its President in 1988. There were obvious signs of change at the time. My predecessor Terry Tamplin tried to gain the interest of the then captains of industry, sadly to no avail. Rationalisation, integration, and competition, not co-operation, were the current word games.

A lower cost study tour to Bavaria was organised during my year to meet the needs of the younger members, although few took the opportunity. However, it was a successful tour and at times an emotional one, as we listened to the almost continual reports of the falling of the Berlin wall by our German guide and driver.

This was one of the highlights of my year. Others were attending meetings in each Section, touring areas of the country to meet and encourage the faithful and hard working members of the Society on their home ground. This however could not stop the momentum of change, could not alter the fact that members were resigning because of redundancies and a shrinking dairy industry.

I recall Peter Lee's ability to type minutes etc. for members of Council on an ancient typewriter. It gave me pleasure to include that piece of antiquity with his farewell gifts. Ros Gale began her term of office when I was President. We quickly achieved a good working relationship that has become a lasting friendship, and that has been an important feature of the Society over the years. Many friendships have been forged through this informal network. The industry has changed and the Society has had to change to correspond with today's perception of a learned Society. I wish future members as happy and as rewarding a period as I have had.

My tenure of office commenced in 1991 and followed on from a very successful year as President by the late Gary Royall. Gary was a stalwart of the Society and I was very honoured to have been his Vice-President. Gary started a number of initiatives during his year to improve relationships between the SDT and other organisations such as the Institute of Food Science and Technology and also the Society of Food Hygiene Technology.

*MR. J. M. BIRD*
*PRESIDENT 1991-92*

Gary and I appreciated that the role of the Society in a contracting UK dairy industry would have to change and that to survive and prosper meant that we needed to extend our membership into related segments of the food industry that, in the past, had only a passive relationship with the Society. Gary was the first President to recognise this and actively seek to promote synergistic relationships with kindred organisations. For my part, I was pleased to continue Gary's work to promote these discussions and attempt to raise the profile of the Society within the general food industry.

I set myself two main objectives for my year of office:
- to complete the arrangements for the 50th anniversary celebrations
- to revive the Northern Section which had sadly lapsed over the years.

I was very fortunate to have the support of Mrs. Ros Gale, Executive Secretary, and Peter Fleming, Vice-President to assist in the formation of a Sub-Committee to organise the celebrations. This Committee was formed from past Presidents and Section officers and we set to work with the objective of organising a central function that would celebrate the past 50 years of the Society and set the scene for the future. The committee also assisted various Sections in organising functions to celebrate the anniversary in their own way. Many ideas were forthcoming and, after many meetings and visits to suitable venues, the committee agreed a two-day residential conference in Cambridge where a select number of senior executives and academics from the worldwide dairy industry were invited to present papers on major topics of interest to the dairy industry.

At the time of the celebrations, the Council of the Society agreed to honour two members of the industry with Gold Medals for their contributions to the Society and the dairy industry in general. Dr. Borg Mortensen, Director of Research at MD Foods, as it was at that time, was awarded his medal for his outstanding contribution to research and product development within the dairy industry internationally, and our very own Dr. George Chambers was awarded his medal for contributions to the dairy industry in general and also to the Society in particular - of which he has been a distinguished member for many years.

It is normal practice for recipients of the Gold Medal to be invited to present a paper at a convenient meeting of the Society and both readily agreed to an involvement at the Conference. Dr. Mortensen duly presented his paper at the Conference and, as a slight departure from convention, Dr. Chambers agreed to give the after-dinner speech at the Conference Banquet. Anyone who knows George will accept that he is renowned for his wit and charm and he rose to the occasion. Many present that night became enthralled by his wealth of anecdotes and there were many sore ribs at the end of his speech. We were fortunate that many of the surviving Founder Members and past Presidents joined with us at Cambridge to make the 50th Anniversary Conference an event to remember.

My main recollection of my year is the excellent Spring Conference organised by the Welsh Section under the guidance of Mrs Hazel Rowlands and Mr John Ladbrooke, and held in the capital city of Cardiff. A delicate balance between work and play left all who attended with happy memories of Welsh hospitality.

During the Conference banquet at Cardiff castle, I was privileged to present Honorary Membership Certificates to Eurwen Richards and Susanne Moore (now Watson) in appreciation of the commitment they have given to the Society over many years. How appropriate it was that Eurwen should receive her award in the capital of her native country!

My year in office as your President was a most moving experience and I was honoured to have the opportunity of leading the Society for a short period, and I trust that I handed over to my successor an organisation at least as strong as I received it from Gary Royal.

## MR PETER FLEMING, PRESIDENT, 1992-1993

*MR. P. W. S. FLEMING*
*PRESIDENT 1992-93*

I was fortunate and honoured to be President of the Society during its Golden Anniversary year of 1993 and was therefore involved in visiting most of the Sections to join in their celebration events.

My abiding memories of the year are the many old friends and new ones whom I met during my travels and this more than anything epitomised what the Society is all about and is a camaraderie which I still miss.

The other feature which I remember was the superb weather which blessed each event which I attended and none more so than the undoubted highlight of the year - the Conference at Cambridge. The papers were of an extremely high quality and delivered by a very distinguished group of speakers.

I was also privileged to make several presentations that year, the most prestigious being the two Gold Medals to Dr. Borg Mortensen and Dr. George Chambers and Honorary Membership to Barrie Powell, Ray Thomson and Paul Young.

One of my most enjoyable visits was to Northern Ireland where I visited a couple of dairies in the morning and played a round of golf with my hosts in the afternoon. As is customary during food plant visits, we were togged out in clear plastic coats and with 'shower cap' hats - not a pretty sight! Unbeknownst to me the local Press covered the visit and a photograph, which I thought was for the local Section, appeared in next day's local paper to be followed by many derisory phone-calls from my *'friends'* in Northern Ireland!

The one underlying recollection of that year, and indeed my entire time in the Society, was the friendliness and good companionship of all the people whom I met and worked with and it is this which makes the S.D.T. such an attractive *'club'* to belong to. Lifelong friendships are also useful contacts building up a network of expertise unmatched in any library or college.

## DR. ANTHONY O'SULLIVAN, PRESIDENT 1993-94

*DR. A. C. O'SULLIVAN*
*PRESIDENT 1993-94*

There were at least three significant features of my Presidential year, as follows:

- I was the last President with an one-year term of office. The next President, Graham Lee, was the first President to have a two-year term. (As a general comment on this development, and based on my own prior experience as Chairman of the Cork Scientific Council in 1968-70, I believe that this has been positive, giving greater continuity in the leadership of the Society. Perhaps the only downside is that it may act as a deterrent to some candidates because of the longer time commitment required)
- I was the first Republic of Ireland-born person to have been honoured by election as the Society's President, even though I was resident in Great Britain at the time. This achievement follows on the desirable precedent created by having three prior Presidents from Northern Ireland: the late Professor John Murray; Dr George Chambers; and Mr George Grier. This precedent was subsequently strengthened by the election of Dr David Armstrong as President for the 2000-02 period. Naturally, I also pleased that this Irish dimension will be maintained by the 2002 election of Mr Michael Hickey, current Chairman of the Southern Ireland Region, as Vice-President.
- The last month of my Presidency, November 1994 witnessed the commencement of deregulation of the GB dairy industry, when the four GB statutory Milk Marketing Boards were disbanded. [The Milk Marketing Board for Northern Ireland (MMBNI) followed suit in February 1995].

Because my Presidential year followed on immediately from all the special and high profile activities of the 50th Anniversary year, under the highly capable Presidency of Mr Peter Fleming, I deliberately adopted a *"technological mode"* as the basis of my Presidency. I decided that the major feature would be to place emphasis on the programme planning facets of the year's work; and these activities were conducted in close association with Mrs Ros Gale, the Society's highly competent Executive Secretary, to whom my personal gratitude is extended.

In retrospect, perhaps I was, inadvertently, laying the groundwork for my subsequent roles as the Society's Programme Planner (from 1995 onwards) and combined Programme Planner/Coordinator (from 2000 onwards).

Furthermore, because Quality Milk Producers Ltd (QMP) was a small organisation, I also tried simultaneously to service the interests of both the SDT and QMP in the programmes which were planned and held during the 1993/94 year. Specifically:

1. I organised the Spring 1994 Symposium at St Helier, Jersey on 25-27 May on the theme of *'THE MANUFACTURE & MARKETING OF NICHE/LUXURY MILK & DAIRY PRODUCTS'*. Wearing my QMP hat, I gave the first paper at this event on *'SUCCESSFUL RETAIL MARKETING OF BRANDED, NICHE MILKS'*; and this Channel Islands milk theme was reinforced by a presentation by Mr Brian Le Marquand, Executive Chairman, Jersey Milk on *'MANUFACTURE & MARKETING OF NICHE DAIRY PRODUCTS IN JERSEY'*. The Symposium also included a visit to Jersey Milk's dairy, a Vin d'Honneur followed by a barbeque on the first evening, and a dinner sponsored by Jersey Milk on the second evening.

2. I also organised, in close association with Mr Denis Prew, Head of the then Dairy (now Food) Technology Department at Reaseheath College, Nantwich, Cheshire, a three-day Residential Course from 7-9 September on *'NEW PRODUCT DEVELOPMENT IN THE DAIRY INDUSTRY.'* As well as a series of lectures, this hands-on course involved a full day of practical sessions, comprising basic new product development (NPD) activities for a range of products which were suitable for inexperienced personnel. The latter entailed actual product development assignments on a team basis; and the end-results were excellent. A third feature was a tour of the then Milk Marque Product Development Centre on the Reaseheath campus, and a joint presentation by G C Hahn & Co and Stephan Machinery (UK) Ltd on NPD examples using stabilisers and specialist processing equipment.

The Annual Conference and AGM, which attracted an attendance of 52, were held in Bristol on 23-26 October, and were well organised by the Committee of the then Western Section, under the successful direction of Stephen Crutchley of Eden Vale Food Ingredients, Staplemead Creamery, Somerset. Attractive technical visits included Anchor Foods' major cheese and butter packing, and aerosol cream manufacturing operations at Swindon, St Ivel at Wootton Bassett and Alvis Bros, organic farmhouse cheese makers, Redhill. I handed on the President's chain of office at that event to Mr Grahame Lee, the first, incoming 2-year President.

GRAHAME LEE, PRESIDENT 1994-1996

In the early 1990s, the Society continued to experience a decline in membership and support for its various activities.

Council had agreed that change was inevitable to contain costs, meet members' needs, stem the decline in numbers and buoy up income. It was felt that the annual change in President and Vice-President was not conducive to effective management. So the Constitution and Rules were amended to prepare the Society in readiness for incorporation, and to provide continuity of management. Accordingly, I was the first President to serve a two-year term of office, being elected in 1994 and carrying on until 1996.

MR. G. LEE
PRESIDENT 1994-96

We were also at the time of probably the most significant changes in the dairy industry: DE-REGULATION. The dairy industry, understandably, had priorities other than continuing to give the Society wide support and we had to realise it was not our automatic right to expect this.

Fortunately, the support from the dairy equipment and ancillary industries was sustained and was appreciated.

Rationalisation of the Society's administration took place with the loss of the Membership Secretary (Joan Lawford) at our Huntingdon office, and an associated extension of the role of our Executive Secretary, Mrs Ros Gale. Committee rationalisation also took place, with the discontinuation of the General Purposes and Programme Planning Committees, and the Scientific and Education Committee was amalgamated with the Editorial Board to form a new Technical & Editorial Board. This certainly led to a reduction in repetitive discussion at Council meetings and travelling costs.

The ongoing review of the Constitution and Rules resulted in the clarification of the previous Custodial/Property trustees; the latter were no longer necessary when Council members were in fact the Managing Trustees.

A review of Society strategy was initiated with widespread consultation amongst members, and eventually resulted in the simplified membership structure we have today.

Whilst we had seen the last of centrally organised spring or autumn Conferences in England and overseas Study Tours, there were still a number of very successful social events demonstrating the will of members to continue meeting - notably, dinner dances in the Northern, Northern Ireland and South Eastern Sections. I also had the opportunity to attend the annual meeting of the Southern Irish Section where dairy farmers were invited, and attended in large numbers.

Instead, one-day Symposia were becoming more in demand and turned out to be an useful means of recruiting new members. During this two-year period, the Society also had its last Residential Course in England at Brackenhurst College, now part of Nottingham Trent University.

It was with great pleasure that I presented the Gold Medal to Dr Donald Muir for meritorious services to the dairy industry and the SDT at the 1966 AGM, when my Presidency ended.

ROBERT MCCANDLESS HAMILTON, PRESIDENT, 1996-98
Picking out highlights of my term in office will undoubtedly offend someone, as the entire two-year period was most enjoyable. However there are some events that stand out in my memory.

My first public speaking engagement was in Kilkenny, where the Southern Ireland Section had arranged a Farmer's Evening, chaired by Dr. Sean Tuohy, its Chairman. Suitably prepared, as I thought, I went to the pre-session meal, which was attended by an eclectic mix of Irish dairy personnel headed up by Minister for Agriculture & Food, Ivan Yates T.D., accompanied by the CEOs of Waterford Co-Op and Avonmore Co-op, Teagasc representatives and others.

*MR. R. MCCANDLESS HAMILTON*
*PRESIDENT 1996-98*

Not surprisingly I found they all had known and done business with Mr. R. W. "Dick" Lawes of L. E. Pritchitt, my employer, who had just passed away a few weeks previously; and their conversation was filled with their anecdotes and reminiscences of him and his stature in the dairy industry in Ireland.

As we made our way on to the platform in the conference room, I was amazed at the size of the audience - Sean Tuohy said there were over 500 farmers in the room. The only thing the audience were interested in was the date of the amalgamation of Waterford and Avonmore (now Glanbia), and then they fully explored the peculiarities of the beef processing market in Ireland. This was a hard-cutting, hilarious debate during which the CEO of Avonmore Co-op, Mr Pat O'Neill, was accused of being lower than a political waffler; he was a professional.

Minister Yates had to leave early and he kept passing me notes urging me to be brief in my closing address. This continued to be a very good meeting until the small hours of the morning, and the Society's thanks to the Southern Ireland Section were repeatedly toasted.

Symposia and Residential Courses were special events at whatever the location, especially Auchincruive (on *'CULTURED PRODUCTS INCORPORATING YOGHURT AND SOFT CHEESE'*) and

Brackenhurst College. The October 1997 Symposium on *'Designer Milks and Dairy Products'*, the April 1998 Symposium on *'Farm Assurance In The Dairy Chain: An Independent Perspective'* and the October 1998 Symposium on *'Extending The Shelf Life Of Processed Milks'* were events that gained for us capacity audiences and were followed with great interest. Even my faux pas at one Symposium was overlooked when I asked the consultant if the dairy industry would ever learn to buy plant which they could operate without the need for a consultant!

However, the main event during my two-year Presidency must of course be the Spring Conference in April 1997 at the Slieve Donard Hotel, Newcastle, Co. Down, Northern Ireland on the subject of *'Food Safety - Protecting & Enhancing Your Business'*.

The V.I.P. guest for the Conference Dinner was Baroness Jean Denton, Northern Ireland Minister for Agriculture. She was a very entertaining lady, who after the dinner asked if she could continue the debate with Dr Verner Wheelock, a Conference speaker, on Government policy re the E.U. Directive on Food Safety. This in itself was totally entertaining as she was able to use all the facts and policy content at will to support her arguments, and the debate was eventually declared a draw. Our M.E.P., Mr Jim Nicholson, closed the Conference with a warning; and he explained the set-up in the Commission as to who was heading up this and how any standard set would be expected to be attained at the highest level.

The Northern Ireland Section must take credit for supporting me in my term of office and a very imaginative Spring Conference.

### J. Westwell, President, 1998-2000

MR. J. S. WESTWELL
PRESIDENT 1998-2000

The irony was not lost on someone who was privileged to be elected as President to the Society and yet discarded from an ailing industry; the timing was coincidental, almost perverse and yet, that's when the reality kicked in - life goes on.

The Society had suffered a major setback when Mrs Ros Gale announced her imminent retirement after almost ten years of dedicated service. The implications were challenging. The move to the Plunkett Foundation at Long Hanborough, Oxford, albeit on a temporary basis, was both frustrating and rewarding from a personal viewpoint. Whilst the Society was able to maintain an administration, but without the expertise of Ros, it called for a 'hands on' approach from members, in particular Officers and Council.

I considered this period to be the *'wilderness years'* as I felt that we had lost contact with the grassroots membership, mainly through the effects of the previous ten or more years, closure of creameries and dairies, and the decline in the old Section activities. There had been a lack of communication from the centre and we had in fact bottomed out!

To survive, the Society had to change and modernise to fit the current climate of the dairy and food industry and indeed the aspirations of the membership. This was achieved through dogged determination, persuasion, and approval from Council without compromising the original tenets of the Constitution. More importantly the flagship of the Society, the Journal, was revamped and is now published internationally, an inspired ambition of the Technical and Editorial Board.

I remember many happy times too, meeting many people within and without the industry. Who amongst the many delegates will forget the tremendous contributions of the Southern and Northern Ireland Sections in not only hosting very successful Conferences; but they were equally kind and generous with their hospitality to all. The spring and autumn Symposia in London were notable successes too, in particular the subject of organic milk and dairy products.

It would be unfair to single out individuals as I could fill a page of acknowledgements, and there were many, but I cannot fail to reflect on the genuine and sincere help, co-operation, and most importantly support given to myself during my term. To those individuals I am grateful.

I have been a member of the Society since 1953, excepting a brief period when I resigned in protest at the Society abandoning a Conference in Northern Ireland because of threats from the IRA. I think I am probably one of the few who has been a member for around 50 years. At various times I have been a member of Council, the Editorial Board, the General Purposes Committee, an elected Trustee, Honorary Secretary, Vice President and President.

My Presidency was a momentous one. It started with the resignation of John Westwell as President and his appointment as Membership Officer with specific responsibility for recruitment. The appointment, a paid post, was for a period of six months.

*DR. D. ARMSTRONG*
*PRESIDENT 2000-2002*

During my time as an officer of the Society, in several capacities, I was much involved in redrafting the rules of the Society, making the case for, and carrying out, the incorporation of, the Society, the appointment of the Plunkett Foundation and subsequently Agri-Food Consultancy (Dr. O'Sullivan) to manage the Society, and the negotiation of the contract with Blackwell Science (now Blackwell Publishing) for the publication of the Journal. These were all time-consuming tasks.

Graham Lee and I conducted the first round of negotiations with the RABDF which almost concluded satisfactorily, and I played a part again when discussions resumed without success. I was also involved in the discussions with the Institute of Food Science & Technology (IFST) over several years. The premises at Huntingdon continued to be a challenge, first with a proposed sale which was sensibly abandoned, and then with the letting of the premises.

The autumn 2001 visits to major dairy centres such as Chadwell Heath and Felinfach were refreshing experiences and had the added bonus of meeting many old colleagues. This is an initiative which I hope will continue.

I record my thanks to Mrs Gale, who was a very professional and efficient Executive Secretary, and to Dr. Tony O'Sullivan, her successor as Executive Director, for their understanding and unstinted help and also to all the other Officers and Council members for their contributions and support in testing times.

Though I am anticipating, I appreciate very much the honour of the Gold Medal which is to be awarded to my esteemed colleague, Dr. E Mann, and me by the Society at the AGM in the House of Lords on 12th November.

I was very pleased when Lord Oulton Wade of Chorlton graciously consented to be the Patron of the Society.

I took over the job as Secretary from Peter Lee in June 1989. He had accumulated a wealth of experience during his time with the Society, and he was a hard act to follow, but my abiding memory of the SDT is the warmth and friendliness of its members, and the help, support and hospitality shown to me by the eight Presidents whom I had the good fortune to know and work with. They all held busy and important positions in their companies, but they gave a great deal of their time during and out of office hours to the Society, and of course their companies supported them and the Society by letting them have time off work to participate in the Society's numerous meetings, Symposia and Conferences.

*MRS. R. GALE EXECUTIVE SECRETARY*
*1989-1998*
*WITH MR. G. LEE*

Eurwen Richards was halfway through her year as President when I took over the job, (in those days the President's term of office was for one year) and my first event was to lead a Study Tour to Bavaria. As it turned out, it was the last Study Tour run by the Society because, despite John Gordon's and the late Garry Royall's efforts to organise tours to America and Denmark respectively, they were not supported by enough members to make them viable.

No 72 Ermine Street, the Society's head office in Huntingdon, was small and inconvenient to work in, and there was very little money to furnish it with the modern machinery and furniture that most of us now take for granted in an office environment, but somehow we managed to get through what seemed to be, at times, an impossible work load with very few hitches over the years.

Membership, which stood at 1,800 in June/July 1989, slowly declined over time due to deregulation and rationalisation in the dairy industry. Joan Lawford replaced Lorraine Attwood as Membership/Journal Secretary; they both worked extremely hard to look after membership and Journal subscriptions and records, helping members with their queries, and liasing with the Society's Sections and Branches. They also helped me with the day-to-day running of the office.

Numerous Committee meetings, Conferences, Symposia, and other events took up a great deal of my time, and administration had to be fitted round these events but I really enjoyed my time as National Executive Secretary and have many fond memories of those years. There are too many people - Presidents, Officers and members - to name in this short space, and too many events to single out for individual comment, but it was a great pleasure and privilege to have met such interesting and dedicated people, who gave their time freely for the good of the Society.

Long may the Society continue to thrive and prosper!

# PAST PRESIDENTS
## 1943-2003

*PROFESSOR H. D. KAY*
*PRESIDENT 1943-44*

*MR. F. PROCTOR*
*PRESIDENT 1944-45*

*MR. G. ENOCK*
*PRESIDENT 1945-46*

*DR. A. T. R. MATTICK*
*PRESIDENT 1946-47*

*MR. J. MATTHEWS*
*PRESIDENT 1947-48*

*MR. E. CAPSTICK*
*PRESIDENT 1948-49*

*MR. J. WHITE*
*PRESIDENT 1949-50*

*MR. SMITH*
*PRESIDENT 1950-51*

*DR. A. L. PROVAN*
*PRESIDENT 1951-52*

*MR. E. L. DOBSON*
*PRESIDENT 1952-53*

*PROF. E. L. CROSSLEY*
*PRESIDENT 1953-54*

*MR. J. C. TAYLOR*
*PRESIDENT 1954-55*

*Mr. H. C. Hillman*
*President 1955-56*

*Mr. H. S. Hall*
*President 1956-57*

*Mr. J. Lewis*
*President 1957-58*

*Mr. F. C. White*
*President 1958-59*

*Mr. J. Ridley-Rowling*
*President 1959-60*

*Dr. R. J. Macwalter*
*President 1960-61*

*Dr. R. Waite*
*President 1961-62*

*Mr. W. W. Ritchie*
*President 1962-63*

*Mr. C. H. Brissenden*
*President 1963-64*

*Mr. J. F. Hunter*
*President 1964-65*

*Mr. J. Glover*
*President 1965-66*

*Dr. J. G. Murray*
*President 1966-67*

*Dr. R. J. M. Crawford*
*President 1967-68*

*Mr. C. S. Miles*
*President 1968-69*

*Mr. T. Idwal Jones*
*President 1969-70*

*Mr. P. A. Hoare*
*President 1970-71*

*Dr. J. G. Davis*
*President 1971-72*

*Dr. E. Green*
*President 1972-73*

*Dr. J. Rothwell*
*President 1973-74*

*Mr. N. L. T. Garrett*
*President 1974-75*

*Dr. G. Chambers*
*President 1975-76*

*Mr. V. C. H. Cottle*
*President 1976-77*

*Mr. S. G. Coton*
*President 1977-78*

*Mr. I. A. M. McAlpine*
*President 1978-79*

*Mrs. S. W. Barron*
*President 1979-80*

*Dr. H. Burton*
*President 1980-81*

*Mr. F. G. Weldon*
*President 1981-82*

*Mr. W. J. Hipkins*
*President 1982-83*

*Mr. A. C. Jackson*
*President 1983-84*

*Mr. R. A. Dicker*
*President 1984-85*

143

*MRS. S. M. MOORE (NOW WATSON)*
*PRESIDENT 1985-86*

*MR. G. McL GRIER*
*PRESIDENT 1986-87*

*MR. T. C. TAMPLIN*
*PRESIDENT 1987-88*

*MISS. E. RICHARDS*
*PRESIDENT 1988-89*

*DR. J. F. GORDON*
*PRESIDENT 1989-90*

*MR. G. R. ROYALL*
*PRESIDENT 1990-91*

*MR. J. M. BIRD*
*PRESIDENT 1991-92*

*MR. P. W. S. FLEMING*
*PRESIDENT 1992-93*

*DR. A. C. O'SULLIVAN*
*PRESIDENT 1993-94*

*MR. G. LEE*
*PRESIDENT 1994-96*

*MR. R. McC. HAMILTON*
*PRESIDENT 1996-98*

*MR. J. S. WESTWELL*
*PRESIDENT 1998-2000*

*DR. D. ARMSTRONG*
*PRESIDENT 2000-2002*

*MR. M. T. WALTON*
*PRESIDENT 2002-2003*

144

## REVIEW OF RECENT SDT SYMPOSIA AND CONFERENCES

*By Dr. Anthony C O'Sullivan\**

The early history of the SDT, as will be gleaned by a perusal of previous Chapters of this publication, witnessed an impressive and extensive range of external activities. As well as many Symposia on specific scientific/technological themes *'of the day'*, there were routine Annual, Spring and Autumn Conferences organised by the many, active Sections of the Society in the earlier days - as well as the normal Annual General Meetings (AGMs). Furthermore, there were annual Residential Courses for more "hands-on" training and an active programme of Overseas Study Tours.

Unfortunately, and increasingly over time, the number and range of external activities organised by the Society have declined. This run-down was caused by the reduction in the organisation's membership and, in turn, resulted in the inability of individual Sections to be able to organise such major undertakings as two- to three-day Conferences - which were the jewels in the SDT's programme crown. In fact, in looking back at the plethora of such activities from today's vantage point, one can only marvel at the commitment and industry of so many members, working on an entirely voluntary basis, to mount such major undertakings.

In more recent years, the increasing consolidation in the British dairy industry has seen more company closures and staff redundancies. Who would have thought that such British dairy giants as Unigate and, now, Express Dairies, would ever disappear from the dairy scene? This inexorable decline has had an adverse, knock-on effect on the membership and, hence, the programmes organised by the Society. This situation will be highlighted by this review of the external activities mounted since 1993, when the Society celebrated its 50th Anniversary with a special Conference at Cambridge on 20-21 April under the successful Presidency of Mr Peter Fleming of the then Scottish Farm Dairy Foods' renown. The illustrious speakers at this 50th Anniversary Conference were Drs. George Chambers and Ian Gordon, and Professors Phillip Thomas and Willie Banks, all from the UK, and Dr. Borge Mortensen and Hanne Werner of Denmark. The Society's Gold Medals were also presented to Drs. Chambers and Mortensen. This special event in the Society's existence attracted a 'heady' attendance of 117 delegates, and with an additional 23 attending the formal dinner on the first evening - bringing the total to 140! Other activities during the 50th Anniversary year were a Residential Course on *'Speciality Cheesemaking'* at Cannington College, Somerset on 1-3 September, followed by an one-day Symposium in London on 27 October 1993, which attracted 35 delegates and which had the Society's 50th AGM immediately after lunch.

*\*Dr O'Sullivan has been the Society's Programme Planner since 1995 and, in early 2000, also assumed the additional role of Programme Co-ordinator when Mrs Ros Gale relinquished that activity. Since his appointment as Executive Director in May 2001, he has continued to execute both these functions on behalf of the SDT.*

This Chapter reviews the Society's programme of external activities from 1993 onwards until today. In turn, objective comments will be made on the advantages, and disadvantages, accruing to the Society from its current activities; and which, from an historical perspective, represent a rather modest, albeit realistic, programme in today's dairy industry environment.

The first year of this Review coincides with the writer's Presidency. Given his past track record in organising scientific/technical events, and his statement at his inauguration that he would focus on technology after the euphoria of the 50th Anniversary year, it is not surprising that he was actively involved with the then, highly capable Executive Secretary, Mrs Ros Gale, in personally organising two of the three main programme events of that year.

The first was a two-day Symposium on *'Manufacture & Marketing of Niche/Luxury Milk And Dairy Products'* which was held in St Helier, Jersey on 25-27 May 1994. The original venue, the Hotel de France, had a serious fire and the event had to be switched to the more attractive

*Dr John Gordon, President of the Society of Dairy Technology, and Mr G Sauvage, Chief Executive of the Guernsey Agricultural and Milk Marketing Board, before the start of the very successful symposium on 'Speciality Cheeses' which was held at the St Pierre Park Hotel, Guernsey from 23 to 25 May 1990.*

Grand Hotel situated on the waterfront. The Symposium was opened by Dr George Chambers, and speakers were Dr Tony O'Sullivan, QMP, Mr Brian Le Marquand, Jersey Milk, Mr Nigel White, Consultant, Caroline Evans, Häagen Dazs, Mr Neil Robinson, St Ivel, Mr Keith Collins, Calthwaite Dairy Products, Mr Steen Reves, MD Foods, Denmark and Mr Charles Hunt, Tesco. The significant mainland attendance benefited from the strong involvement by, and sponsorship of, Jersey Milk, which entailed a morning visit to its multi-product dairy and a dinner on the closing evening.

*Delegates at the 'Purple Suite' Jersey.*

A highly successful Residential Course was held at Reaseheath College, Nantwich on 7-9 September 1994 on *'New Product Development in the Dairy Industry'*. As well as $1^1/_2$ days of relevant lectures, a tour of Milk Marque's Product Development Centre (subsequently closed, unfortunately) and a joint presentation there by G C Hahn and Stephan Machinery, the middle day was taken up by practical sessions. Denis Prew, Head of the then Dairy Technology Department (now Food Technology) at Reaseheath and his colleagues sub-divided the 16 delegates into four groups, and each group was charged with developing a new product formulation within a specific dairy product category. This hands-on approach was greatly welcomed by the participants, who were stimulated by the product development briefs given to them and responded accordingly.

The third event of the 1993-94 year was a three-day Annual Conference at the Grand Hotel, Bristol from 23-26 October 1994 on *'Developments in Dairy Processing and Management'*, which was successfully organised by Stephen Crutchley, Chairman, and Committee members of the then Western Section of the Society. The Conference had three main themes:

- an update/overview of some aspects of developments in milk and milk products' processing (including *'Developments in Pasteurisation & Standardisation'*; *'Advances in Spray Drying'*; *'Application of Membrane Systems'*; and *'Manufacture of Fresh Cheese Under High Quality Condition'*)
- practical management techniques presented with emphasis on dairy industry application (including *'HACCP - A Manager's Guide'*; *'Practical Aspects of Quality Systems'*; and *'Computerised Process Control And Management Systems'*)
- techniques to control/reduce wastage within dairies, whether losses to effluent, in overfill or detergents (including *'Management Control of Detergents'*; *'Waste management*

*Development'; 'Waste Management - How To Reduce Wastage And Proven Solutions'; and 'A Factory View Of Wastage Control')*

The Society's 51st AGM was held late on the opening Monday afternoon and was enlivened by a debate on the respective roles of the Management and Custodial Trustees at that time. The Tuesday of the Conference offered Local Study Tours to St Ivel, Wootton Bassett, Alvis Brothers, Redhill (organic & farmhouse cheesemakers) and Anchor Foods, Swindon. This was the last occasion when the Society's autumn AGM coincided with an Annual/Autumn Conference.

Another activity where the SDT was partly involved was a Symposium organised by University College, Dublin and Moorepark's Dairy Products Research Centre **in association with** the Society. This two-day event occurred on 8-9 March 1994 at Silsoe College, Bedford; and the writer, the President at the time, chaired one session and also spoke at the evening dinner.

### 1994-96 YEARS

These and subsequent periods will be discussed on a two-yearly basis, as these coincided with the President's term being extended from the previous one-year norm to a biennial basis. The first President to serve such a two-year term was Mr Grahame Lee, a Regional Manager with Milk Marque, the successor voluntary Milk Group to the statutory Milk Marketing Board for England & Wales.

*Mrs Suzanne Watson with Dr Ernest Mann (right) looking after the Society's interests at the 24th International Dairy Congress in Melbourne in September 1994.*

The first event during Mr Lee's Presidency was an one-day Symposium on **'Dairy Ingredients In Food'**, which was held in London on 1 March 1995; and this also coincided with this writer assuming full responsibility for the Society's programme planning function. Speakers at this Symposium were Dr. Eamonn Pitts, The National Food Centre, Dublin; Michael Forrester, Foodtec Services; Kristian Albertsen, MD Food Ingredients, Denmark; Marie Relihan & Maurice Lapthorne, Golden Vale Products, Ireland; and Associate Professor Daniel Mulvihill, University College, Cork. The event attracted 42 delegates, four of whom were from overseas.

The 1995 Residential Course took place on 6-8 September 1995 at the University of Reading on the subject of **'Heat Treatment of Milks And Juices'**. Understandably, it drew heavily from the University's staff expertise, and speakers included Mr Andrew Wilbey, and Drs. Jukes & Lewis. SDT members Messrs Norman Critchett and James Carnell spoke on ingredients for flavoured milks. The second day was fully given over to practical sessions giving hands-on experience with HTST pasteurisation, UHT and 'in-container' sterilisation, and supporting laboratory analysis. The Friday morning presentations were given by Messrs Andrew Wilbey and John Rigarlsford, and by Miss Eurwen Richards, Richards Consultancy and an ex-President of the Society in 1988-89.

The very attractive Annual Conference scheduled for 29-31 October 1995 outside Oxford had, unfortunately, to be cancelled.

A significant internal event took place on 3 April 1996 at Brackenhurst College, Nottingham when the Society's Officers, Section Chairmen and Secretaries held a day-long meeting "on the problems currently experienced by Sections, possible solutions and the way forward". Constructive ideas were ventilated during the initial overview, discussion groups and summary sessions resulted in the formulation of appropriate strategies.

The 1996 Residential Course on **'Basic Microbiology and HACCP: A Practical Introduction'** was held at Brackenhurst College, Nottinghamshire on 4-6 September and attracted 9 delegates. Most of the lectures were given by Brackenhurst staff, including SDT member Dr Elizabeth Whitley, and were supplemented by other Society members, Mr Ralph Early, Harper Adams College and Miss Eurwen Richards, discussing HACCP. The practical sessions involved basic microbiological techniques, including rapid and confirmatory testing.

Another activity undertaken by the writer in his capacity as the Society's Programme Planner (and perhaps not well known to members) was responding to requests from the National Dairymen's Association (NDA) to organise one of three Conference Sessions at its annual *'Dairy Industry Show (DIS)'* held in various northern (e.g. Blackpool) and southern (e.g. Bournemouth) locations in early September every year. As well as organising this two-presentation Conference Session, the Society's President also chaired the meeting. This SDT input started in 1996, and continued for four successive years up until September 1999. The latter occurred in Buxton, Derbyshire on 7-8 September and, to give readers a flavour, the SDT Session was deliberately geared to the liquid milk interests of the NDA members by having the overall theme of *'Innovation in Liquid Milk Processing & Marketing'*. Two excellent contributions were made by Lisa Palillo, MD (now Arla) Foods on *'Adding Value to Milk'* and Richard Hogg, Dairy Crest on *'The Flavoured Milk Market'*.

The Autumn 1996 one-day Symposium on *'The Current Status of Bio-Fermented Milks: Focus on Europe'* attracted only 26 delegates and was held at the Society of Chemical Industry (SCI) on 29th October. The attractive panel of speakers was Dr. Judy Buttriss, National Dairy Council; Mr Carsten Hansen, MD Foods, Denmark; Dr. Driessen, Mona/Campina Melkunie, The Netherlands; Dr. Ian Gordon, GIRACT, Switzerland and SDT member; Society members Professor Valerie Marshall, University of Huddersfield & Dr Adnan Tamime, Scottish Agricultural College, in a joint presentation on the starter cultures for bio-fermented milks; and Mr Lars-Ebbe Nilsson Tetra Pak, Sweden on recent developments in fermented milks' technology, including ultrafiltration.

## 1996-98 YEARS

This two-year period represented the Presidency of Mr Robert Hamilton, who worked with Pritchitt Foods' operation at Newtownards, Co. Down, Northern Ireland. The *piece de resistance* of his period in office was undoubtedly the well attended (79 delegates, with 21 from Great Britain) and highly successful Spring Conference held at the Slieve Donard Hotel, Newcastle, Co. Down on 20-23 April 1997. This Conference was sponsored by Smurfit Corrugated Cases, and introduced the desirable concept of third party financial contributions to Society events. The topical theme of this well organised Conference, under the Conference Committee Chairmanship of Mr Andrew White, was *'Food Safety - Protecting & Enhancing Your Business'*.

The eminent speakers included Messrs John McKee, Stan Brown, Mike Johnson, Dairy Council for Northern Ireland, James Bryant, Dr David Jervis, Unigate European Food Group and Professor Verner Wheelock. A local politician, Mr Jim Nicholson, MEP, spoke on *'What does the future hold for Food Safety? A personal opinion'*. This political input was continued by Baroness Jean Denton, Northern Ireland Minister for Agriculture, who was the Conference's entertaining and knowledgeable after-dinner speaker.

The 1997 Residential Course on *'Cultured Products Incorporating Yoghurt and Soft Cheese'* took place at the Scottish Agricultural College (SAC), Auchincruive on 3-5 September and featured major contributions by SDT members Dr Adnan Tamime, Lecturer at SAC, and Professor Donald Muir of the nearby Hannah Research Institute (HRI). Other lectures were given by Messrs John Bird, Tetra Pak (UK) and an ex-President of the Society in 1991-92, Peter Schmidt of G C Hahn and Co, Gerry Larner, APV Dairy and Dr Jim Bruce, SAC. The practical demonstration on the manufacture of different fermented milk products included the production of Greek-style yoghurt and Fromage Frais by ultrafiltration.

The Autumn 1997 Symposium was held in London on *'Designer Milks And Dairy Products'* on 29th October, and attracted an attendance of 32 delegates. The seven presentations were sub-divided into (i) *'Designer Milks at Production Level'*, which had Miss Rosemary Mansbridge, ADAS, Dr John Newbold. BOCM Pauls and Mr Anthony Hall, R Keenan UK discuss dietary manipulation of milk fat and milk protein inside the farm gate and (ii) *'Designer Milks & Dairy Products at Processing Level'*, which featured Robert Bouhon, Corman SA Belgium, Professor Donald Muir, HRI and Dr Colette Shortt, Yakult UK covering in-plant technological manipulations. The final presentation on *'Retailer Perspectives on Designer Milks and Dairy*

*Products'* was by Mr Victor Willis, R&D Manager for Safeway and SDT member, and he gave a commercial view of this area.

The Spring 1998 Symposium was held in London on 21 April at the new venue of the Royal Institution of Great Britain (RIGB) on the very topical theme of *'Farm Assurance in the Dairy Chain: An Independent Perspective'*. Because of both the topicality and relevance of the Symposium theme, it was successful in attracting 54 delegates. This Symposium broke with tradition by commencing with the consumer, through discussing the *'Perspectives of the Multiple Retailer'* as seen by Jane Perkins of Sainsbury's and SDT member. This consumer orientation set the stage for all the subsequent presentations, which provided perspectives from producer to consumer. Mr Brian Pocock, Unigate spoke on the National Farm Assurance Scheme, and was followed by a dairy farmer (William Kilpatrick), Breed Society (Tim Brigstocke of the Holstein Friesian Society), Milk Seller (Mike Hurst, Milk Marque) and Milk Processor (Ian Cameron, MD Foods). The final paper giving the Microbiologist's perspectives was presented by Dr Jim Bruce of SAC.

There was no Residential Course in 1998, while the Autumn 1998 Symposium took place on 21st October at the Royal Institution on the overall theme of *'Extending the Shelf-Life of Processed Milks'* and in association with the United Kingdom Dairy Association (UKDA), the UK's National Committee of the International Dairy Federation (IDF). The latter venture was an attempt by the Society to broaden the potential attendance base by pre-circulating the programme to members of the other organisation associated with the particular event. In return for this support, Mr Brian Peacock, Vice-Chairman of the UKDA (and also a Society member and Chairman of the Milk Development Council) chaired the afternoon session of the Symposium. Speakers were Drs Mike Lewis (SDT member), University of Reading; H Ranjith, Milk Marque's Product Development Centre; Sean Tuohy, Moorepark Technology Centre, Ireland; Hans Engall, Tetra Pak, Sweden; and Debra Henyon, Elopak, USA and Rapporteur, IDF Group of Experts on Extended Shelf-Life (ESL) Milk. The final presentation was given by Mr Jonathan Middlemiss, Farm Produce Marketing, Cheshire on *'Commercial Aspects of Marketing ESL Milks'*. The combination of the theme's relevance and the UKDA association resulted in a high delegate attendance of 73, of which 8 were from outside the UK.

### 1998-2000 YEARS

This two-year interval coincided with the Presidency of Mr John Westwell of Associated Co-operative Creameries (ACC), Newcastle-upon-Tyne, and with the second year being the Millennium and a period of enhanced activities. The Spring 1999 Symposium was on the highly topical subject of *'Organic Milk And Dairy Products'* and also attracted an excellent attendance of 73, similar to its previous autumn's counterpart. The Symposium examined in detail the domestic potential of organic milk production, processing and marketing, and drew on the most relevant UK experiences at each of these three stages. The embryonic UK situation was complemented by a presentation on the Danish dimension, where, at that time, organic milk production was of much greater significance to its dairy industry. The panel of speakers embraced Messrs Patrick Holden, The Soil Association; David Wilson, Dutchy Home Farm and Mrs Sally Bagenal, Organic Milk Suppliers' Co-operative. Messrs William Best (liquid milk), Nick Reaks (cheese) and Mrs Rachel Rowlands (yoghurt) spoke on their experiences as UK organic processors; and were followed by Messrs Christian Mortensen, Milk Allocation Company, Denmark and David Jones, Central Buyer, Dairy for Waitrose. Subsequently, the writer published a detailed account of this Symposium in the Society's journal (Vol. 53, No. 1 February 2000 pp 20-27).

The Autumn 1999 Symposium on *'Generic Marketing of Milk: Why And How?'* had to be cancelled due to poor registrations. This was unexpected, as dairy farmers were about to vote during that time on whether they were willing, for the first time since deregulation of the domestic milk market in November 1994, to agree to co-fund generic promotion of milk in association with the liquid milk trade.

The Society decided to mark the significance of 2000 by mounting **a major** Spring Conference rather than the normal one-day Symposium. This Millennium Spring Conference on the overall theme of *'Dairy Technology 2000'* was organised by the Southern Ireland Section, under the capable Chairmanship of Dr Sean Tuohy, and with support from SDT centre. It was held in Kinsale, Co. Cork, an historic and beautiful maritime town in West Cork, over the weekend of 20-23 May. This timing was a major break with SDT tradition, and was suggested by Mr Paul Moody, a member of the Organising Committee and was based on American practice. In the event, the combination of excellent week-end presentations and an attractive recreation/social programme succeeded in attracting 38 attendees from the UK, together with 62 from the host country, a total of 100 delegates - a significant achievement. Another feature of this outstanding event was that the Organising Committee expanded on the concept of sponsorship introduced at the 1997 Conference in Northern Ireland by obtaining significant contributions from a number of Irish dairy companies.

UK contributors to the programme included Mr Neil Davidson, Express Dairies, Dr David Roberts, Dairy Research Consortium Ltd., the SDT stalwart, Professor Donald Muir of the Hannah Research Institute, and Dr Mike Gasson, Institute of Food Research, Norwich. Continental European speakers were Mr Barry Wilson, Dairy Industry Newsletter, Mr Jorgen Iversen, APV Nordic Systems, Dr Ejnar Refstrup, NIRO A/S and Mr Jaco Baron of Tetra Pak Tebel. The very strong Irish dairy technology base was reflected in contributions from Drs Liam Donnelly and Seamus Cross of the Dairy Products Research Centre, Moorepark, Nicholas Simms of the Irish Dairy Board and Eamonn Pitts of the Rural Economy Research Centre of Teagasc (the Irish Agriculture & Food Development Authority). Another new feature was a Breakfast Meeting on the Tuesday morning, when another SDT stalwart, Dr George Chambers, gave a thought-provoking presentation on *'Dairy Technology 2000 and Beyond'*. Subsequently, the Society's Technical & Editorial Board decided to publish the papers delivered at this Conference in full as a special November 2000 issue of its *'International Journal of Dairy Technology'*. Coincidentally, this was the last issue of the journal published by the Society before Blackwell Science (now Blackwell Publishing) commenced operation in February 2001.

Highlighting the significance of Agriculture and Food in the booming Irish economy, the Southern Ireland Section was able to muster the combined attendance of the Minister for Agriculture & Food and his Minister of State at the Conference's two social events. Mr Ned O'Keefe, Minister of State for Food and whose constituency embraces Moorepark, welcomed SDT guests at the Saturday evening Drinks Reception, while the Minister of Agriculture & Food, Joe Walsh - a native of West Cork and an ex-researcher at Moorepark - was the Guest of Honour at the Sunday evening Conference Dinner. His impressive 'off-the-cuff' speech extolled the Irish Government's major investment in its agricultural and food sector.

The Southern Ireland Section's initiative in organising a postgraduate Dairy Research Poster Competition confined to student members and in memory of the late Eamonn McCormick, who had been a founder member, long-serving Secretary and past Chairman of the Southern Ireland Section, was a great success, attracting 25 entries of high calibre. The award of the **E P McCormick Gold Medal** to the winner, David O'Sullivan, was made by Eamonn's widow, Anne McCormick, during the course of the Conference Dinner. Another initiative undertaken by the Organising Committee was the incorporation of a Trade Exhibition, which attracted seven exhibitors; and their contribution added to the overall profitability of a Conference format unique in the annals of the Society since its establishment in 1943. Furthermore, four full and twenty-five student members joined the SDT as a direct result of participating in the Kinsale activities.

As well as a welter of spectacular, outdoor activities, technical tours were laid on to both the Dairy Products Research Centre, Moorepark and the Faculty of Food Science & Technology, University College, Cork. Those who visited Moorepark were greatly impressed by the uniquely versatile, and commercially-significant, pilot plant facilities of Moorepark Technology Ltd. - and reinforcing Minister Walsh's message.

The main conclusion which the Society drew from this unique Kinsale experiment was that the combination of a week-end timing, allied to a location of high and attractive social amenities such that delegates' wives and partners are well catered for, is one that the SDT must allow for in its future activities' planning. The unknown that was Kinsale may well be the blueprint for success in the future.

A second Millennium year event was a Residential Conference dealing with all aspects of Cleaning-in-Place (CIP) technology. Entitled *'CIP - For a Clean Future'*, it was mounted by a Conference Organising Committee chaired by Dr John Crawford in close association with the Northern Ireland Region's Committee, and took place at Loughry College - The Food Centre, Cookstown, Co. Tyrone, Northern Ireland on 10-12 September 2000. This successful activity, which combined a Residential Course and a Conference, attracted 74 delegates, of which 37 were from the Province, 31 from Great Britain and 6 from the Republic of Ireland. Speakers included Messrs David Turner, Paul Lindsay, David Lloyd, Ian Herron, Bob Wogan, Harry Moss, Alan Stack, Miss Lynn Whan, and Drs Ian Campbell and Michael Rowe. Relevant trade displays/presentations were also made by companies in the CIP area, including Dunlop Engineering, Ashland UK, Wedeco UV Systems, Henkel Ecolab, DiverseyLever/Nalco, Biotrace, and Testec. This Residential Conference ended with a tour of the Loughry facilities, with its new Food Technology Centre being the largest, purpose-built food technology facility in the UK.

2000-2002 YEARS

These two years represented the Presidency of Dr David Armstrong, a dedicated member who has given a lifetime of service to the Society. Complementing the Millennium Spring Conference and Loughry College's Residential Conference, SDT centre organised a high-profile Millennium Autumn Symposium on *'Milking More From Milk: Using Milk Ingredients To Add Value'* at the Royal Institution's London location on 29 November 2000. This third Millennium year event attracted 43 delegates, of whom seven were from overseas - including representatives from Pakistan and Australia. An array of seven outstanding speakers examined milk proteins (Professor Pat Fox, University College Cork), milk fats (Dr Ken Burgess, Dairy Crest) and milk sugars (Dr Matti Harju, Valio, Finland) as major sources of ingredients; and these were followed by specific contributions on *'Milk ingredients for nutraceuticals'* (Dr Jan Steijns, DMV International, The Netherlands), *'Ultra Whey 99: a whey protein isolate case study'* (Mr James Neville, Volac International), *'Probiotic milk ingredients: opportunities'* (Professor Kevin Collins, University College Cork) and *'Nutritional value of milk'* (Dr Anita Wells, National Dairy Council). Six of these seven contributions were subsequently published in the journal during 2001.

This Symposium also marked the first occasion that the Society sought, and received, external financial support for a *'local'* Symposium activity, through Yakult UK Ltd sponsoring one speaker and Volac International hosting a very successful Wine Reception at the end of the event.

The Spring 2001 Symposium had the highly topical theme of *'Safety And Assurance of Milk And Dairy Products'*, and covered the safety/assurance aspects of milk as a raw material, of dairy processing and dairy/food retailing, followed by the role of information technology (IT), genetic engineering/modification, implications for the dairy industry of *Mycobacterium paratuberculosis* and a contribution on the role and functions of the Food Standards Agency. Again, this event achieved even more external support, in that one speaker was sponsored by the RABDF and Qualtrace Solutions, Cork hosted the lunch. A total of 37 delegates were attracted to this event.

Whilst these two Symposia were very worthwhile for those present, it was unfortunate that the attendance was less than expected, with the Spring 2001 event being less well attended than its previous autumn counterpart (37 v. 43). This outturn is, unfortunately, a reflection of a shrinking marketplace, in that there are fewer technical people now working in the dairy industry, and there is increased competition as a result of new organisations now offering similar events. Because of these developments, it has becomes imperative that the Society expands sponsorship support of its external Conferences/Symposia in order to make them more cost-effective.

It was not possible to organise a Symposium in Autumn 2001, as the writer had taken over the Society's management from the Plunkett Foundation in May 2001 and was totally committed in coming to grips with this new role.

The Spring 2002 Symposium maintained the Society's commitment to safety initiated in the Spring 2001 event by selecting *'Aspects of Responsible Dairy Processing'* as its theme; and it was held in association with the Dairy Industry Association Limited (DIAL), with the latter's Director General and SDT member, Mr Jim Begg, chairing the afternoon session. Whilst the programme approached safety and efficiency issues from the processor's perspective, it also widened the scope to embrace environmental aspects as well. The basic aspects of *'Integrated Production and Pollution Control (IPPC)'* were given by Mr Mark Maleham, Environment Agency, whilst its dairy processor application was covered by Mr David Hopkin on behalf of Robert Wiseman Dairies, the first UK dairy to obtain IPPC approval. Similarly, the background aspects of *'Climate Change Levy (CCL)'* were treated by Mr Gareth Stace, DIAL and its dairy processor's application was discussed by Dr George Plemper, Express Dairies. Other aspects of responsible dairy processing covered included papers on *'Operational Cost Savings in Dairy Plant Water Usage'* (Dr Paul Williams, Ondeo Nalco), *'Code of Practice for HTST Pasteurisation'* (Dr Ed Komorowski, DIAL and SDT member) and *'Likely Future Developments'* (Dr Ken Burgess, Dairy Crest and SDT member). A total of 44 delegates attended this Symposium, and sponsorship of £2,500 was generously provided by Tetra Pak, DiverseyLever and DIAL.

In another departure from Society tradition, no Autumn 2002 Symposium was held, as Council opted instead for the innovation of an autumn series of domestic dairy plant visits, based on the *'Responsible Dairy Processing'* theme of the Spring 2002 event. Four visits to five dairy plants with leading-edge technologies were organised in the October - November period. These were to Dairy Crest's Chadwell Heath liquid milk superdairy, Essex; Volac International's whey processing and Aeron Valley Cheese's manufacturing and pre-packing operations at Felinfach, Lampeter, Mid-Wales on the same day; Robert Wiseman's latest liquid milk dairy, Droitwich; and United Milk's butter/powder factory complex, Westbury, Wiltshire.

A total of 59 members signed up for these visits, although only 51 actually participated in them. In fact, three members saw all five plants: Dr David Armstrong, and Messrs Colin Young and Angus Wielkopolski. Because the visits were only open to members, the Society gained 9 new members by virtue of their involvement in the visits - giving a similar recruitment result to a Symposium.

Two of the dairy plant visits took place during the final days of David Armstrong's Presidency, whilst the final two plants had Mr Maurice Walton installed as President.

2002-2003 YEAR

Because this Chapter is being written in late September, it will cover only most of the first year (2003) of the two-year Presidency (2002-04) of Mr Maurice Walton, a senior Manager with the detergent company, JohnsonDiversey. As mentioned in the paragraph immediately above, half of the four autumn 2002 domestic dairy plant visits occurred at the very beginning of his Presidency.

The Spring 2003 Symposium focussed on *'Chilled Yoghurts And Desserts' (CYD)*, fresh, short-life, added-value dairy products showing impressive, annual growth and a key driver of retail dairy sales in the multiples and supermarkets. Despite a surprising lack of co-operation from the major UK manufacturers, which necessitated even greater organisational efforts than normal, a very attractive programme was finally arranged. The 7 topics under the overall CYD theme were: *'Health Benefits of Yoghurt'* (Dr Michelle McKinley, The Dairy Council); *'Labelling - The Key To Successful Marketing'* (Dr Ed Komorowski, DIAL); *'Latest Developments in Starter Cultures'* (Dr Anne Skriver, Chr. Hansen A/S, Denmark); *'Recent Trends in Yoghurt/Fermented Milks Technology'* (Dr. Adnan Tamime, Consultant and SDT member/Technical Series Editor); *'Organic Yoghurts & Desserts'* (Mr Karl Tucker, Yeo Valley Organic Company); *'Use of Stabilisers in Manufacture of Dairy Desserts'* (Mr Ian Tregaskis, G. C. Hahn & Co.); and *'Preparation & Processing of Viscous Dessert Products'* (Mr Alan Stack, Tetra Pak, SDT member and Honorary Treasurer).

The 44 attendees, coincidentally the same number as the Spring 2002 Symposium, were privileged to hear 7 presentations of uniformly high standard and consistency; and these, in turn, sparked off extensive and animated debate. The registration fees and sponsorship from both Tetra Pak and DIAL, combined with the organisational costs now being absorbed in the Executive Director's annual consultancy fees, resulted in an encouraging surplus of £4,104.72. Given the current UK dairy industry climate and the rather specialised, and highly competitive, nature of the Symposium, this financial contribution is a very acceptable outturn; and it must be associated with the 9 new members gained from their attendance at the event. However, one obvious lesson is that it is essential for the Society to select Symposium themes which are both topical and, critically, of wider relevance to the general dairy industry, thereby facilitating the attraction of a potentially larger attendance.

SUMMARY: ADVANTAGES & DISADVANTAGES

The main advantages accruing to the Society from organising its continuing programme of Symposia, Conferences and similar public events are manifold, as follows:

- meeting the SDT's *primary* objective of advancement of Dairy Science & Technology through fostering scientific and technological developments and information transfer on them
- contributing to the SDT's finances by running them at a surplus. In fact, surpluses under this heading represent the fourth main income source for the Society, after membership subscriptions, Huntingdon office rental and journal royalty in decreasing order of magnitude. However, it has to be accepted that, given the current attendance levels, this consideration will become less significant in future years - although this has been counterbalanced by sponsorship, with £2,500 being the target for each individual event
- attraction of new members, in that such public events give an essential interface with non-members - and, especially, new entrants - in the UK dairy industry. For example, over the two Symposia in 2000/2001, albeit not that well attended, the Society still gained 22 new Members and two defaulters returned to membership - a gain of 24 member subscriptions. This is a constant feature of all Symposia/Conferences and, on average, at least 10 new members are routinely recruited
- visibility, credibility and publicity/PR for the Society and its activities - impacting on both members and non-members alike. This facet has been augmented by dairy industry-wide publicity for the SDT, especially through articles summarising such events in both *'Milk Industry'* and *'Dairy Industry International'*. These Symposia are also summarised for our Members by Mrs Pauline Russell, a member of the Technical & Editorial Board, in a subsequent issue of our *'Members' Newsletter'*; and the Society's appreciation is extended to her for this activity
- networking potential, given that attendance at a particular event gives one an ideal and unrestricted opportunity to network with the speakers and other delegates
- source of rigorous articles for the Society's *'International Journal of Dairy Technology'*, a consideration that will become much more critical as Blackwell Publishing expands the size of future issues of this publication, the SDT's flagship. For example, six of the seven papers presented at the November 2000 Millennium Autumn Symposium were published in the Journal in the May and August 2001 issues - and these also had the advantage of being review-type rather than research articles and, hence, of greater interest to our members.

The two main disadvantages of organising these external events are:

- increasing difficulty in obtaining speakers. In the past, when technology played a much more important role in the UK dairy industry, potential speakers were very pleased, even honoured, to be invited to speak at a Society event; they also had the possibility of publishing a paper in the Society's peer-reviewed journal. Unfortunately, in today's more pressurised workplace, people are much less interested in accepting speaking commitments - and the necessity of converting a verbal presentation into a subsequent written format is even more off-putting. In fact,

153

some people do not even respond to an initial invitation and have to be chased up - just to get a rejection!

- associated difficulty of attracting SDT members. Like all others working in the dairy industry, our members are not insulated from the time pressures, and associated stresses, of today's workplace. In fact, if the Society were to depend solely on its members, then its external activities would run at a loss. The only way of making these events viable is aggressively marketing them to non-members, tailoring this effort to those working in the area covered by the particular event's theme. But a plus from this additional input is that it facilitates the attraction of new members. A consequence of a reduction in the number of delegates, whether member or non-member, is obviously that the contribution to the Society's financial resources is concomitantly lowered.

In summary, the advantages accruing to the SDT from its programme planning activities currently greatly outweigh the disadvantages. However, this balance will have to be carefully monitored in the future to make sure that this positive relationship still holds. Finally, it should be recognised that the Society has built up an enviable expertise in the organisation of scientific/technical events; and this capability could be used in collaboration with other organisations, both nationally and internationally.

# CHAPTER EIGHT

## PAPERS OF SPECIAL SIGNIFICANCE

Over the years since 1947 many papers of particular interest or significance have been published in the Journal. In this chapter I reprint an abridged selection.

These papers are a mine of information and in the light of our situation to day of startling facts e.g. Crossley reported that TB had been eliminated from the national herd though Brucellosis was still a problem. Incredible though it may seem pasteurisation was opposed by many for reasons as invalid and emotional as the opposition today to G.M. Technology. It is also refreshing that reference could be made without apology to the soldering dexterity of "Teams of girls."

Crossley made the point that while whey was of high nutritional value it was "financially worthless" and a burden. He did, however, recognise the value of new processes which now make whey a valuable raw material. The papers by John Inglis remind us of the days when milk was collected in churns or cans. This was a tedious and costly operation and involved much physical work. No one regrets the passing of the milk can.

M Bessey and F Harding show again how much milk producers in the UK have lost as a result of the destruction of the milk boards. Bessey makes the point which is important in today's context that the milk boards were set up because cooperation enterprise did not exist. The argument sometimes heard that the milk boards prevented cooperative development is, therefore, untrue.

J Gleeson in his paper has highlighted the huge benefits gained by Irish farmers from the European Union. Professor Kay sets out the birth and growth of the society in the context of the contemporary dairy scene.

Macwalter made this inimitable and devastating comment on fast food "People Petulantly Picking at Plastic Packages of Pitiful Provender" and other pertinent remarks about modern management practices.

I would emphasise that there are many other papers of merit which could be selected.

These papers I have selected show the development of the Society, changes in research and education, the chequered growth of the milk industries of the UK and Ireland, and overall changing attitudes, good and bad, over a period of six decades.

They also demonstrate the eclectic interests of the Society even the environment is not neglected. They record the improvements in technology and significant developments in the economic framework, including the consequences of joining the EEC (now the EU). Dr Armstrong gives a snapshot of the horse dealing in Brussels over the CAP which is still decades later, unreformed.

Two watersheds are apparent: entry into the EEC and deregulation. The latter was a consequence of the former.

## SIGNIFICANT PAPERS

1.  Economic Aspects (of Milk Transport) J. G. Inglis, Dairy Industries July 1945.
2.  Progress and Trends in the Dairy Industry in N. Ireland 1924 – 49 D.A.E. Harkness, JSDT Vol 4. 24-29, 1950
3.  Milk Transport in England & Wales, J. G. Inglis, Milk Industry Jan 1951.
4.  The early history of the Society, Professor H. D. Kay, JSDT Vol 11. No 1 1958.
5.  Some Recent Developments in the manufacture of milk powders, R. Waite, JSDT No. 3, July 1964
6.  A half century of dairying: from Tradition to Technology, Professor E. L. Crossley, JSDT Vol 29, No 1 Jan 1976.
7.  Changing Attitudes. R. J. Macwalter, JSDT Vol 29. No. 2 April 1976.
8.  Recent developments in milk production in Ireland P. A. Gleeson, JSDT Vol 35. No. 1 Jan 1982.

9.  The role of the Milk Marketing Board in marketing M. E. Bessey, JSDT Vol. 35 No. 1 Jan 1982.
10. Milk Quality Central Testing, F. Harding, JSDT No. 1 Jan 1982, p.24
11. Time, Technology and the Dairy Industry, T. R. Ashton, Vol. 36 No. 3 July 1983, p66.
12. The importance of Agriculture in N. Ireland, J. A. Young Vol. 36 No. 3 July 1983, p68
13. The European Econ Community and UK Dairying with special reference to N. Ireland - a personal view, D. L. Armstrong, JSDT Vol. 36 No. 3 July 1983.
14. Protection of the environment: Friends of the Earth view of the role of the dairy industry, L. Stupples. Paper given at symposium 'The Dairy Industry Clean & Green, 22 Oct 1991.

## ECONOMIC ASPECTS (OF MILK TRANSPORT) DAIRY INDUSTRIES JULY 1945
### By J. Inglis

"This is in no degree a scientific paper......
In considering this subject it is seen to fall into two sections – the transport of milk in small containers, known variously as churns, kits, cans or tankards on the one hand, and the transport in large containers on the other. The latter cannot be manhandled and usually consist of tanks of 1,000 gallons capacity and upwards. They are generally insulated.

*Standard vehicle with upper deck for farm collection of milk.*

Although it is by far the most common method of transport it is far from being ideal for such a susceptible commodity.... It is necessary at the present time to use these small containers for collection of milk from individual farmers. It is also necessary today to deliver treated milk to retailers in small churns when their requirements are below the capacity of a small tank. The method of delivery in insulated tanks, however, should be further extended if necessary by the use of smaller tanks. In this connection far too many distributors do not make the simple necessary adaptations to allow their plant to receive milk in bulk.

In discussing the transport of milk in churns the units to be considered are the churn, the type and dimensions of the lorry body and the type of the chassis and power unit. It is good to learn of the present attempts to standardise churns. There has been a great deal of waste of transport in the past, due to the provision of churns that cannot be loaded economically. The 17-gallon taper churn, for example, is anathema to a haulier... With suitable conveyors and steel decks or cast iron grids it is questionable if handling is greatly aided by the cylindrical churn. It is interesting to note that if a squared churn 12ins. X 12ins. was used, 105 of them could be loaded on the same size lorry which would hold only eighty-eight cylindrical churns of the same capacity, a 19.5 per cent saving of space. The eighty-eight churns would have the 13$\frac{1}{2}$ ins diameter which is now proposed and which is, in itself, a great improvement on many of the churns in use today. However, if cylindrical churns continue to be used, the main point is that the smaller the diameter the better they load.

*Lowering churns by means of a steel churn slide.*

Whilst on this subject of churns, I hope that your Society will consider the introduction of new materials. The 10-gallon tinned steel churn of approximately 34 lbs. weight means that a high percentage of dead weight is carried, particularly with farm collection, when the churns are seldom full. Assuming the churns are 85 per cent full, the weight of tinned steel churns averages 29 per cent of the total load. If we take, on the other hand, an aluminium-magnesium alloy churn of the same dimensions but less than half that weight, the dead weight of the churns is only 16 per cent of the total load. The problem to answer is whether it is cheaper to carry 13 per cent of unnecessary dead weight in order to avoid charges on any additional capital which might be incurred by providing alloy churns.....

### Lorry Bodies

Collection of milk from the farms alone requires the equivalent of 7,000 3 – 5 ton lorries to be engaged daily on collection. Comparatively few of these lorries have bodies which are ideal for the job. Milk in churns is conspicuous for its bulk rather that its density. The maximum width permitted today is 7ft. 6ins. overall. The legal length is governed by the amount of overhang permitted. Overhang is now conveniently defined as being 50 per cent of the wheelbase and it is, therefore, obvious that long wheel base vehicles should be used. A flat platform gives the best results…..

### Loading the Lorry

This is a point that needs close attention and one to which little heed is commonly paid. It is difficult to achieve economical loading unless the churns are of a uniform diameter. There are three ways of arranging churns on a lorry: (1) in straight rows; (2) in rows which are staggered sideways; (3) in rows which are staggered lengthwise. Usually the least effective method is the straight row.

This matter is of particular importance to those despatching milk in churns. Unless the maximum number of churns which can be loaded is ascertained and the churns loaded accordingly, drivers have been known to re-arrange the load so as to include a few "buckshee" churns unnoticed.

The second factor to be taken into consideration in loading is that of weight….

From these various details of dimension and weight it will be seen that when churns are reasonably full, when their diameter does not exceed $13^1/_2$ ins. and when a suitable platform has been provided, it is possible to get a full pay lead of ninety churns on the platform body, and there is no need in such cases to have recourse to an upper deck. However, when the churns are not well filled the upper deck is needed in order to get a full payload…..

In the case of almost all milk haulage the seven-day week is the predominant factor to contend with….

The whole must be controlled and progressed by a master chart. Any fault which arises due to negligence should be traced back.

The immediate supply of guaranteed reconditioned components on an exchange basis at a reasonable price will be the best arrangement for covering major overhauls…..

There are two small but important points which come under this heading of organisation. Firstly, as far a possible, arrangements should secure that there is only one driver per vehicle. If a relief driver is employed on days off, and does not use a spare vehicle for this purpose, he should be in the nature of a rotary man or charge-hand and senior to the regular driver.

The second point is the importance of keeping up the appearance of the vehicle…...

In the case of seats, I have yet to find one which will last for more than a fraction of the life of a lorry. Men cannot be expected to take a pride in a dirty, battered lorry….

At present time lorries fall roughly into two categories, known colloquially as "lights" and "heavies." Both appear to do their job very well and both depend on efficient maintenance for good results. Comparative costings reveal no significant overall difference and certainly not enough to justify the inflexible views which the advocates of either side put forward….

It looks as if the Diesel engine, with its great fuel economy, will have an increasing part to play….

### Routeing Vehicles

Sometimes the organisation for loading and maintaining vehicles is first-class but the organisation of the routeing (that is the time when vehicle and man are not under immediate supervision) is left rather to chance….. The first essential is to have an accurate record of the location of each point of call. The 1in. Ordnance Survey maps, on which are superimposed the military grid, will be found very convenient for this purpose. A six-figure reference number recording each point of call can then be filed with other particulars. Change of route can then be worked out or checked in the office and checks made for unnecessary mileage….

*Costs*

Lorry operating costs can be conveniently grouped into time costs and mileage costs. There is sometimes an advantage in splitting the first group into standing charges and other charges as seen in the attached example (Table 1V), but the division between time and mileage charges is clear. I append a typical analysis of costs for the year ending September 30th, 1944.

No particular merit is claimed for the figures shown. Other operators, in fact, are known to operate on lower figures.

TABLE 1V

ANALYSIS OF COSTS OF OPERATIONS OF
THREE MAIN FLEETS

M.M.B. Twelve Months ending September 30th, 1944

| STANDING CHARGES | Cost per hour d. |
|---|---|
| Drivers' Wages | 24.48 |
| Vehicle Licences | 2.58 |
| Vehicle Insurance | 1.07 |
| TOTAL | 28.13 |

| RUNNING COSTS: | Cost per mile d. |
|---|---|
| Depreciation (Mileage Basis) | 0.55 |
| Mechanics' Wages | 0.29 |
| Repairs | 0.73 |
| General Maintenance | 0.10 |
| SUB TOTAL MAINTENANCE | 1.67 |
| Petrol | 2.24 |
| Oil | 0.11 |
| Tyres | 0.76 |
| Tubes | 0.03 |
| TOTAL RUNNING COSTS | 4.81 |

| | Cost per vehicle |
|---|---|
| TOTAL ESTABLISHMENT CHARGES | £54  1  0 |

| TOTAL COST PER MILE | 8.48d |
|---|---|

It is possible to operate economically with smaller loads when the distance is short. It follows that smaller lorries may prove more economical for short distance work, and so a 3-tonner carrying a 650-gallon load may be more satisfactory than a 5-tonner carrying 900.

Milk is worth today approximately 2s. per gallon and distributing costs nearly 1s. and transport costs about 1d. Any saving in the 1d. which involved waste of food or increased the distributive costs is obviously false economy. Nevertheless, the cost of transport from the farms alone amounts to £3½ million per annum at the present time. Many economies are still essential before the present arrangement can be considered satisfactory....

## Haulage of Milk in Tanks

The main points submitted under the heading of Organisation and Costs apply to the haulage of milk in tanks by road just as much as to operation of milk lorries. The tank, when full, is a heavy unit which has to be transported either by the larger type of motor chassis or by rail. The largest single factor in road tanker costs is labour and, from this point of view, the larger the tank the more economical it is to operate. The railway companies also recognise the importance of large-sized units and allow increased rebates for the larger sizes.

In dairies too, the largest item of cost is again labour. The most efficient use of labour will continue to be sought by increasing the amount of continuous flow and decreasing the amount of batch handling. One of the first steps is the elimination of the churn in the dairy, and the tendency must be for more tanks to be used in place of them. Milk, owing to its perishable nature always seems to lag behind other liquids in adopting the technique of bulk handling. It is unlikely that we shall see, in the light of present knowledge, pipeline technique for

*Bulk collection of milk from a farm in Scotland.*

handling milk, and the increased use of tanks probably represents the limit of progress for some time to come.

In deciding whether the tanks should go by road or rail the main factor is distance….. In the case of rail tanks, weight does not appear to matter, but this is not so where the road is concerned…… Tanks are not so far adapted for the delivery of milk in small quantities. Milk could, however, be pumped direct from the road tank to the retailer's holding tank. The problems to overcome would be the provision of sufficient agitation to keep the fat content of the milk uniform, and measuring and delivering arrangement on the lines of the electric petrol pumps now in common use, but which can be readily sterilised……

It can only be a matter of time until the marked economies of handling by tank forces a solution to minor technical problems such as these.

In the matter of collecting milk from farms, the day may come when instantaneous quality tests will permit the acceptance of milk at the farm. Milk could then be measured qualitatively and quantitatively, accepted at the farm and immediately tipped. When the specially equipped vehicle is en route to the next call the milk could be pumped through a filter and brine-cooling unit into an insulated tank. The tank might be detachable and various ways could be found for transporting it to its final destination, the milk in the collecting tank being delivered without further handling.

PROGRESS AND TRENDS IN THE DAIRY
INDUSTRY IN NORTHERN IRELAND, 1924-49

**D. A. E. HARKNESS, C.B.E., M.A.**
**Permanent Secretary, Ministry of Agriculture, Northern Ireland (JSDT 4  24-29 (1950))**

*"The dairy herd*
The basis of the dairying industry in any country obviously lies on the farms and the dairy stock that are kept on the farms.

In 1924 there were 256,739 milch cows and 14,171 heifers in calf recorded on farms at the June Census - total of 270,910 dairy cattle. Fifteen years later, in June 1939, the numbers were 244,973 milch cows and 24,716 heifers in calf-a total of 269,689 - i.e. almost identical with the figure in 1924. Last year, in June, 1949, the Census showed 309,560 milch cows and 66,080 heifers in calf on farms - a total of 375,640 dairy cattle - i.e. about 105,000 more than in either 1924 or 1939, representing an increase of about 39 per cent.

It is apparent, therefore, that despite all the efforts by way of subsidies and improved marketing conditions that were made in the 1930's to improve the position of dairying in

Northern Ireland - and considerable efforts were made by the Government with this object in view - we only just succeeded in maintaining our dairy cattle population. On the other hand the ten years since 1939 have shown a very considerable increase in the size of the dairy herd.

In 1925, at the time of the first Northern Ireland Census of Agricultural Production, it was estimated that the average yield of milk per cow was 407 gallons per annum. In the two years preceding the war it was estimated that the average yield per cow was 391 gallons - a decline of sixteen gallons compared with 1924/25. The estimated yield fell further during the war years - to as little as 328 gallons in 1941/42. This, of course, was largely due to difficulties with regard to feedingstuffs and during the past eight years the tendency has been upwards. In 1948/49 the yield was estimated at 403 gallons - an increase of 75 gallons from the wartime low figure and 12 gallons above the pre-war average but still just below the figure estimated at the time in respect of 1924/25.

There were various factors which accounted for the decline in milk yields in the fifteen years prior to the war. As I have indicated, the dairy cattle population held its own during these years but over a great part of the country farmers were more interested in the quality of the calf given by their cows than in the yield of milk. In the early 1930's the price paid for milk delivered to creameries fell to absurdly low levels. After the passage of the Milk and Milk Products Act in 1934 certain financial assistance was made available to improve the price of milk for manufacture but the cattle subsidy of 5/- per live cwt. which was also introduced in 1934 had even more influence and the swing to beef continued.

The number of Shorthorn bulls was maintained but there has undoubtedly been much more attention given to the milking quality of the Shorthorn. The fall in the popularity of Aberdeen Angus and Galloway bulls is apparent and equally the great increase in the number of Friesian and Ayrshire bulls standing. Nevertheless, the yield is still no higher than 403 gallons and although the shortage of feedingstuffs undoubtedly constituted a handicap during the war years the past few years have seen a substantial improvement in supplies. When it is found that in 1948/49 the estimated yield per cow was 526 gallons in Scotland and 592 gallons in England and Wales it is apparent that there is great scope for the improvement of yields in Northern Ireland.

Before leaving the dairy herd I must refer to one outstanding development of the past decades the great increase in autumn calving and the proportion of milk produced in the winter months.

In 1939 only 34.2 per cent. of the total sales of milk were made during the months of January, February, March, April, November and December. In 1949 no less than 41.7 per cent. of the total sales were in these months.

*Quality improvements*

In 1925 the responsibility for the supervision of cows, cowsheds and dairies rested with the local authorities who were responsible to the Ministry of Home Affairs. His position was governed by the Cowsheds and Dairies (Ireland) Order of 1908. The first major legislation dealing with milk that was passed by the Northern Ireland Parliament was in 1927 when the Sale of Milk Act (Northern Ireland) was enacted and came into operation on the Ist January, 1928. It provided that no person should sell milk under a special designation as "Certified" "Grade A" "Tuberculin Tested" or "T.T." except under and in accordance with a licence granted by the Ministry of Home Affairs. This was the start of the Grade A movement and although the Ministry of Home Affairs was the licensing authority a great deal of work in the encouragement of producers to take out licences was done by the Ministry of Agriculture.

In 1934 the Milk and Milk Products Act was passed and the Ministry of Agriculture became the licensing authority for producers of Grade A, Grade B or Grade C milk. Every producer who sold milk for liquid consumption required to have one or other of these licences but producers who sold Grade D milk for manufacture did not require to be licensed. Grade A and Grade B milk had to be sold in bottles and this represented a great improvement in the cleanliness of milk. Grade A milk had, of course, to be from cows that had passed the double intradermal test, while Grade B and Grade C herds were subject to a clinical examination by the Ministry's own veterinary officers to ensure that the cows were clean and healthy.

Bacteriological standards were prescribed for Grade A and Grade B milk while the standard for Grade C milk was the methylene blue test.

*War-time developments*

In the years preceding the war sales of milk off farms were around 40 million gallons a year, of which 15 million were for liquid consumption and the balance was used for manufacture - almost entirely for butter and cream because it was only in 1938 that one small factory was established for the drying of milk by the roller process. The war brought about a great change in the market for milk. Previously the liquid market had been confined to a comparatively small radius round Belfast where about 1,500 producers had annual contracts, on terms fixed by the joint Milk Council created by the 1934 Act, for the sale of milk to distributors. Outside this area-and also to a fair extent in Belfast - any sale of milk for liquid consumption was effected by producer/retailers who found their own customers among consumers. Certain levies were imposed on the sale of milk for liquid consumption. These went into a Milk Fund created by the 1934 Act together with licence fees, levies on the sale of butter and margarine, and certain Exchequer contributions.

The Milk Fund was used to defray administrative costs and to pay certain subventions to producers who sent their milk for manufacture. They were assured of a minimum average price of 5d. per gallon in summer and 6d. (later 6½d.) in winter, together with a bonus of 2d. a gallon if they had a Grade A or Grade B licence and Id. (later raised to 2d. in 1937) in the case of a Grade C licence-holder.

In Great Britain the English and Scottish Milk Marketing Boards operated a pooling system. It was impossible to operate a pooling system in Northern Ireland because the weight of manufacturing milk would have depressed the pool price to a level that would have been quite unfair to, and quite unremunerative for, the liquid milk producers who had to maintain a relatively level rate of production all the year round. Consequently there were two markets for milk in Northern Ireland - a liquid market centred round Belfast, protected by the annual contract negotiated by the joint Milk Council, and an unsheltered competitive market for milk for manufacture. As I have indicated, however, the producer who had to sell on the manufacturing market got some financial assistance.

The war greatly increased the demand for milk for human consumption. Sales of milk for the liquid market increased as follows:

SALES FOR LIQUID CONSUMPTION

|         | In Northern Ireland |         | For shipment to Great Britain |         |
|---------|---------------------|---------|-------------------------------|---------|
| 1939/40 | 15,471,934          | gallons |                               | gallons |
| 1940/41 | 17,545,573          | "       | 864,334                       | "       |
| 1949/50 | 41,325,324          | "       | 3,428,246                     | "       |

There were various factors accounting for this remarkable increase in liquid consumption in Northern Ireland - higher employment, higher wages, the presence of large numbers of troops (British and American), the shortage of other foods and last but not least, the National Milk and the Milk in Schools Schemes. In addition there was a big increase in the requirements for processed milks - evaporated milk for military and civilian use, Household Milk and National Dried Milk for infants.

Milk became agricultural priority Number 1. Greatly enhanced prices were paid to farmers for milk and in addition in November, 1940, the Government became the purchaser of all milk just as it had already become the purchaser of all fat cattle, sheep and pigs and either then or later became the purchaser of many other agricultural products, e.g. eggs, wool, potatoes and flax. An Order was made by the Ministry of Food requiring every person who produced milk in Northern Ireland to place that milk at the disposal of the Ministry of Agriculture and the Ministry became responsible for the payment of producers for all the milk they sold.

This led, in 1942, to a uniform price being paid to all producers of milk in Northern Ireland, irrespective of whether their milk went for liquid consumption or for manufacture, provided that it complied with a reasonable standard of cleanliness and quality. As from the 1st October, 1942, all producers of milk (other than producer/retailers) were directed to deliver their milk at some 30 grading centres at which there were milk grading officers appointed by the Ministry of Agriculture....

The concentration of all milk supplies at a relatively small number of grading centres involved the closing of a number of the smaller creameries - principally the auxiliary ones at which cream had been separated and sent on to central creameries for manufacture into butter...

Every churn of milk received was examined by the grading officer to ensure that it was satisfactory. Platform examination was supplemented wherever possible by the short time resazurin test. Milk which was graded as up to liquid standard was paid for at a uniform price regardless of the class of licence held by the producer or whether he had a licence at all. In addition however, holders of Grade A milk licences received a premium of $3^1/2$d. per gallon (now raised to 4d per gallon) and the holder of a Grade B licence premium of $1^1/2$d. per gallon. Milk which failed to pass the test as up to liquid quality was paid for as of manufacturing quality. In 1943/44 some 6.693 per cent. of the milk supply was graded as manufacturing: in 1949/50 the percentage was only 1.644. There can be absolutely no doubt that this system of platform grading had a most satisfactory result in effecting a marked and rapid improvement in the general quality of the milk offered for sale by Northern Ireland pro-ducers.

The condition of cows and byres was, however, not ignored. As soon as possible legislation - the Milk Act of 1944 - was passed requiring all producers who sold milk to be licensed by the Ministry.... The improvement effected under the 1944 Act in the standard of cow byres was quite remarkable.

Another important development that took place in October, 1942, was the rationalisation of milk collection, so as to avoid overlapping and to save petrol, and the institution of a flat trans-port rate for the collection of milk. For 1.6d per gallon his milk was collected and a clean churn returned to him the following day. This meant a double set of churns and we had to obtain tens of thousands of additional churns to enable this service to be provided. They were sold or hired to farmers. The return of a clean sterile churn was to large numbers of small farmers an immeasurable boon in helping them to supply clean, sweet milk....

In Northern Ireland we have never under-rated the importance of beef production and we want to see all the calves reared that it is possible to rear properly. But neither do we want to exclude the small producer from the milk market and that is one of the reasons why we attach so much importance to the <u>development of a good class of dual purpose Shorthorn</u> - an animal giving a good milk yield and also a good quality calf for rearing.

*Milk processing and utilisation*

The increase in the demand for liquid milk - both for Northern Ireland consumption and for shipment to Great Britain - meant that milk had to be drawn from a much wider area than was usual pre-war when it was a case of collecting from a 10 or 15 mile radius around Belfast and the morning milk was in the hands of consumers within a few hours. Milk had now to be bulked and transported for considerable distances - involving the elapse of a considerable period of time before it reached the consumer. In the interests of safety this called for the past-eurisation of the milk and consequently in 1942 the Ministry of Food agreed to the Ministry of Agriculture erecting and operating a number of Government Milk Depots - seven in all were established. Dromona, Dunman Bridge, Tassagh Lisnaskea and Ballynure were completely new erections; Killyman and Enniskillen involved large extensions to existing premises. All were equipped with H.T.S.T. pasteurising plants and also extensive refrigeration capacity. There was also a great improvement in the equipment and facilities at the creameries which were recognised as grading centres. Thanks to these developments Ulster milk sent to Great Britain earned the highest praise for its good quality despite the testing experience of up to 28 hours spent on the journey.

We were fortunate during these years to get a large supply of pasteurising and other dairy equipment: so much so that in 1945 we were able to schedule Belfast and district as a safe milk area in which the only milk allowed to be sold was Grade A milk bottled on the farm of production, and pasteurised milk. Provision for scheduling safe milk areas had been made in our 1944 Milk Act – in accordance with the policy laid down in the Imperial Government's White Paper on Milk Policy published in 1943 – and Belfast was the first area in the whole United Kingdom to be so scheduled. In 1949 the area was extended top include all urban districts and seaside resorts. Today approximately 85% of the milk consumed in Northern Ireland is pasteurised.

Then the war provided the opportunity for Northern Ireland to enter the market for high category milk products. The great increase in sales of milk off the farms meant that despite a 150 per cent. increase in liquid milk consumption and the shipment of up to $3^1/2$ million gallons of milk a year across the Channel, we were able to maintain our production of butter and yet have available up to 21 million gallons of milk for classes of milk production virtually unknown to us pre-wart, viz. condensed and evaporated milk, National Dried Milk, spray dried milk, chocolate crumb, as well as a number of proprietary milk foods. We have also undertaken a large programme of drying and condensing skim milk resulting from butter manufacture. We have seen the establishment of milk factories in Northern Ireland by such famous firms as Nestlé's, Benger's, Bovril and the C.W.S.

In the years ending 31st March, 1940-50 milk available for manufacture has been utilised as shown below:

|         | BUTTER | CONDENSED MILK | MILK POWDER | OTHER PRODUCTS |
|---------|--------|----------------|-------------|----------------|
|         | Gallons | Gallons | Gallons | Gallons |
| 1942/43 | 19,578,460 | 1,579,621 | 1,196,048 | 1,997,443 |
| 1949/50 | 22,788,825 | 10,906,793 | 8,234,472 | |

It has been possible for me to touch only very briefly on several war-time developments which had many features of interest - in particular I have had no time to deal with the organisation of milk transport to Great Britain by both sea and air which played an important part in easing the milk shortage in Britain during the difficult years of the war and post war period. Nor have I been able to deal with the course of milk prices.

*Conclusion*

I must now try to sum up the direction in which the developments I have outlined are leading us. I am going to suggest that we must concentrate our efforts on the following lines in the period upon which we are now entering - the period following the war and the first five years of post war reconstruction that have now passed.

1. In the field of production we must concentrate on securing a substantial increase in the yield per cow through better breeding and equally better management.

2. In regard to quality it is significant that in England the Wales the Milk (Special Designations) Act, 1949, envisages the elimination of accredited milk (roughly the equivalent of our Grade B) within a short period of time so that the only milk on sale will in effect be tuberculin tested milk and pasteurised (or safe milk). It is clear that public interest in the safety and purity of milk will not for long tolerate the sale of milk which is neither tuberculin tested nor pasteurised. We have, however, lagged far behind Great Britain in this respect and must do all we can to make up leeway in our efforts to eliminate tuberculosis.

Consideration should be given to the introduction of a scheme for the payment of milk on the basis of quality.

As regards pasteurisation it is clearly important that even greater care shall be taken to ensure that pasteurised milk is not contaminated after pasteurisation. This means that pasteurising and bottling should be part of a continuous process and that milk should not be pasteurised and then carried to other premises for bottling.

3. Economic policy today is concentrating on disinflation and the restraint of wages and personal incomes. The rapid expansion in milk consumption that occurred during the war and immediate post war years has slowed down or is coming to an end. Other foods are becoming more plentiful. If we are to maintain milk consumption then we must satisfy the consuming public not only that milk is nutritious but that milk is safe.

4. As production of milk increases in Great Britain the need to ship it from Northern Ireland will decline and perhaps disappear. This, however, will set free large quantities of high quality milk suitable for the production of high category milk products.

5. The possibility of the restoration of cream production has been canvassed recently. That is, a development we should watch carefully to see whether it provides a suitable outlet for our milk in post war conditions.

I am sure that the formation of a Northern Ireland Branch of the Society of Dairy Technology is a very promising augury for that future and that it shows that those interested in dairying in Ulster are alert and anxious to play their part in placing the industry upon the highest possible basis of efficiency.

## THE EARLY HISTORY OF THE SOCIETY
## PROFESSOR H.D. KAY
### (JSDT Vol. 11 No. 1 1958 – Response to Gold Medal Presentation)

"When you, Sir, as President of the Society of Dairy Technology wrote to me some months ago to tell me that the Council of the Society had honoured me by deciding that I should be awarded the Gold Medal of the Society, and that it would be presented to me at today's meeting, you also, in your persuasive way, indicated that something was expected of me at that meeting beyond mere blushing acceptance, and that a short address to the Society on some topic to be chosen by myself would be expected.' Said Professor Kay.

As many here will remember, the Society came into being on 2nd March 1943, in the middle of the war. But this was merely the official birthday.

The idea of an organised dairy technological group or club or association dates back, in this country, to before the outbreak of the war in 1939. In fact, but for that event it is likely that the Society might, today, have been two or three years older. During the tension preceding the outbreak of war and in the early days of the war, there was undoubtedly the feeling that any move to form a Society could wait for better times. I well remember a talk I had with my friend, Dr. J. G. Davis, then a colleague and head of the section of Bacterial Metabolism at Shinfield, early in 1941, in which we argued the situation. He was then in favour of making a move, despite war conditions, towards the initiation of an association of dairy scientists.

Other colleagues with whom I discussed the suggestion were against such a move at such a time. I myself felt that with our whole national future at stake, as it certainly was in 1941, and with the great amount of additional war work to be done and the claims on the time and energies of everyone in the dairy industry, the time was far from auspicious. Not only were we all concerned to maintain the supply and quality of liquid milk, which was, in a very real sense, our key human foodstuff during the war, and to maintain it under tremendous handicaps of shortage of staff and equipment and transport, but we had also A.R.P. or Home Guard duties which occupied whatever time we had left. One night in three or four, at a Wardens' post or a Report Centre, in addition to one's daily task, did not leave an overplus of either time or energy to encourage new enterprises, and some of us, in all the three categories of which the Society is now made up, felt that we should be courting a real flop if we tried to launch a new society in the midst of all "the shouting and the tumult."

But a new hope began to arise in the autumn of 1942. We were no longer alone against the enemy. The war itself had brought us together in the sense that it made us realise how much we depended on each other and how much pooled knowledge on the one hand and personal acquaintance and friendship on the other were of basic value in dealing with the severest of wartime handicaps. These were two of the main reasons, in my opinion, that began to put a new complexion on the problem as to whether a new Society might be founded. During the autumn of 1942, I had some quite informal chats as to the ripeness of time both with colleagues like Mattick, Davis, Sutton and others, at Shinfield, and with several scientific, technological and other men in, or closely associated with, the milk industry and also in dairy engineering, and found quite a volume of opinion in favour of going ahead rather than of waiting.

It seemed, therefore, that the time had come to send out a feeler to a much larger number of individuals, and the following letter was drafted:

9th November 1942

Dear Sir or Madam,

As you may know, there has been much informal talk recently about the need, even in war time, for closer contact between those who deal with the day to day technical aspects of dairy work (includ-ing milk production, processing and manufacture, milk plant manufacture, quality in milk and dairy products) and those engaged in the more purely scientific, research, and control sides of dairying. The general objective that has been in mind is the furtherance of the application of scientific knowledge to the dairy industry, both as regards the war effort and in *post-bellum* developments.

As a result of these talks, a purely informal meeting has been arranged for Monday, 23rd November, at 11 a.m., at the Waldorf Hotel, Strand, W.C.2, to consider whether any specific course of action is warranted. It is hoped that you will be able to attend.

Coffee will be served from 10.30 to 11 a.m.

Yours truly,
(Sgd.) H. D. Kay.

I still have a list of just over eighty people to whom this letter was sent, and they are indeed a representative cross-section of the industry dairy scientists, technologists, progressive directors, dairy managers, dairy engineers, even one or two concerned with legislation and control. The quite astonishing result was that three-quarters of those written to turned out for this meeting, an amazingly high proportion when one remembers how overworked most people were at this time. In addition, several letters were received from those unable to attend, to express their support for some further course of action.

Many here present will remember this meeting though our number is, in the nature of things, slowly getting smaller. After coffee had been served (even in the direst period of rationing there was still some coffee) I explained in more detail why the meeting had been called. Whilst technical organisations existed which served the production side of the milk industry, there had been little or no common meeting ground for those engaged in the technology of milk handling, distribution or manufacture, apart, perhaps, from the Dairy and Ice Cream Equipment Association which dealt very competently with one section only of this field. An independent organisation, the main object of which was to facilitate technical and social intercourse between those engaged in dairy science and technology, seemed to several members of the meeting whom I had previously consulted to be worth serious consideration, even then in the turbulence of war-time, though a few seemed to regard the project as one that should wait for happier times. It was true that the war would almost certainly limit the scope of the activities of any new organisation that might possibly be decided on, and very few of those present had spare time to devote to the successful development of any ambitious new society. I suggested, nevertheless, that a start, if made in the near future, might be useful, even in helping us to meet some of the war-time problems of the industry. It would in any case help to confirm the sense of unity in an industry whose outstanding national importance the war had already underlined. We should aim, if it was decided to found an association of some kind, at one which would embrace dairy and creamery managers, directors and technical staff, teachers, advisers and research workers in dairying, dairy bacteriologists and control staffs, dairy engineers and equipment makers, civil servants connected with the dairy industry and also milk producers and farming advisers whose interests took them beyond the farm gate.

Though it was already known that several present at this meeting were in favour of forming some such organisation, I think many of us were amazed at the unanimity and enthusiasm of those who spoke. Though in my notes of the meeting I have not, unfortunately, the names of all those present, I have the names of the speakers. I think it might be of interest to mention them, and I give them in alphabetical order, not in the order in which they spoke:

T. R. Ashton, A. L. Barton, A. E. Berry, G. W. Scott Blair, E. Capstick, F. J. Clarke, H. B. Cron-shaw, Ben Davies, A. G. Enock, Graham Enock, J. C. F. Fryer (Secretary, Agricultural Research Council), A. Lauchlan, W. A. Lethem, J. Mackintosh, A. T. R. Mattick, F. Procter, A. L. Provan, W. W. Ritchie, E. A. Shepheard and S. B. Thomas.

Most of these people are still with us; many have held high office in our Society.

All but one of these speakers were in favour of starting an association as soon as possible, and many valuable suggestions were made at this meeting and noted by my secretary. I had also received in support of the proposal, letters from several people who were not able to attend the meeting.

The meeting came to the decision that an *ad hoc* Committee be set up with the following terms of reference: "To consider what organisation should be formed to meet the needs and views expressed in today's discussion and to draw up proposals." It was further decided that these proposals should be submitted to a general meeting of those interested to be held in London on or about 1st March 1943.

The *ad hoc* Committee consisted of the following (again in alphabetical order): E. B. Anderson, R. Arnison, T. R. Ashton, A. L. Barton, E. Capstick, J. G. Davis, Graham Enock, H. C. Hillman, H. D. Kay, J. Mackintosh, J. Matthews, A. T. R. Mattick, A. McBride, F. Procter, A. L. Provan, H. G. Robinson, E. A. Shepheard and W. G. Sutton, with power to co-opt. Whether or not it is legitimate to call this group the Founding Fathers of the Society, I will leave to others to decide. It will be noticed that practically every one of these people not only contributed much time and thought at the very beginning of the Society, but held office during the critical early days of the very young Society, or has done so since.

At the first meeting of this *ad hoc* Committee which was held on 7th December, 1942, the following were co-opted: H. B. Cronshaw, Ben Hinds, S. B. Thomas and T. G. Wilson. These four names must also be added to the roll of the Found-ing Fathers.

A chairman for the meeting, and an honorary Secretary (W. G. Sutton) were elected; the latter, a most fortunate choice, was at that time Director of the Commonwealth Bureau of Dairy Science and was a tower of strength in all the early secretarial and administrative work of the Society.

A lively discussion then took place, in which many diverse views were launched, on the objects of the proposed association, its management, eligibility for election, classes of membership, possible subscription rates and so on. One suggestion was that the new organization should be an Institute of Dairy Technology, somewhat analogous to the Royal Institute of Chemistry, with high professional standards, an examination for entry, a high annual subscription and a building of its own in London, the last provided, presumably, by an enlightened Government! But eventually some kind of uniformity of view was arrived at, at least on the general principles. It was decided to meet again about a month later, and at that meeting to thrash out details, so that really well-considered and down-to-earth proposals could be put on paper and brought, with the unanimous backing of the *ad hoc* Committee, before the General Meeting to be held in March 1943.

The Sub-Committee duly met and made some quite specific recommendations. Sutton and I were asked to put these into shape well before the General Meeting and circulate an appropriate document to all those likely to be present at that meeting. This was done, and 272 invitations were sent out, each with a copy of the document, for a meeting to be held at the London School of Hygiene on 2nd March 1943.

Unfortunately the exact numbers who turned up for this historic meeting were not actually taken, nor, I am sorry to say, have we the names of all those present. An estimate made shortly after the meeting was that well over 100, possibly as many as 130, people were present. The first part of the document, also historic, that had been circulated read as follows:

***"Improved technological co-operation in the dairy industry"***
***"Report of ad hoc committee"***

"The committee appointed by the general meeting at the Waldorf Hotel on November 23rd 1942 'to consider what organization should be formed to meet the needs and views expressed in today's discussion, and to draw up proposals' has met twice. The committee co-opted at its first meeting Dr. H. B. Cronshaw, Mr. Ben Hinds, Mr. S. B. Thomas and Mr. T. G. Wilson. Professor Kay acted as chairman and Mr. Sutton as secretary.

*"The committee recommends:*
"(1) That a Society should be formed under the name of the Society of Dairy Technology (the committee's first choice) or The British Dairy Science Association (the committee's second choice).
"(2) *The objects* of the Society should be: (a) To improve practice in all branches of the dairy industry by the application of knowledge gained from experience and experiment; (b) To provide opportunities for discussion and collaboration between persons interested in improving the technical practices of the dairy industry; (c) To encourage technical education for persons working with milk and its products; (d) To encourage scientific enquiry into problems arising in the dairy industry; (e) To encourage the distribution and use of technical information among persons working in the dairy industry.
"(3) That it should be clearly understood that the Society should not engage in political activities and should not infringe on any of the duties and responsibilities of existing organisations.
"(4) That there should be *two types of membership* in the Society: (a) *Ordinary Membership,* which would normally be open to men and women holding responsible positions of a technological, scientific or educational character in, or directly connected with, the dairy industry; (b) *Associate Membership*, which would be open to men and women interested in the technological, scientific or educational side of the dairy industry, but not qualifying for Ordinary Membership."

And so on, twenty-nine clauses in all.

With only a few amendments - one very useful one was sent in in writing from Dr. Seligman, our first Gold Medallist - these twenty-nine recommen-dations were accepted by this General Meeting. A good many of those here today, and especially those who are or who have been at some time members of Council, will realise how little these original regulations have had to be amended since that time. In fact, the *ad hoc* Committee had done a very satisfactory job.

This meeting, as you have seen, made the important decision that the name of the proposed Society should be that recommended as its first choice by the *ad hoc* Committee, namely, the Society of Dairy Technology. Another important (in fact one of the most important) recommendations of the Committee had been the method of electing mem-bers of Council to ensure that no single interest out of the three major interests represented in the Society (1, milk utilisation, 2, dairy engineering and equipment, 3, milk production dairy education, research, administration and control, could obtain control of the Society, even in the unlikely event of such an attempt being made.

Further, the formation of local sections was approved, and two days only after the general' meeting of 2nd March, the first section - the Midland and South Wales section - not only applied for sectional status within the Society, but submitted a complete list of draft rules for their section! I believe Mr. John Lewis who is, as you know, our President-elect, had much to do with this speedy and enthusiastic move.

Other important matters decided at this 2nd March meeting were that, in future, voting for membership of the Council should be made by postal ballot, and that the President of the Society, who would also be Chairman of the Council, should be elected by the Council from amongst its own membership. The 2nd March meeting then proceeded to appoint a Provisional Council, intended to hold office for more than a year until an election, according to the rules just accepted, could be held about the end of June 1944. Names were proposed for the Provisional Council, including the officers other than the President, and after a ballot the following were the first officers and other members of Council to be elected:

*Officers* - Joint Honorary Secretary: A. L. Barton and W. G. Sutton; Honorary Treasurer: E. A. Shepheard; Honorary Assistant Treasurer: H. C. Hillman; Barton was to be the Meetings Secretary and Sutton to deal with other business.

*Members of Council* (Group 1) E. L. Crossley, R. C. Home, A. McBride, F. Procter. (Group 2) R. Arnison, H. B. Cronshaw. Graham Enock, J. Matthews. (Group 3) E. Capstick, J. G. Davis, H. D. Kay, D. M. Smillie.

It was decided to have the first General Meeting of the newly-formed Society in about two months' time, at which meeting the official constitution and rules could be formally agreed. In the meanwhile, it was urged that prospective members should send in their subscriptions - in fact, one or two had done so at the meeting on 2nd March. But it was only *after* this meeting that subscriptions really began to come in.

When the preliminary informal meetings were being arranged, finance for the cost of rooms, coffee, stationery and postage, &c., had been provided by one or two of (shall I say) the *prospective* Founding Fathers. The more optimistic of them had doubtless hoped that, if and when a Society was founded, these early payments might be reimbursed to them. But any financial difficulties of this kind in the pre-subscription era were relieved when an anonymous well-wisher (who asked that his anonymity should be preserved) gave £100 to the embryo Society to cover just these early expenses before subscriptions came in. This was a very present help in time of need, and saved the acting-secretary-cum-treasurer of the prospective organisation the invidious task of having to make a whip round.

Up to today, I think that only two persons beside myself have known *who* this donor was. I have recently persuaded him to allow me to reveal his name and he has rather reluctantly consented - it was our fellow member, W. A. Nell. His gift showed characteristic generosity and thoughtfulness. This was, indeed, the spirit in which the Society was born; people not only gave their time and work entirely free of charge to the Society but were prepared to put their hands into their pockets in addition, if need be. That spirit still remains, I am glad to say, to this day, though in view of the great increase in secretarial work we had fairly soon to appoint a paid secretary.

The actual birthday was then 2nd March, 1943, right in the middle of the war. The Provisional Council met as soon as possible. Questions of importance had to be tackled before the next General Meeting; one was to prepare a new edition of the official Constitution and Rules, as amended at the 2nd March meeting, and to make sure they were such as to allow the Society to be considered a charity, for taxation purposes, within the legal meaning of that word which is, I assure you, Sir, entirely dissociated from faith and hope! Sutton and I took legal advice on this matter before the Constitution and Rules were put into the form to be submitted to the General Meeting. Another question which the Provisional Council dealt with at its first meeting was to decide who should be the first official President of the Society.

In the meanwhile, Barton was arranging for the first *general* meeting to be held on 24th June, 1943, and got together a distinguished group of five speakers to take part in what was, in effect, a symposium on Trends and Problems in the Dairy Industry. The speakers were F. Procter, E. L. Crossley, J. Matthews, N. C. Wright and E. Cap-stick. Each gave a 25-minute paper and there was a 20-minute discussion on each, opened by W. A. Nell, E. Capstick, Richard Seligman, A. T. R. Mattick and S. B. Thomas respectively. Why we overworked Capstick in this way, I don't know, except that he was a very willing as well as an able horse!

Before the papers, important business was transacted - the constitutional, and now legally vetted, framework which had been circulated beforehand to all members in good standing *was approved without alteration*. To enable the Society to hold property and funds, the meeting authorised the Council to nominate up to six Trustees. These were Graham Enock, A. McBride, W. A. Nell, F. Procter and Richard Seligman, to whom E. White was added shortly afterwards. Professional auditors were also appointed. The names of no less than 240 foundation members were posted in the Hall at this meeting.

I think at this stage (June 1943) I must draw my short history to a conclusion, though perhaps I ought to mention that about the time of this meeting we lost our first Treasurer, E. A. Shepheard, who was transferred to work in Australia. His successor was Graham Enock, again a most fortunate choice, through whose work during six years of office the finances of the Society were put on a sound foundation.

May I just add a word about the future? The Society is, I think, now in a position to bring even greater influence to bear in advancing the future of dairy technology in Britain. One of the quite outstanding directions in which this advance could be strengthened is in developing higher education in dairy technology. When one considers the very serious way in which this problem is being tackled in countries like Denmark, Holland, Norway and New Zealand - to mention only a few dairying countries-one realises how far behind we are lagging in this country. Even the relatively small facilities we have are not being made full use of. I am not referring to dairy research where we are in a fairly good position, but to the supply of the next generation of dairy technologists of high calibre - our younger successors in this Society.

Is it that working conditions in the industry are relatively poor compared with those in many other industries? Is it that salaries and emoluments are worse in our great dairy industry, with its annual turnover in milk alone of over £300,000,000, than in many others? What is the reason for the astonishingly low entry that has occurred in recent years, of young men with good brains into dairy technology?

It seems to many of us in this Society that this problem is perhaps the most serious one that the dairy industry is facing, whether it realises it or not, at the present time. If one is permitted to extrapolate into the future a short message from today's account of the past, I should like to exhort all members of our Society, individually and collectively to address themselves with all speed to its solution.

Well, Mr. President, I have come to the end of this brief sketch of our beginnings and delivered my short message for the future. Our Society, today nearly 1,750 strong, is now a very viable organisation, an established power in the land, well known throughout this country and abroad. Its reputation is high in the industry not only as a platform for scientific and technological discussion of all dairy Problems, but also as a centre of good fellowship for the great majority of those in Britain who are interested in the advancement of dairying.

## MILK TRANSPORT IN ENGLAND AND WALES
### J. G. Inglis
### (Milk Industry January 1951)

"Mr Inglis gave this paper to a meeting, held under the auspices of the Food and Agriculture Organisation of the United Nations and the British Council, at Reading

It is obvious that the pattern for transport must follow the marketing process and the best transport arrangements can only be achieved if milk is marketed successfully.

The aim of the Transport Manager is to streamline the route to market so that no unnecessary time elapses between production by the farmer and consumption by the ultimate customer......

We in this country are subject to a number of limiting factors when planning our transport. In the early part of the century, when many of our depots and dairies were built, the movement of milk depended on horse and rail. Handling centres were established which conformed to the limited radius of the former and the relatively slow speed of the latter. Today the provision of new handling centres at the best strategic points for integration with modern road transport fails to keep pace with the rapid improvement in the latter.

Every depot or collecting centre within 30 or 35 miles of a large consuming centre tends today to be redundant unless it is performing some special service. Nevertheless we are forced by present conditions to develop at a pace that lags some way behind possibility.

Other limiting factors which we experience in this country are:

(a)    The revolution in the hours and working conditions of agricultural workers, which means that milk is seldom available for transport from the farm before 8 a.m.

(b)　The town dairy works approximately a $7^{1}/2$ hour day and likes the early part of the day devoted to sending the milk out on the retail rounds. In order to allow the staff to continue straight on with the reception of processing of milk, these dairies would choose to start receiving milk by 8 a.m. at the latest.

(c)　Our lorry drivers are bound by statute to have a rest period after $5^{1}/2$ hours of duty and they must not exceed an 11 - hour day.

(d)　Lorries over 3 tons unladen weight and almost all tankers are limited to a maximum speed of 20 m.p.h. which means they must he scheduled to run at 17 m.p.h. The effective radius for operations with one driver is therefore well under 100 miles.

It may be seen, therefore, that the existence of the established interests of both capital and labour tends to slow down progress, however desirable that may be....

Milk is not only a bulky commodity to transport, it is also highly perishable....

Quite apart from these disadvantages, the size of the milk haulage industry in this country is seldom appreciated.

*The beginning of bulk collection.*

In England and Wales milk has to travel from 140,000 farms to over 7,000 separate destinations. Half of the milk has usually to make a second and sometimes a third journey. It has to do this seven days a week in all weather conditions. It is estimated that the gross weight of milk including containers carried on primary and secondary haulage is nearly 20,000,000 tons per annum and the ton-mile figure is astronomical. Some 10,000 vehicles are employed in collecting and delivering milk to the first destination. Approximately half are vans and small conveyances and the balance are lorries, mostly of the five-ton type.

It will be seen that rather more than half of the milk goes direct from the farm to the town distributor or from the farm direct to the retail customer. The remainder goes to depots which balance the liquid milk market and manufacture any surplus to the requirements of that market.

The chart [omitted] indicates that the transport of milk sold by wholesale is controlled by the Milk Marketing Board as far as its first destination. Beyond that point purchasers make their own arrangements under direction from the Ministry of Food. Nevertheless this secondary transport is still paid for by producers at a "frozen" charge.

*The Control of Milk Transport to the First Destination*
The reasons for placing the control of the collection of milk from the firms under the direction of a single authority were as follows:

1.　The elimination of cross transport, of which a great amount existed prior to 1942.
2.　An acceleration of the routes to market.
3.　The use of fewer and better loaded lorries.
4.　Improved arrival schedules causing less congestion at the receiving dairies.
5.　To secure that the rate paid was appropriate to the job.

It was appropriate that the producers' Board should be chosen for carrying out this task, because it is the producer who pays for getting his milk to market. During the war the Ministry of Food, the distributors and the hauliers worked together with the Board to achieve the maximum economies in vehicles and manpower when forwarding the milk to market. The Ministry of Food later calculated that some 40,000 lorry miles per day had been saved. Both before and after that period substantial savings have been made and this work is estimated to have reduced the cost of transport to the producer by over £1,000,000 per annum....

*Vital Stage*

The most vital stage in the transport of milk is the journey to the first destination. The reason, of course, is that the greater part of the milk cannot before this stage be submitted to mechanical refrigeration, neither can the necessary low temperature be maintained by insulation before the first destination is reached. Only after chilling and insulating against temperature increase does the perishability of milk diminish and its transport become less exacting....

Broadly speaking, milk cannot be conveniently and economically transported from the farms to first destinations from a radius much in excess of 35 miles ....

All sorts of vehicles are used where small consignments are involved and there can be no standard type.

Provided the load to be carried is sufficiently large the most economical type of vehicle for milk collection appears to be the mass-produced five-tonner....

We do not in this country make much use of covered vehicles for transporting milk from the farms. Provided it is cleanly produced, reasonably cooled on the farm and kept in the shade until loaded, there is no reason for it to become unmarketable during transport from the farm.

Any consideration of the present cost of milk collection must be influenced by the considerable rise in costs of operating road transport which has taken place since 1942. In 1944 in a paper read to the Society of Dairy Technology (included in this chapter) figures were quoted which showed that milk-collecting lorries operated by the Milk Marketing Board cost 8.84d. per mile run. The latest cost for operating the vehicles of the Milk Marketing Board for milk collection is 11.30d. per mile....

Reliable figures for the period prior to 1944 are not available but it may he assumed that the increase in the cost of operating the Board's vehicles between 1942 and 1950 is not less than 35 per cent. Table No 3 attempts to show a comparison of the position as it existed in 1942-43, with the position today, and with the position as it might well have been without rationalisation.

The very large saving indicated in the table can fairly be claimed to result in a large measure from the reorganisation of milk transport which has taken place.

When the Board assumed responsibility for ex-farm transport, an endeavour was made to discover a formula which would automatically decide the payment to be made to any particular haulier for any particular job. Up to now it has not been found practicable to employ such a formula because there are so many variable factors.

## TABLE 3
### *The Cost of Milk Collection from the Farms in England and Wales*

*A Statement of Total Cost per annum and the Cost per gallon for the years 1942 – 43 and 1940 – 50. Also the Estimated Cost without Rationalisation*

Average collection rate 1942-4 ..................................................... 858d. per gallon
       (The average rate is obtained by dividing the total
       cost £2,895.000 by the 809,747 gallons carried)
Increase in operating costs 1942-43 – 1949-50 .......................... 30%
      .858 + 305 equals ......................................................... 1.115d. per gallon
Average collection rate actually paid 1949-50 ............................ 830d.
      (i.e. total cost of £4,306,000 divided by
      1,245,675,000 gallons)

      Difference ................................................................. 285d. per gallon

      —

.285 X 1949-50 gallonage of 1,245,675,000 gallons
represents an economy of £1,483,000 per annum.

The custom has always been to pay milk hauliers a sum per gallon of milk carried rather than a rate per mile, and this practice has the merit of encouraging a contractor to load his vehicle as fully as possible. The arrangements which operate today are that each contractor is paid a rate which fits as closely as possible the job he is called upon to do. The time and mileage that will be involved for the year are ascertained and agreed and the latest available annual gallonage, which will provide his revenue, is used when computing the rate. This system, which makes use of the information gathered during rationalisation, means, in effect, that the cost plus contract is used, but in view of the inevitably varying conditions, there does not seem to be any practical alternative. The Board who hire the transport also operate pilot units at various centres throughout the country with a wide variety of conditions. From these units operating costs are carefully and accurately compiled. This information guides the Board in their negotiations with hauliers.

Each haulier signs a contract with the Board, and there is no doubt that the system has worked well over the past eight years. If any disagreement on rate arises which cannot be settled by direct negotiation, reference is made to the Joint Haulage Committee in accordance with the terms of the contract. This Committee is composed of representatives of both sections of the milk haulage industry and the Board. It has met several times a year since 1942 and has reached unanimous agreement on the rate to apply in each of the cases in dispute referred to it for settlement.

**Summary**

The transport of milk from the farms of England and Wales is in itself a large industry with a turnover of nearly £4,500,000 per annum.

1.    The Milk Flow Chart [omitted] illustrates how the milk produced in the year ending March 31, 1950, flows to its market. It shows also that milk from wholesale producers is controlled as far as its first destination by the Milk Marketing Board.

2.    Detailed knowledge is necessary if the maximum economy in milk haulalge is to be achieved.

3.    The reorganisation of milk haulage along these lines can result in economies both at the receiving dairy and in the total cost of transport.

The estimated saving due to rationalisation and control of milk transport are shown in Table 3, as amounting to well over £1,000,000 per annum and this constitutes a small but nevertheless worthwhile reduction In selling costs…..

## SOME RECENT DEVELOPMENTS IN THE
## MANUFACTURE OF MILK POWDERS
(JSDT Vol. 17 No. 3 July 1964)
R. WAITE - The Hannah Dairy Research Institute, Ayr

"Before I discuss recent developments in the manufacture of milk powder, I would like to spend a few minutes considering the mode of occurrence of the various milk solids, since this may affect the physical and chemical properties of the resultant powder. The fat in freshly drawn milk is present in small globules, ranging in diameter from 2-5um, which exist as an oil-in-water emulsion stabilised by a layer of absorbed protein-phospholipid around each fat globule. The membrane protein differs from the other milk proteins in amino acid composition and is the carrier of protein-bound copper. Alterations in the membrane structure during milk powder manufacture may therefore affect the keeping quality of the powder.

All except 4 to 5 per cent of the nitrogen in milk is present in the form of proteins dispersed in the aqueous phase, with casein accounting for about 80 per cent of the total nitrogen. Casein is relatively heat stable, bulk milk requiring about 20 min at 135°C to become denatured; but very considerable differences in coagulation time exist between the milk from different cows and even between the milk from the four quarters of an individual cow if any quarter has been affected by sub-clinical mastitis. The albumins are the next largest group of proteins, containing 13 to 15 per cent of the total nitrogen and are much more heat sensitive, particularly the blood serum albumin which forms part of this group. The globulins, although present in the milk of healthy cows in only small amounts, 1 to 2 per cent of the total nitrogen, are also heat sensitive and both these latter groups gain in importance because, when denatured, they may lessen the heat stability of the casein.

The lactose in milk is in true solution and one might think it would not play any great part; but under some heating systems it can contribute a certain amount of acidity, and it is quite surprising to find that an increase in acidity of as little as 0.02 pH units (a very slight change), is enough to cause a decrease in the heat stability of the casein. The minerals again would appear to be quite inoffensive, occurring as they do only to the extent of 0.7 to 0.8 per cent in liquid milk; but toward the end of the drying process, when the proteins are particularly sensitive, these salts will now be present in a concentration of some 5 to 6 per cent and could cause local protein denaturation if this phase of the drying was for any reason prolonged. When milk is heated, and in milk powder when the moisture content is high, there occurs an undesirable chemical reaction between lactose and protein, mainly through the amino acid lysine. It is undesirable because it gives rise to a cardboardy flavour and can ultimately cause browning of the product. More importantly, this lysine-lactose compound is not broken down during its passage through the gut of monogastric animals, like ourselves, and consequently some of this essential amino acid is no longer nutritionally available and such milk powder has a lowered biological value. There is considerable danger of this interaction taking place during some methods of 'instantising' milk powder.

Turning now to my main subject it can be said that for many years the manufacture of milk powder has meant the removal of the water either on steam heated metal rollers, or by hot air after the milk has been reduced to droplet size by one of the several methods of spray drying. These two systems, producing material excellently suited for many purposes, still account for almost all the world's production of milk powder, but during the last 10 years or so different methods, or variations of existing methods, have been tried experimentally and some have reached commercial operation.

*Ancillary equipment*
In the spray drying process of milk powder manu-facture the raw milk is usually clarified, homo-genised, preheated and concentrated before drying. Nearer home, the APV plate evaporator would seem to have similar desirable qualities in reducing the time involved in raising the milk total solids to the 40 to 50 per cent required for spray drying. If the recommendation to Ministers of the Milk and Milk Products Technical Advisory Committee that direct heating of milk by steam injection, followed by vacuum removal of the added water be accepted, we can expect to see this principle applied more widely than it is at present in the production of concentrates for milk drying.

*Spray drying*
The spray drying process, in which water is removed from milk droplets by large volumes of heated air could, in theory, result in dried milk solids which had received little or no additional heating. This is, of course, because the latent heat of vaporisation requires the application of a certain amount of heat to a liquid simply to change that liquid to vapour without any change in temperature. As we shall see, the latest spray-drier is designed to come close to this ideal state, but most spray driers operating during the last 40 years have used inlet air temperatures which cause appreciable and unwelcome heating of the almost dried milk particles. This has been necessary in order to obtain the desired rate of powder production. Several of the inventions patented in the last decade, reviewed recently by Bullock (1962), relate to mechanical or pneumatic devices to prevent such nearly dry particles from touching, or from adhering to the walls of the drying chamber. Others endeavour to reduce overheating in the chamber by reducing the inlet air temperature and using supplementary infra-red heat within the chamber, either in a single or double drier (Jehlicka, 1955). More violent atomisation of the milk (Coulter

& Townley, 1959) and rearrangement of the relative positions of the hot air inlets and the liquid spray (Barzelay, 1958), are among other suggestions for greater efficiency of drying. From tirne to time there are reports of spraying methods which produce solid particles of powder. This would have obvious advantages in reducing the fat oxidation in spray-dried whole milk powder and it is a pity that, after the original report, very little is heard of these methods again (Hallquist, Bergstrom & Nordh, 1949; Zhilov, 1958). There is no lack of patented methods dealing with spraying problems but it remains to be seen how, many of these newer ideas will prove to be commercially acceptable.

*Foam drying*
Passing now to different methods of producing dried milk, the main drive has been to produce more readily wettable powders with a better taste on reconstitution and this has been achieved in several ways. Perhaps the simplest, and certainly the one that involves least structural change to existing commercial spray driers, is the method of foam drying developed at the Eastern Utilization Research and Development Division of the United States Department of Agriculture by Hanrahan and his colleagues (Hanrahan, Tarnsma, Fox & Pallansch, 1962). The principle is very simple. It Is to change the normal stream of concentrated milk to the drier into a vigorous foam immediately prior to atomisation in the drying chamber. This is done by injecting nitrogen at 2,000 lb/in$^2$ into the concentrated milk supply line 3 ft from the spray J nozzle, which has an orifice diameter of 0.04 to 0.05 in.

*Puff drying*
Another development, from the same Division, is the puff drying method described by Sinnamon, Aceto, Eskew & Schoppet (1957). Puff drying is defined as 'the formation of a highly expande-sponge-like structure of dried material from a thin film of liquid concentrate under conditions of low temperature and high vacuum.' Not any foam will do; it has to be a very closely knit affair formed by small uniform bubbles so that the amount of unexpanded intercellular material is kept to a minimum. The powder was completely soluble, with undamaged whey proteins and had excellent dispersibility in cold water, a dispersibility that it retained when stored for 12 months at 73'F. Gas packing in nitrogen was advantageous to prevent staling. About this time Kielsmeier in Canada (Bullock, 1960), who had been working along very similar lines, had constructed a pilot plant vacuum belt drier to do a similar job continuously and the Eastern Regional Laboratory people have used this machine to put their process on a continuous basis (Aceto, Sinnamon, Schoppet & Eskew, 1962). This development must ring a bell in the minds of the older members of this Society, because George Scott Ltd., of Leven, Fife, manufactured vacuum band driers nearly 50 years ago. Apparently nobody thought of feeding them gassed milk concentrate, or if they did, did not appreciate the future large potential of retail sales of readily wettable milk powder.

*The BIRS process*
Another new drying process, already in commercial production, has been designed to work at about normal atmospheric temperatures and, because its almost universal application in the food industry, is attracting a great deal of attention (Lang, 1963). This is the BIRS process, developed by Dr Peter Hussmann, which has been variously called 'Cold drying' or 'No heat' process. I am greatly indebted to Messrs. H. J. Heinz Co. Ltd., who in this country are the associates of the BIRS company, for giving me a copy of a lecture delivered by Dr. Hussmann in Germany in 1961, in which he described his process in some detail.

   According to Dr. Hussmann the cost of production is, if anything, slightly less than conventional spray drying....

*Vacuum freeze drying*
My last manufacturing process proper concerns the well-known system of accelerated freeze drying. I propose to say very little about it because there appears to be one major snag so far as milk drying is concerned. You will all be familiar with the method whereby a liquid is quickly frozen, placed in a vacuum chest at very low pressure and sufficient heat supplied to cause

sublimation of the ice.  This system has produced excellent dehydrated meat, fish and vegetables but when milk is frozen there is rupture of the fat globule membrane.  (It has been suggested that the crystallisation of the lactose may be responsible).  This leads to a very high percentage of free fat in the powder (i.e. fat easily extractable by solvents), as much as 35 to 95 per cent compared with 5 to 15 per cent for ordinary spray-dried powders (Nickerson, Coulter & Jenness, 1952).  When the powder is reconstituted this causes oiling-off, fat churning and poor wettability.  Commercial equipment is now available for freeze drying milk which has been freeze concentrated and theoretically the system has much to recom-mend it.

*Butter powder*
I thought it might be of interest to mention a recent development in powder manufacture that is not, strictly speaking, milk powder.  This is the butter powder, containing 80 per cent fat, that the C.S.I.R.O. at Melbourne has produced (Hansen, 1963).  Pasteurised cream, casein and skimmed milk together with a little alkali and emulsifier are mixed, emulsified and pasteurised before being spray-dried in a conventional drier.  The product is cooled and packed as a stable, non-greasy, freeflowing powder which has good keeping properties.  It is intended primarily for the bakery trade, although it could presumably be recombined with reconstituted skimmed milk powder to give a whole-milk.

*'Instant' milk powders*
I will conclude my talk with some reference to the manufacture of powders, mainly skimmed, which possess a greatly enhanced wettability and are known generally as 'instant' powders.

Originally, in the Peebles (1956) method this was a two-stage process and many of the patented methods that followed Peebles kept to the same principle. Latterly the trend has been to the production of the same type of powder in a single process.

Despite the high solubility of all properly made, properly stored spray-dried milk powders when vigorously agitated in water, their wettability and consequent dispersion is normally very poor.  So long as milk powders are used mainly by other food industries where mixing equipment exists to overcome low wettability, this property is not of major importance, but as milk powders gain popularity with the housewife, who is not so equipped, then a high rate of wetting and dispersion is required.

Peebles found that by raising the moisture content of ordinary spray-dried skimmed milk powder to 10 to 20 per cent and at the same time giving the powder some mechanical agitation, the fine particles could be caused to clump together.  During this process the lactose, which exists as an amorphous glass in ordinary powder, crystallised to the monohydrate form.  A second drying process removed the excess moisture, and it is at this stage that great care has to be taken to avoid the formation of the undesirable lactose-lysine compound I mentioned earlier.  These treatments prcduced a granular, free-flowing, extremely wettable powder completely soluble in water. The product was so commercially successful that many other patents have since been filed that claim to achieve the same end by different means. Steam or moist air could obviously replace water and the changes can be rung on the agitation and re-drying processes.  As these variations have recently been well reviewed by Bullock (1962), I will not list them here.

It is neater and no doubt cheaper to substitute one operation for two and this was the next step in 'instantising'.  The changes have followed two main lines.  The first of these is more a cleaning-up of the Peebles process than anything intrinsically different.  Using a conventional spray-drier the powder is removed with a moisture content of about 18 per cent and an agglomerating and re-drying section are tacked on to the original drier.  A more novel method is to use a larger drying chamber (120 ft high) and to manipulate particle size either by the concentration of the milk and the rate of flow (Coulter & Townley, 1959) or to mount subsidiary spraying jets some 3ft below the main jets so that the partially dried particles receive a further coating (Amurdson, 1960).  The latter method affords a very flexible system and in one variation of it concentrated skimmed milk is sprayed from the main jets and butter oil from the secondary jets to produce what is claimed to be an 'instant' whole milk powder.

The 'instantising' of whole milk powders has, so far, had rather less success but in the last year or so at least one really 'instant' whole milk has been produced in America and is now also being manufactured in Australia for retail sale. Messrs. Cow & Gate Ltd. have patented a process for the production of a puff-dried whole milk powder very similar to that which I described earlier...

## A HALF-CENTURY OF DAIRYING: FROM TRADITION TO TECHNOLOGY
## PROFESSOR E. L. CROSSLEY

(JSDT Vol. 29 No. 1 Jan 1976 – Response to Gold Medal Presentation)

"First of all I wish to thank all my friends in the Society for the honour accorded to me today, which will always be a source of pride and a reminder of times good and bad, but always rewarding, with an inner feeling of guilt that really it should have been shared with numerous colleagues who have helped so much in whatever I have tried to do......

For these I have chosen a title which seems to me to embody the changes many of us have seen in milk production, dairy technology and education.

Dairying is a highly traditional activity and every reputable textbook gives references going back at least to Biblical times. This means at once that it is difficult to change anything other than by slow stealth, which is what has happened over the past 50 years. Only recently have heretics seriously considered taking it apart and recombining the constituents in forms suitable for humans rather than calves, but the future will see much more of this.

Apart from condensed and dried milks, all dairying was originally a farming activity, traditionally enshrined in the image of the milkmaid, but with the gradual development of large-scale operations dairy work other than milk production was transferred to dairy factories, although much of this still remains recognisable as primarily a scaling-up of essentially traditional methods. Somehow tradition has got mixed up with chemical engineering. In the years up to 1939 labour was readily available because the industry offered security of employment, and coal was also cheap, so there was little real interest in labour saving, mechanisation or conservation of energy.

Equally drastic changes have occurred in milk production technology, but I will refer only to one of which I have personal recollection. In the 1920s the chemical composition of milk was possibly better than it is now, but hygienically it was grossly contaminated and frequently referred to in the trade as 'liquid dynamite.' Hand milking was almost universal and modern methods of production were virtually unknown. Milk carried a high risk of pathogenic infections and its keep-ing quality was extremely short - indeed a good housewife always boiled it. It was commonly uncooled, or at best only to about 65°F with the available water supply.

A major landmark was the development of clean milk production techniques in a very poor cow-shed at the NIRD by one of our former presi-dents, the late Dr. A. T. R. Mattick. The methods were simple - clean overalls, clean milking, washing of utensils (little more than a bucket and a milking stool), using more elbow grease than is popular now and completed by steam sterilisation. The problems of milking machines were still to come.

The results were dramatic and extremely low levels of bacterial numbers were frequently achieved. Special grades of raw milk were introduced and the production of TT milk began. But most milk remained a possible source of tuberculosis and also of brucella infections, the latter at the time difficult to diagnose and frequently unrecognised by the medical profession...

A curious blind spot remained. It was generally believed that provided the bacterial population of raw milk could be reduced to a low level by hygienic production, a satisfactory pro-duct could be distributed without low temperature refrigeration or heat treatment. Even the NIRD was strongly opposed to pasteurisation – I make no comment except that all is now for-given! Possibly it could have been done whilst milk consumption was small and distribution local. But as population and milk consumption increased and transport was required over ever increasing distances, this became manifestly impossible. Heat treatment developed rapidly, combined much later with efficient refrigeration at the farm. Now we are perhaps in danger of the reverse error of masking poor production methods by refrigeration.

Milk was purchased, by direct contract between dairies and farmers, usually an annual contract made in October. This was a very hectic period in the dairy year, when dairy managers accompanied by a reserve driver toured the farms in their area to secure milk supplies. Although farmers were free to sell their milk to any buyer, in most cases both parties knew that the farmer was going to sign up. But this required long preliminary bargaining, finally completed by appropriate refreshment. This, combined with temperamental motorcars, rendered milk buying a rather hazardous occupation.

*Birth of the Milk Marketing Boards*
The birth of the Milk Marketing Boards was an event of major importance. This ended the jungle of milk buying and created an orderly marketing system, with immediate benefit to milk producers. The policy of, developing the liquid market as a first priority has revolutionised the economic structure of the industry, and indirectly slanted research and development in this direction rather than to manufactured products.

The level of milk consumption was quite low and its high nutritional value largely unknown to the general public; in some city communities it was only used from a tin. The major concentration on increasing milk consumption, school milk schemes, and above all the food crisis during the Second World War have doubled per capita consumption. However, we may always have to counter conflicting pressures between consumer demands for cheap milk as a vital food and a reasonable living for producers and distributors; neither should be reduced by politicians and economists to the status of underpaid social workers.

*Science in the dairy*
Scientific work early in the century was largely chemical, directed mainly to milk composition and the prevention of adulteration, which had been widespread in the 19th century; this formed the basis of legislation which has changed very little to this day. Mental suspicions remained for some time that producers often watered milk, if only they could be caught out! Many years passed before it was realised and accepted that quite genuine milk produced by honest cow could fall below the presumptive legal standard as today we know to our cost. The landmark here was the introduction of the freezing point test and its acceptance by the Courts…..

Of course heat treatment had been practised for some time, for 20 years almost entirely by a flash heater known as the Danish Scalder, which subjected milk to very brutal treatment and produced a mixture of rather doubtfully heated milk and froth. But with the legal sanction and definition of pasteurisation, the holder process became compulsory, for reasons valid at the time. The first small batch heating vats, with crude instrumentation or none at all rapidly grew up into large multicompartment holders. During this period there also occurred one of the major advances of this century - the invention of the plate heat exchanger, by our first Gold Medallist, the late Dr. Seligman, which was to solve many of the technical problems of heating and cooling milk.

The bacteriologists having determined the heating temperature and time required to render milk safe and ensure adequate keeping quality, the engineers had to make sure that every drop of milk received the correct treatment. In their time the large holder plants were very fine machines, although they had their problems and 20 per cent of the working day was devoted to cleaning and sterilising them - by hand of course. At this time too the NIRD produced the original phosphatase test - one of many major contributions of Professor H. D. Kay. Although less sen-sitive than the modern plant control instruments we have now, it provided a laboratory control test of correct heating previously missing, and even convinced sceptical engineers that their valves really could leak. At the same time bot-tling of milk on a large scale was also developing, and eventually post-pasteurisation contamination became the major cause of poor keeping quality.

It was, too, a period of major change in con-structional materials. Wooden butter chums were to survive for a long time, but otherwise most equipment was made of tinned copper or steel. Patches of exposed copper were a common sight, and the NIRD had still to demonstrate the importance of copper as a catalyst of fat oxidation. This was followed by extensive use of aluminium until it was superseded by stainless steel for many purposes.

177

Slowly and quietly work was proceeding which led to the HTST process we know today, although it was against the law. During the war years the potential advantages became obvious, although legend has it that personal intervention was needed by Winston Churchill himself to change the law. The HTST process started a new generation of improved control instruments, on which many modern processes are completely dependent.

The holder dinosaurs faded away, but in retrospect it should be remembered that they established pasteurisation as a successful process and were the main factor in the elimination of milk-borne tuberculosis. Since then we have made doubly sure by eradicating tuberculosis from the national dairy herd, but have still to complete this task for brucellosis. Pasteurised milk had to overcome much early prejudice and pseudo-scientific distortion, but received strong support from the medical profession; now it is regarded as normal milk with which alternatives are compared.

*Microbiological tests*

This was an exciting period in bacteriological laboratories. It is strange to remember now that there was very little formal teaching of bacteriology outside medical schools, where there was naturally little interest in non-pathogenic organisms; dairy courses provided one of the few exceptions....

In a period of rapid and vigorous expansion, with close co-operation between research institutes, the advisory bacteriologists and industrial laboratories, a whole range of new techniques was developed for application to milk and milk products, accompanied by vigorous argument about the interpretation of results. This was the time of innumerable colony counts and presumptive coliform tests, which laid the basis of hygienic control and in some quarters was still strangely regarded as the basic standard of comparison. Eventually it was found essential to introduce simpler, cheaper and quicker tests which could be applied more frequently; serious wastage of milk during the Second World War led to regular testing of all farm supplies by dye reduction tests and ultimately to payment for hygienic quality.

However, we may be in need of a new look at our microbiological methods in the changed context of the modern industry. The tests applied to raw milk supplies are a relic of the past and have only limited relevance to milk which will be processed, as nearly all milk has been for many years. In general, we need methods slanted away from counting inoffensive bacteria and towards those that really matter. The coliform group has probably been subjected to more detailed scrutiny than any other, and remains a valuable indicator of plant hygiene, although not an automatic danger to public health. But in modern technological processes the role of public enemy number one is probably passing to the spore-forming species, which merit much closer atten-tion, including their hiding places at the farm.

*Chemistry and nutritional value of milk*

In contrast, after a fallow period discoveries in the fields of chemistry and nutrition have been racing ahead. Perhaps the major factor in this renaissance has been the development of chromatography. The secrets of proteins and amino acids - in casein more complicated than we imagined - are being unravelled together with new knowledge of the minor but important constituents, including flavours.

In the nutritional field, most of our present knowledge has been accumulated in quite recent times, including the discovery of nearly all the known vitamins. The high nutritional value of milk is even more widely recognised in the popular mind and increasingly the possible nutritional effects of a new process are taken into account; only the economists remain impervious to nutritional considerations.

With increasing complexity the need for co-operation and discussion between the varied interests linked ultimately to milk becomes more pressing. This need became acute with the difficulties and stresses of war, and a major step for the future taken during those dark days, against all the odds, was the foundation of this Society.

## Condensed milk

Sweetened condensed milk now finds its way to the confectionery industry rather than the domestic kitchen, but evaporated milk has retained a market of its own.

## Evaporated milk

Evaporated milk was the first industrial attempt to produce a near-sterile dairy product. With this product we learned a great deal about destruction of heat-resistant spores and the control of can leakages, incidentally well in advance of the canning industry. Evaporated milk manufacture involves considerable skill in maintaining the physical stability of the canned product during storage.

## Dried milk

Proceeding a stage further, we arrive at dried milk. The roller process has been used since early in the century, mainly for production of infant foods or drying the skim milk resulting from cream or butter manufacture. The introduction of spray drying has been one of the most important technological developments. Although spray drying had been used on a small scale, the major development of large plants in this country occurred after 1930. In spite of high capita and thermal costs, spray drying expanded rapidly as a large modern industry and dried milk is now an important international product and a major source of first class protein for developing countries. Moreover, it may be a starting point for some new dairy foods.

Nearly all new processes involve fresh microbiological problems, often because they provide suitable living conditions for species which never gave trouble before. This happened with spray drying due to the particular operations which precede drying. The combination of heat treatment, continuous evaporation at relatively low temperatures for many hours, and holding the warm concentrate for more than one hour before drying all built up a specialised thermoduric flora. So although the process worked quite nicely in the technical sense, many early spray dried powders contained astronomical numbers of bacteria. By introducing modern methods of plant cleaning, some changes in plant design and methods of operation, more instrumentation and strict laboratory control, the high bacteriological standard we know today was achieved. But there were chemical problems too.

As dried milk was immune to microbiological attack, it could be stored for long periods and transported in small weight and bulk over wide distances. This provided ample time for chemical oxidation of the fat and development of the well known tallowy flavours…..

Basic research on the chemistry of fat oxidation was intensified at the Hannah Institute and in Cambridge. The practical solution adopted was removal of the oxygen and replacing it with nitrogen by gas packing in sealed metal containers. The process proved highly successful and achieved keeping qualities up to seven years; during the war years it was operated on a very large scale and used to rescue starving populations when the war was over. An interesting sidelight was a need for perfect soldering of containers, as the smallest pinhole leak could be disastrous. This was achieved by teams of girls, who developed amazing skill and dexterity. In fact their failure rate of can leaks was about a quarter of that achieved by professional tin-smiths; of course the trades unions were never told!

Shortly afterwards, a keeping quality long enough for normal commercial purposes was achieved by high temperature preheating to produce antioxidant sulphydryl groups. In addition, the different problem of stale flavours, discolouration and loss of solubility was traced to a Maillard reaction triggered off by high moisture contents, particularly in dried separated milk. This in turn led to the use of more satisfactory packaging and the adoption of alternatives to expensive metal containers….

In theory dried milk is the logical form of milk distribution, and may be the ultimate practical solution. It would reduce transport costs and storage space, and could be stored for long periods without refrigeration in shops or the domestic larder. This would need a full cream powder, equal in flavour and nutritional value to pasteurised milk, having a long storage life and instantly soluble in cold water.

So for the present this role seems likely to be filled by UHT milk in conjunction with aseptic packaging, particularly for milk distribution in hot climates. The process is likely to extend to other dairy and food products. We are now entering a new era of development in which micro-biological problems disappear almost entirely; instead they are replaced by an entirely new set of chemical and physical problems which we have not experienced before.

*Progress in dairy hygiene*

Another striking change has also happened. From sheer necessity, strict hygiene has always been a major requirement - indeed almost the dairy religion. After use all equipment had to be dismantled completely, thoroughly cleaned by hand and then sterilised by steam, or at least hot water..... So we should be grateful for the chemical studies of milk plant deposits, the formulation of detergents to remove them, and particularly the replacement of steam by chemical sterilants for many jobs in the dairy and on the farm. Young dairy workers today do not realise their good fortune, with time and labour saving in-place cleaning, all built into the plant and automatically controlled.

*Cream*

Cream in pre-war days was a luxury product for special seasons and occasions; for all in the cream trade the strawberry week in June and Christmas week were a nightmare. The use of preservatives was not prohibited until 1928, amid forebodings of disaster. Sterilised products in cans and glass jars were quickly produced - one of my first assignments in industry. Pasteurisation, based essentially on methods used for milk, developed rapidly with modification of the design and opera-tion of plant, and keeping quality risks are now associated mainly with trade conditions of distribution which render the slightest post-pasteurisation contamination a potential disaster.

*Butter*

Internationally speaking, butter is an extremely important product - economically much too im-portant. But in this country it is given a low priority when milk is short and we depend on imports for the bulk of our butter requirements.

I will only venture two comments. Nutritionally speaking butter is not the most important fraction of milk, nor is buttermaking the most laudable objective of dairying.

Second, in comparison with its competitor margarine, it starts with a flavour advantage and a certain social cachet. But it has the major weakness of uncertain spreadability, which the margarine industry has exploited ruthlessly. Why should we not imitate the technical methods of the margarine industry, by fractionating butter fat, modifying the individual glycerides and re-combining them to secure the desired properties? An urgent research programme is needed, with a changed attitude of mind and pressure to clear away any outmoded legislation. This will be, more productive than trying to protect butter by dubious legislation. Of course the first hesitant steps have been taken, but not in this country.

*Cheese*

And now cheesemaking, the acme of tradition, combining a high degree of individual skill with almost frightening scientific complexity. This is no place for the research worker who likes to see quick results.

It is true that the crude basic chemical changes are now known and the major microbiological progressions and enzyme actions have been established, but many details remain to be elucidated. In fact the present legal standards give little indication of real quality. Further progress seems most likely to result from the technique of aseptic cheesemaking developed at the NIRD....

'In recent years cheese manufacture has undergone spectacular changes, with the successful grafting of large-scale mechanisation on to the traditional art....'

Cheese whey presents the paradox of a product, of high nutritional value which is financially almost most worthless, even a burden. Of course the utilisation problems are very difficult and have been discussed for 40 years. Some whey products are now used to feed humans instead of bacteria in effluent plants, but a great deal is still used very inefficiently, in animal feeding. The potentially exciting feature is that it has provided an opening for the new techniques of ultra filtration and reverse osmosis, which promise to recover the valuable protein. Full realisation of the theoretical possibilities awaits further developments in membrane technology; once this happens there will be new applications in many fields of dairy and food technology. With whey there may be a problem with the residual lactose. Here we ought to think seriously of converting it to biomass and obtaining still more first-class, protein. The expertise to undertake the research programme required is probably available in the dairy industry; surely it should not be beyond.

The dairy industry has reached a stage when it will no longer be safe to rely on traditional products, as it has done throughout this century, including the recent boom in yoghurt - a revival of an ancient product. Fortunately this attitude is changing and there is now a flurry of welcome activity to develop new dairy products and to seek wider co-operation with the food industry.

## DAIRY EDUCATION OVER FIFTY YEARS

Every individual is his own educational expert, but whatever one's opinion of our educational system may be, at least the quiet dairy revolution of the past 50 years could not have happened without it. The dairy industry, in the widest sense, has a long tradition of support for education in the UK, and in some respects has provided an example for the rest of the food industry to follow. Naturally it began as a branch of agricultural education, with the simple objectives of teaching the particular agricultural aspects of milk production combined with the practical arts of making butter and cheese. The first real school for specialised dairy teaching was the British Dairy Institute founded 90 years ago by the BDFA (now the RABW) at Aylesbury. It remained there only a few years before moving to Reading to join a young college which was to grow up into the University of Reading. In suc-ceeding years the other well-known dairy schools at agricultural colleges were established in other parts of the kingdom. All these courses remained relatively unchanged for 25 years; in fact they were closely attuned to both farm and dairy practices of the period.

They included a great deal of practical work, particularly traditional buttermaking and cheese-making to high standards, and above all an emphasis on superlative hygiene achieved by scrubbing and polishing imposed under almost military discipline. Besides the equipment, this included personal cleanliness, special uniforms long before protective clothing became general, and everything else from the ceiling to the drains.

*Diploma courses*
Many early courses were of short duration, but the need for longer courses was visualised from the start. A major historic step was the introduction in 1896 of the National Diploma in Dairying as a joint effort by the BDFA, the Royal Agricultural Society, and the Royal Highland and Agricultural Society, incidentally paid for entirely with their own money. As the sciences of dairy chemistry and microbiology developed, these were incorporated into the diploma courses. There is no need now to stress the important contribution of the NDD, originally the only professional qualification in dairying.....

At Reading some bits of research work were started, as it happened well before their time, but at least this led to designation in 1912 as the official national centre of dairy research. This activity grew rapidly and necessitated a move to the more spacious site at Shinfield in 1923.

But as new technological, large scale processes developed and the sciences to go with them, the emphasis changed. Educationally the whole field could not be covered properly in the time available, and financially the cost of modern equipment (even on a pilot scale) with the buildings to house it, presented daunting obstacles. All educational progress was halted during the Second World War, but in the post-war era there have been remarkable changes, not least in the mental attitudes of Whitehall. The war had revealed serious inadequacies in the educational system, and for the first time some real state finance became available. From 1949 onwards the University of Reading and other colleges were able gradually to re-equip their dairy departments and laboratories. Moreover student grants were introduced and potential students were no longer de-barred from courses by low parental income. Before the war there were very few grants available and even then their value was far from generous - £10 a term plus fees was a common figure.

The NDD course survived until quite recently; inevitably its passing has aroused sentimental regret, but we must accept that it has served its purpose in this country. Personally I believe that a course of this type, which combines milk production with simpler technology, is still the real need in many developing countries. It has been succeeded by our new national diplomas, which are orientated towards science and technology and away from agriculture. First results are very encouraging from the point of view of milk processing and manufacture. A serious educational gap remains in milk production technology, where almost the only surviving course is the one-year National Certificate in Dairying.

*Degree courses*

At Reading the need for more advanced teaching of dairy science and technology had been realised by the College for a long time, and as soon as it was able to do so, with the grant of its Charter in 1926, the first degree course in dairying in the UK was inaugurated. This course, and later a second course at Nottingham, ran very successfully up to 1965, when the greatly increased costs of advanced teaching created economic difficulties. These troubles were relieved, temporarily as it has turned out, by merging dairy courses within newly established general courses in food science and technology which had long been overdue. The use of expensive laboratory and pilot plant facilities was spread over a larger number of students, to the advantage at least of the taxpayer. Much of the basic teaching in science and technology is common to all the students, and a wider knowledge of food science is an advantage to dairy students, increasingly so in present conditions.

The major difficulties arise with those unique aspects which are not common, but are specific to the dairy industry and moreover represent important sectors of it both technically and economically. It must not be forgotten that milk is still the most important agricultural product in this country. Time alone will tell whether adequate instruction in these specialised but essential fields can be achieved within a general course. At Reading a post-graduate course in dairy science is provided to make sure that the necessary teaching remains available, and similar solutions have been adopted in New Zealand and other countries with important dairy industries. In the USA, which was the first country to merge dairy and food science courses, experience has ap-parently not been altogether satisfactory and some universities are reverting to dairy science degree courses. Our new diploma courses also operate within a general context of food technology, but it was soon found necessary to provide a dairy technology option.

Perhaps the most difficult aspect of dairy teaching is the problem of integration. The basic sciences can be taught quite well as the normal academic disciplines of pure science courses, but understandably these rather tend to become isolated in separate compartments of the student mind. The teaching problem, once the basic science foundation has been laid, is to weld the whole together and create some awareness of this. There will always be a need, in research and industry, for specialists in particular sciences. Equally there is a need for individuals with a more general outlook and a practical turn of mind, mentally able to cross the various scientific boundaries but still able to cope with new scientific and technical developments as they come along.

*Technician courses*

There remains the equally important and quite essential group of people who do a great deal of the work - the plant operatives and other floor workers, and junior laboratory staffs. Apart from various short courses, usually agriculturally biased, not a great deal was done for them until the early 1930s, when the City and Guilds of London Institute established part-time courses in milk processing and distribution subjects. These were purely vocational courses at technical colleges, open to all without any entry requirements and at purely nominal fees. Originally they were evening classes and could extend over four years for the full course. Their value was soon recognised by the industry and they remain an essential part of the educational system.

These courses have their own problems. Compared with some other industries the number of students is relatively small, and dairy technicians are thinly spread over the country rather than concentrated conveniently in local areas. In the eyes of officialdom this is a deadly sin. Fortunately very useful support has been forthcoming from the new Industrial Training Board.

When this Society was founded, one of its first acts was to set up an Education and Research Committee. In many respects our members are the eyes and ears of the dairy world, and this Committee has in various ways helped to establish courses, find specialised teachers and recruit students in local areas.

But a final step was needed to integrate the whole system, and particularly to involve the industry more directly. This was achieved 1958, when the industry itself set up the Training and Education organisation now known as DITEC. This Committee includes members from all sectors of the dairy trade, the MMB, Minis of Agriculture and the Industrial Training Board and is able to speak for the whole industry on educational matters. It has a full-time secretariat and its activities cover all levels of education and training, including establishment of courses, provision of finance for recruitment of students by advertising and careers films, student grants clearing house for placing students in industry during their training and periodic surveys of personnel needs of industry.

## CONCLUSIONS

At last we have an organised, coherent educa-tional system, covering all academic levels and all dairying careers, which marks the end of the training era commonly known as 'sit by Nellie.'

It would, however, be foolish to avoid two pessimistic features of the present scene. The serious economic situation of the country has created a crisis in the educational world, with panic decisions being taken on purely financial grounds with little reference to the educational consequences. The other disturbing thought is that in contrast to the past, the majority of new technological developments in this country in the last decade have been introduced from abroad. It may well be time for a completely new look at our whole national system of research and development."

CHANGING ATTITUDES<br>R. J. MACWALTER<br>(JSDT Vol. 29 No. 2 April 1976)

"Changing attitudes in the world at large on important issues have effects on the direction of research and development in the dairy industry.

One of the delights of retirement is that one can afford to be difficult and 'ornery'. It is about the only thing one can afford, so one makes the most of it. Thus it was that when your Conference Secretary asked me to speak on the subject of 'New uses for milk ingredients', I said 'No'. I had given a talk on this subject in 1948 and nothing had happened since to cause me to change what I said then. By the time I had replied it was too late to ask anyone else, so my suggestion of a talk on 'changing attitudes' was accepted with barely concealed alarm. But I must tell you, ladies and gentlemen, that this Conference is really already over, and if you feel inclined to fade away now, you will have missed nothing - I shall go maundering on regardless.

The keynote of the Conference has been 'the changing scene' and all the papers have responded to that note. I think though that we should recognise, that to the technologist change means 'progressive development' because our method of progress almost invariably, is to make logical extensions of existing knowledge rather than to introduce revolutionary concepts. Neither are we concerned overmuch with the social consequences of our work beyond its application to our industry.

On the other hand, in the wider world of industry, change is not always brought about as the result of logical technical development within industry, but quite often it is the result of the impact of people on each other, with all the emotional overtones which move people - like envy and ambition. One of the most potent factors is that of imitation, or more politely, 'the industrial climate'. This produces waves of innovation which seem to run through the whole of industry in the western world. For instance, do you remember computers? Not so long ago computer was a word which could strike terror to the uninitiated in the same way as the word 'Kaiser' was used by mothers to frighten their children into submission. Yet now it has little more power to affect us than (say) the word 'chastity'. Think of the importance of the principle of centralisation 30 years ago, introduced to avoid duplication of effort and to assist in corporate thinking. Contrast that with the modern view of the value of decentralisation to arrest the dead hand of Head Office and to stimulate independent and responsible thought at the perimeter. Yet changes of this kind have a profound effect on the direction of the work we undertake as technologists. Again there can be sudden changes of direction due to the pressure of public opinion. The recent upsurge of feeling against pollution is an example of this. The growing dangers of pollution have been well recognised by technologists for many years, but I suspect that the force of the new movement will be far more effective than all our reports.

We have had in this Conference, examples of two important changes in attitude of the kind I have in mind. One, the impact of the Common Market, an economic and political concept not arising from our own dairy industry, but clearly likely to stimulate considerable change in our technical operations in the future. Another is the paper of Professor Yudkin on the changes in the nutritional significance of milk.

Almost all businesses began with a single-minded man with ambition, a skill to exploit and confidence in his ability. The successful businesses absorbed others and became larger. Those who founded the business were at this time in control, and applied their skills in the product or service which formed the company's *raison d'etre*. Such a company directed by specialists in its business, carrying a large proportion of the wealth of the founders and personally supervised, was a well-knit organisation yet could draw all the economic benefits of size. The question arises 'Is there a limit to the advantages of size?' The first problems to appear are those concerned with communications and the gulf separating the governors and the governed. So the personnel department emerges. This is followed by others like market research and management services as, well as the specialist groups like architects and property men. Now we are into a structure which is so complex that it needs to be controlled by men trained in the techniques of managing businesses, in contrast to the men who controlled by virtue of their knowledge of the business and their stake in it. We have now entered a new phase of company development.

As men mature - all right - as men become older, they raise their eyes from the immediate foreground and those who have realised their ambitions seek less personal rewards. Thus we have men like Carnegie, Nuffield and Wolfson who directed their wealth and organising skills to the benefit of the community at large. Our industry, however, with its social importance, presented its leaders with the opportunity on their doorsteps to promote projects of importance to the industry generally and to the public it serves and there are many examples of work being undertaken without much expectation of direct personal return.

It seems to me that the situation becomes rather different in a large modern enterprise such as I have described which is controlled by those skilled in management techniques. Such leaders with their knowledge learned in a variety of different types of company, cannot have the experience and tradition of those whose whole lives have been spent in one industry. Indeed their measure of success must be the balance sheet for they have no other. They cannot escape from the need to show better and better financial results each year, and if they do not, they have failed in their own eyes and in the eyes of all like-minded people. This attitude has important effects on the reactions of staff at the lower level, for when they see that the acquirement of management skills is regarded more highly than skills in the production side of the business, these are the goals they set themselves. So the best men tend to be drawn from the other activities on which the company's progress depends, into this field. There they may learn the esoteric language and techniques of man-management which are common to all companies and serve to isolate their disciples from the rest of mankind. Having learned them they are now ready to move to other companies and other industries….

However it is the short-sighted view of the supreme importance of next year's profitability which imposes the greatest restriction on the vital work of development technologists, and we see the effects of this in many undertakings.

Let us now look at our own industry where the trends I have referred to have not been entirely absent and are likely to grow. If short term profitability is the touchstone by which technical achievement is to be judged, how, for example, would we rate the efforts made by many dairy companies over the years to improve hygienic standards of milk, to devise methods for monitoring antibiotic residues, to find better ways of purifying dairy effluent? These can scarcely be justified in money terms except very broadly and over a long period, condi-tions which do not impress accountants. I fear that this type of work would not be approved under the conditions I have postulated. More serious still, our industry has devoted a great deal of effort to improving the nutritional safety of its products and in monitoring production. Much of this has been embodied in legislation and over the years, the regulatory authorities have learned to trust responsible manufacturers to bring all their technical expertise to bear on matters of this kind. In the up-to-date view of business efficiency these matters do not qualify for investigation for they do not contribute to profitability. More concisely the view is 'it's OK if it's not illegal.' It may be said that if there are problems of this kind to be investigated, let the legislators discover them and impose what safeguards they wish. I suggest that if this view were to prevail, we should end up with a set of cumbersome and unworkable regulations, and a serious loss of confidence between the two protagonists involved the manufacturer and the legislator.

We would expect a more conservative attitude towards imaginative research and development, for the more far-sighted work is the least susceptible of evaluation in terms of short-term profitability. Yet of all times, this is the moment when every original idea must be mobilised. It will not be enough to leave it to Government organised research bodies who themselves are suffering restrictions.

You may have by now received the impression that I have misgivings about the effect of sheer size on progress. These misgivings extend also to size in relation to production units themselves, as distinct from the size of the organisation which owns them. I appreciate that there are advantages in unit cost to be gained by a large throughput, but when troubles occur the consequences are correspondingly large…. I was very interested to see that Sir James Barker in his talk to the Society last October, thought that there was a case for smaller bottling plants, but not for creameries. Well, he is moving in the right direction.

If I am told that it is fanciful to suggest a reversal of an economic trend which has served for many years, my answer is that the last few years have demonstrated that none of us understands the principles (if any) of economic theory. I even suggest that the worship of continuous annual growth may well be a cause of much of our trouble. 'Faster, faster' said the Red Queen in 'Alice' - create a market and then satisfy it - there's no end to progress. Isn't there? I find that commercial philosophy (if I may use such a word) is very puzzling, where clever notions become basic principles overnight. I suspect that we would, benefit from a little more scepticism and less solemn worship of what happens in the United States when the quality of life seems not to be markedly better than elsewhere. If Gilbert and Sullivan were with us still, they would find excellent material for one o their gentle satires. Of course, Gilbert would be about 130 years old, so we could perhaps forgive him if he wrote a libretto something like this.

*I'm an organisational man,*
*Fast growth is the nub of my plan,*
*Not technological,*
*Very pragmatical,*
*Give 'em all ulcers young man.*

*I'm a dab at the marketing game,*
*At finance all say I'm the same,*
*Growth logarithmical,*
*By methods statistical,*
*Pays better than public acclaim.*

*When expansion's beginning to drop,*
*Capital spending I'll dock,*
*Utmost austerity,*
*To hell with posterity,*
*R and D's first for the chop.*

Having destroyed the whole fabric of modern commercial practice, I am now free to turn my attention to another changing attitude - that of the dairy worker. Our industry, during my time in it, has been the embodiment of the Victorian belief in the sanctity of hard work. I suppose those of us who could see the daylight at the end of the tunnel were able to regard this maxim with tolerable equanimity in the hope that we might, in the fullness of time, and from a comfortable office, make sure that this virtue continued to be practised. But the worker himself, who by reason of lack of opportunity or education, or most probably, lack of that divine discontent which spurs men on, tended to take a different view. Particularly if he could see elsewhere far more comfortable and easier jobs available, without the need for week-end working. Since the last war the industry has found the need to try to reduce the amount of hard graft in dairy work, and you are all familiar with what has been done, but I want to discuss the price which has to be paid for these efforts, which have, without question, improved the lot of the operator and made possible the continued recruitment of staff.

In the field of in-place cleaning of dairy equipment, it is my experience that results are, on average, better than before, because the bad results achieved by bored and unskilful workers have been eliminated. But so also have we eliminated the very good results produced by dedicated and conscientious workers. Similarly where we have improved conditions of butter and cheesemaking by continuous and mechanised techniques, very poor quality has given place to uniform average quality. Of course, if you want to make butter and cheese of prize quality you still make it in small units of plant lovingly maintained by craftsmen.

Similar observations apply in other areas than the production floor. For example, the mechanisation of many laboratory tests is justified if the routine steps in the work can be eliminated or speeded up thereby reducing costs and making possible perhaps a more comprehensive control system. Some of these new procedures involve the replacement of analysis based on the measurement of a characteristic of the constituent in question, by measurement of property not specific to that constituent. An example of this is the measurement of moisture content. Classically it is done by finding the weight of water evaporated by heating a sample. Modern methods measure the electrical properties of the sample. This is justified if check tests are frequently made to ensure that instrument drift or extraneous causes have not invalidated the relationship of the property measured with its moisture content. I am concerned that when procedures of this type are fully established, there may not be available the skilled analytical staff to recognise the hazards when they occur. The overall benefit may justify the risk, but does not eliminate it. So I assert that we may sacrifice excellence for acceptable average quality, justified by the belief that thereby a greater number of consumers may be served at tolerable cost than would otherwise be possible. This view is accepted over a wide range of products beyond the food and dairy field. The question arises 'How far can we allow this sacrifice to go?' It will continue because pressures will grow. It is likely for instance that the need to introduce a five-day week all through the industry will become insistent. Then we shall have a lengthening interval between the cow and the consumer. This will call for improvement in the microbiological quality of milk and dairy products, or some more effective means than at present to arrest microbial action. Or, a general falling off of quality at the end of the food chain?

Meantime, we have an in-built mechanism for accepting the standards which prevail, since each succeeding generation bases its criteria of quality on its own immediate past. The old, who look back to a more remote past see greater changes, and are more apt to criticise. Those of us in that age group must be careful not to fall into the error of assuming that food quality is a static thing which can change only for the worse. This is the view of the antique collector who has no praise for work less than 100 years old. On the other hand one has only to look at some modern architecture to demonstrate that modern techniques can be aesthetically unacceptable. So I say that we must be continually on our guard, so as to delay any acceleration of loss of excellence which seems to be the common consequence of unrestrained technological advance. There are rumblings in the air from the consumer already; and this brings me conveniently to my third example of changing attitudes - the attitude of the consumer and his relationship with the supplier.

I visited recently a railway station restaurant where the items on the menu were numbered so that it was possible to communicate with the waitresses who had no English beyond the phrase 'Service not included', a piece of unnecessary information for it was quite clear that no-one was getting any service. I ordered a roll, butter, cheese, a glass of milk and coffee. I doubt if Professor Yudkin would quarrel with that? What did I get - eventually? A tiny rectangle of aluminium foil, and after the labour of opening it there was a mass of oxidised butter oil which I had to scrape off the wrapper. The cheese was a similar but slightly larger rectangle of green curd swimming in more butter oil. The roll was a sphere of cotton wool encased in what seemed another wrapper of cardboard (the crust). The milk was in a vandal proof carton and was curiously flavoured to confer upon it eternal life. The coffee was a cup of hot water to which had been added granules of a brown powder and topped up with a white fluid with which no cow could have accepted even distant relationship. Why was I submitted to this travesty of a meal - doubtless nutritionally adequate but aesthetically revolting? For my part, if the whole lot had been mixed together into a pap and put into a tube, I should have been saved the frustration of sitting down in hope. I could have caught an earlier train with ample time to buy some indigestion tablets, the waitress could have been declared redundant and spent her time and dole in the fresh air. I could even imagine the label on the tube expressed in that brash travesty of our glorious language peculiar to prepared foods.

'Eat the new quick exciting way, the safe way without unhygienic knives and forks - just like the astronauts do. Save hours a day'.

Up and down the motorways of this land, in its cafés, in its aircraft, there are thousands of <u>People Petulantly Picking at Plastic Packages of Pitiful Provender,</u> and cynically reading the description on the packs telling them of the oven-fresh, crunchy, creamy delights within. One of the most pleasant activities in life not yet classified as a major sin, is thus debased into an unattractive fuelling operation. I am told that the consumer prefers these developments. I think it may well be that 'the consumer' that disembodied statistic does react in this way; but real people, the ones you and I know, do not.

I would advise marketing executives, were I so bold, to pay some attention to the tastes of their own families, and less to experts in merchandising. I am sure their own kith and kin are but little different from the majority of their customers. I doubt whether the erudite terms like 'merchandise' and 'consumer trends' do more than isolate the producers from the eaters. Having been so bold as to raise one point, I may as well raise another, and suggest that producers should examine their products not so much as it leaves the production line as when it arrives on the plate or in the larder.

I said in another context that succeeding generations accept standards which would not have been tolerated two generations before, but this is true only up to a point. History is littered with examples of great surges of feeling which change the logical direction of events when ordinary people begin to ask where they are going and do not like the answer.

I will try to draw together the threads of this rambling talk by making the following points.

The influence of the policies of large organisations and of technological developments, on the lives of people are very powerful. But so is the power of people to resist, if the changes that are brought about are seen to produce unwelcome effects on their lives. This resistance may be slow to develop. It seems to me that at this time when changes can take place very rapidly we must be more alert than ever before, to their long term effects… We cannot leave our business and our lives in the hands of arrogant and insensitive people with narrow views of their obligations.

I realise that I shall not find many in this audience to agree with what I have said, because our views on matters of this kind are conditioned by our particular experience, for this is the datum line from which comparison begins. Our experience differs with our age group and is distorted by our memory. For instance, I have a picture of a period of perpetual sunshine, when I lunched on new crusty rolls, fresh butter and mature Cheddar cheese; at work my Chairman consulted me on every step and always took my advice; when everyone called me 'Sir.' 'Ichabod, Ichabod' I say with the poet 'the glory is departed."

# RECENT DEVELOPMENTS IN MILK PRODUCTION IN IRELAND
## P. A. GLEESON

Agricultural Institute, Moorepark, Fermoy, Co. Cork, Ireland

(JSDT Vol. 35 No. 1 Jan 1982)

"The past decade and Ireland's accession to the EEC in 1973 has brought about a marked increase in investment and output in Irish agriculture.

*Progress in Irish agriculture*

The pattern of growth in the period 1953-79 and the volume of gross and net agricultural output and farm materials are shown in the Figure (Kearney, 1980a). The average annual growth rate over this period was 2.6 per cent, but in the sub--periods 1953-59, 1960-69, and 1970-79 the growth rates were 0.79, 2.25, and 3.17 per cent, respectively (Table 1). The corresponding growth in net outputs in the three sub-periods concerned were 0.19, 0.89, and 2.34 per cent, respectively. As gross agricultural output increased it was accompanied at all times by a faster expansion in the volume of farm materials, for example, feeds, fertiliser, and seeds. The annual growth rates in milk output were 2.37, 7.43, and 6.10 per cent for the three periods and arise from a combination of both increased cow numbers and increased milk yield per cow. This is illustrated in Table 2 which shows the changes in total cow numbers, creamery and beef cows from 1953 to 1980 (Kearney, 1980a). In addition, the milk delivered for manufacturing purposes and deliveries per cow are also given. In this paper the data presented relate to manufacturing milk only, because in Ireland the liquid milk trade is generally separate from the milk manufacturing industry.

**TABLE 1**

Annual growth rates in gross output, farm materials, net output, and manufacturing milk (%)

|  | 1953 - 59 | 1960 - 69 | 1970 - 79 | Overall |
|---|---|---|---|---|
| Gross output | 0.79 | 2.25 | 3.17 | 2.61 |
| Farm materials | 3.38 | 6.89 | 5.01 | 5.71 |
| Net output | 0.19 | 0.89 | 2.34 | 1.68 |
| Manufacturing milk | 2.37 | 7.43 | 6.10 | 5.27 |

Volume indices 1975 — 100

188

TABLE 2
Trend in cattle numbers and milk output, 1953-80 [Abridged]

| Year | Total cows ('000) | Beef cows ('000) | Cattle output ('000) | Creamery cows ('000) | Milk deliveries (M gal) | Delivery/cow (gal) |
|---|---|---|---|---|---|---|
| 1953 | 1,174 | 120 | 955 | 754 | 232 | 308 |
| 1960 | 1,284 | 201 | 1,064 | 798 | 280 | 351 |
| 1970 | 1,713 | 387 | 1,449 | 1,084 | 510 | 470 |
| 1980 | 2,035 | 448 | 1,679 | 1,377 | 842 | 611 |

*Factors in influencing milk supplies*

Cow numbers, milk deliveries to creameries, and delivery per cow are shown in Table 2. Milk deliveries increased from 510 M gal in 1970 to 864 M gal in 1979. However, milk delivery declined to 842 M gal in 1980 and the present prediction for 1981 is a decline of approximately, 5 per cent on the 1980 figure. The rapid expansion in milk deliveries can be partitioned between increased delivery per cow (60 per cent) and extra cows (40 per cent).

The increased use of concentrates, especially in the period after 1975, might be a reflection of the earlier calving of dairy herds. There was evidence at the end of 1978 of a growing reliance on the use of concentrates for milk production even though there was continuing evidence of under-utilisation of grassland either as grazing or as quality winter feed.

*Income and related factors*

In the 1950s farm incomes in Ireland, expressed in terms of income for a family labour unit, had been increasing at about 5 per cent while inflation had been running at around 3 per cent. During the 1960s incomes rose annually by about 8 per cent while the average yearly rate of inflation was 5 per cent. During the 1970s the per capita income increased by an average of 24 per cent from 1970 to 1978, and the inflation rate during this period averaged 13 per cent, giving a real per capita increase in income from farming of approximately 8 per cent per annum (Kearney, 1980a). In the subsequent two years however, incomes have fallen by 19 per cent in nominal terms while the real per capita incomes are estimated to have declined by about one-third.

Both output and input prices have risen dramatically during the 1970s. In the period 1975-78 farm prices outpaced the increase in input costs due, in the first instance, to the dramatic rise in cattle prices and, in the second, to green pound revaluations, transitional adjustments, and relatively large annual increases in Community prices. It was during these years that a spectacular increase occurred in farm incomes, a peak occurring in 1978 as indicated by the incomes of the self employed (Table 5). All of these indices have been in decline since that year. Over the past two years agricultural prices have increased by only 3 per cent, input costs have increased by 29 per cent, volume of gross output has declined by 2 per cent, while real per capita incomes have dropped by one-third. Currently, the output/input price relationship is less favourable than at any time in the 1970s except 1974.

From 1975 to 1979 output price index increased each year, and most of the gross output increase from £859 M to £1,677 M can be attributed to produce price rises rather than increased output. The index volume of gross output increased from 100 to 110 during that period.

*Structure of the Irish dairy industry*

Agriculture is one of Ireland's most basic industries and accounts for approximately one-fifth of total output and two fifths of all exports. Although milk accounts for less than one third of farm output it has a major influence on the beef industry. Irish dairy exports increased from just under IR£10 M in 1960 to over IR£600 M at the end of the 1970s. This latter figure represents 16 per cent of total exports and makes a significant contribution to balance of payments due to the low import content.

According to *EEC Dairy Facts and Figures 1980* (MMB) agriculture is a major source of employment in Ireland. In comparison with the EEC and, in particular, the UK, the percentage employed in agriculture is high even though it has dropped dramatically in the last 20 years. Agriculture contributes over 16 per cent: of the gross domestic product based on 1978 estimates, considerably higher than the 2.6 per cent reported for the UK. While there has been a rapid decline in those employed in agriculture in general from 1960 to 1979, there is a similar trend in the number of creamery milk suppliers (An Bord Bainne). In 1960 there were 97,533 milk suppliers, falling to 89,875 in 1970, and to 69,500 in 1980; this represented a decrease of 29 per cent over a 20-year period. These changes in recent years have resulted in heavy expenditure in both improved buildings for labour efficiency and increased machinery purchases to replace hired labour.

**TABLE 6**

Agricultural employment and contribution of agriculture to gross domestic product

|  | *Employment in agriculture 1978* | | Contribution of agriculture to GDP 1978 |
|---|---|---|---|
|  | '000 | % of total labour force | % |
| Ireland | 223 | 21.5 | 16 |
| EEC | 8,229 | 7.6 | – |
| United Kingdom | 737 | 2.8 | 2.6 |

**TABLE 7**

Percentage of civilian labour force employed in agriculture (Ireland)

| *1960* | *1965* | *1970* | *1975* | *1978* | *1979* |
|---|---|---|---|---|---|
| 34.2 | 30.4 | 25.3 | 22.3 | 21.5 | 20.4 |

The size structure of the dairy herd in Ireland is compared with the UK and the EEC averages for 1979 in Table 8 (MMB). The percentages of cows in each size group are very similar for both Ireland and the EEC average. However, herd sizes are much larger in the UK. The data relating to Ireland in Table 8 include herds for both liquid milk and milk for manufacture. The average size of dairy herd in Ireland increased from 10.4 in 1975 to 12.4 in 1977 and to 14.2 in 1979.

**TABLE 8**

Size structure of dairy herds, 1979 (% of cows in each size group)

|  | *Size of herd (cows)* | | | | | | |
|---|---|---|---|---|---|---|---|
|  | *1 -4* | *5 - 9* | *10 – 19* | *20 - 29* | *30 - 49* | *50 and over* | *No. of cows ('000)* |
| Ireland | 5.3 | 8.0 | 19.9 | 17.6 | 24.1 | 25.1 | 1,503 |
| UK | 0.4 | 0.8 | 3.4 | 6.0 | 17.6 | 71.8 | 3,288 |
| EEC | 6.0 | 10.6 | 23.4 | 18.2 | 19.2 | 22.6 | 25,080 |

*Size and structure of the Irish dairy processing industry*

The size and number of the processing units has altered considerably over the last 20 years (MMB). The number of creameries decreased from 169 in 1960 to about half that number in 1976 (Table 9). This figure had fallen still further by 1980 to 52 cooperative intake points representing 20 manufacturing companies. Corresponding with the reduction in the number of creameries, the average intake per dairy increased from 1.66 M gal in 1960 to 9.37 M gal in 1976. In 1970 the five largest co-operatives handled 29 per cent of the total milk for processing while in 1980 the largest six cooperatives processed 57 per cent of the total milk for manufacture.

190

## TABLE 9
Number of dairies* and average daily intake

|  | 1960 | 1965 | 1970 | 1973 | 1976 |
|---|---|---|---|---|---|
| Ireland | 169 | 164 | 162 | 117 | 82 |
| Average intake/dairy (M gal) | 1.66 | 2.39 | 3.17 | 5.69 | 9.37 |

*1960-70 data represent the number of creameries while 1973 and 1976 data represent the number of organisations.

Milk for manufacturing purposes in Ireland is produced on a seasonal basis, two peak months accounting for 30 per cent of the annual supply and, in recent years, 2.7 per cent being produced in December and January (Keane, 1980). The national creamery milk intake pattern for 1973-80 is presented in Table 10. This table also shows the utilisation capacity of the industry and the peak to trough delivery ratio. This utilisation figure shows the potential capacity for the processing of increased milk production in Ireland without any additional investment in processing plant provided that the milk is produced in months other than the peak months.

## TABLE 10
National creamery milk intake patterns (%) 1973-80 [Abridged]

| Month | 1973 | 1980 |
|---|---|---|
| January | 1.3 | 1.5 |
| February | 2.3 | 3.4 |
| March | 5.5 | 7.2 |
| April | 10.0 | 11.4 |
| May | 15.3 | 15.8 |
| June | 16.2 | 15.5 |
| July | 14.9 | 14.4 |
| August | 13.1 | 12.3 |
| September | 10.6 | 9.4 |
| October | 7.1 | 5.8 |
| November | 2.7 | 2.7 |
| December | 1.0 | 1.1 |
| Total | 100 | 100 |
| Utilisation %* | 51.5 | 52.7 |
| Peak/trough month | 16.2 | 14.4 |

$$*\text{Utilisation } \% = \frac{100}{\text{peak month} \times 12}\%$$

### Milk assembly
The major changes that have taken place in milk assembly in the last 15 years were due in part to the development of the new technology of bulk collection but were also linked to more general developments in the dairy industry. These developments include the virtual ending of the return of skimmed milk to suppliers, the centralisation of processing, the rapid increase in milk output per farm, and the decrease in the labour availability on farms. The first refrigerated bulk milk tanks were installed on some Co. Waterford dairy farms 15 years ago, and the changes in milk assembly procedures in the industry have been dramatic since that time. By the end of 1977 over 50 per cent of all milk in the country was assembled by bulk tanker direct from farm-refrigerated bulk tanks to the factory. This involved about 20 per cent of milk producers. While the milk assembly rationalisation programme began over 15 years ago the pace of rationalisation

has varied considerably between co-operatives, resulting in complete transition to bulk in a number of co-operatives. By 1980 an excess of two-thirds of all milk was collected by road tanker while close on one-third (25,000 approx.) of milk suppliers have refrigerated milk tanks. Thus a spectacular transition from the traditional to an ultra-modern milk assembly system has taken place in a period of 15 years, which reflects the rapid changes that have occurred in the Irish dairy industry in recent years.

*Farm investment*

The past decade and Ireland's accession to the EEC in 1973 have brought about a marked increase in investment in Irish agriculture. Agricultural credit expanded rapidly with the prosperity in Irish farming. It is estimated that over £1,300 M was invested in farm buildings and machinery in the last decade (Hickey, 1980). Lending by banks and the Agricultural Credit Corporation (ACC) to farmers increased from £87 M in 1970 to £1,082 M in 1980. Farmers invested primarily in extra stock, associated buildings, and feeding facilities. Crop acreage increased, and this resulted in additional machinery and storage facilities, for example: forage harvestors, tractors, and combined harvesters. Considerable investment was also made in new farm dwelling houses and modernisation of existing ones.... Capital investment in the latter years was also reduced owing to the unprecedented levels of interest rates and the reduced availability of credit.

*Funding of capital investment*

The make-up of this capital investment in Irish agriculture varied over the years, building and machinery accounting for about 40 per cent of the total investment in 1973, about 80 per cent in 1975, and 65 per cent in 1978.

While farmers have borrowed heavily in recent years to finance increased capital investment the reality is that state grants and farmers' own contributions still account for at least 40 per cent of the net capital investment on Irish farms.

*Change in farm incomes, 1978-79*

The data from the 1979 Farm Management Survey indicated that about 31 per cent of full-time farmers had a family farm income of less than £2,000 as compared to 24 per cent in 1978. Based on an estimate of 120,000 full-time farmers in 1978, between 8,000 and 9,000 more farms, had incomes of less than £2,000 in 1979. In 1978, 40 per cent of full-time farms had incomes in excess of £5,000 but this declined to 30 per cent in 1979. Thus, an additional 12,000 farms fell below the £5,000 income level in 1979.

The average family farm income per farm on full-time farms was £4,363 in 1979, a decline of 18 per cent on the 1978 figure. The income was derived from gross output of £11,702 and costs of £7,339. Gross output per farm increased by 2.5 per cent from 1978 to 1979 but total costs rose by 20 per cent. By far the greatest increase in costs was related to extra concentrate feeding, which increased by 36 per cent, and fertilisers by 14 per cent. Both of these items combined accounted for 70 per cent of the increased direct costs.

The data shows clearly that dairying is the most profitable enterprise in Irish farming. On all farm enterprises there is a major opportunity for expansion when the figures related to forage acres per livestock unit are examined.

*Potential of dairy farming*

Despite the decline in farm incomes in the last two years the data highlight the divergence between the better farmers, that is, those with high income relative to farm size and who were influenced by the high product prices obtained since joining the EEC, and those who have been unresponsive. The adoption of dairying as an enterprise has long been advocated as the most profitable enterprise in Irish farming particularly for small holdings.

High production must be obtained with comparatively low costs based on maximum use of grass and conserved silage for winter feed. Feeding accounts for a large proportion of cost in milk production (Gleeson, 1980). The major ingredients the diet of a spring calving cow are grass (66 per cent), silage (25 per cent), and concentrates (9 per cent). The relative cost of these feeds expressed on a digestible dry matter basis.

*Role of research in improving incomes in dairy farming*
In the future it is the efficient management of high-yielding cows at high stocking rates that will protect and expand incomes from dairying, combined with efficient processing and marketing. Production and processing research has a major contribution to make in achieving these objectives.

The Agricultural Institute Dairy Research Centre at Moorepark was established in 1959. The research programme involved both production and processing. Before its establishment there was little research activity ongoing for the dairy industry, particularly in relation to production. The major emphasis in production research related to efficient production and utilisation of grassland on different soil types, improved yield and quality of silage, and the role of concentrate supplementation in efficient profitable dairying. In relation to animal management, the research priorities related to breeding, replacement heifer rearing, herd fertility brucellosis, mastitis control, lameness, and calf mortality. Other areas of research involve testing of milking equipment, farm building design, farm layout, cleaning milking equipment, and milk assembly.

The processing research has been involved in improving both the chemical and bacteriological quality of milk and dairy products and the development of new products. The research to date has removed many of the constraints that existed in the 1960s and 1970s at both farm and processing level to enable the industry to expand and develop to its full potential in the future.

# MILK QUALITY: CENTRAL TESTING
## F. HARDING
(JSDT Vol. 35 No. 1 Jan 1982)

The effect of raw milk quality on product quality is discussed in explaining the reasons for quality payment schemes. In October 1982 the Milk Marketing Board of England and Wales will centralise its testing for the quality payment of its 43,000 dairy farmers in six laboratories. The sampling, sample identification, sample transport, testing, recording, and reporting systems which are being developed for use are described.

The quality and yield of products, whether cheese, butter, powder, or short, shelf-life products such as yoghurt, can be seriously affected by the quality of the milk. Consequently, the financial return to the processor and to the dairy farmer can be seriously affected if the quality of the raw material is poor. Impairment of the hygienic quality of the raw material may be affected by bad milk production practices, faults in storage during transport, or faults in handling by the processor of the raw material. This paper summarises the effects that poor quality milk has on financial returns to the processor and the dairy farmer and emphasises the quality of the raw material at the production stage in order to explain the reasons for the use of quality payment schemes.

One has considerable sympathy for the dairy farmer who has to produce milk under difficult conditions and at unsocial hours…. The yield and quality of the final products depend on the composi-tion and hygienic quality of the raw milk supply; hence the quality of the raw milk supply can have a significant effect on returns to the producer and the processor. Even the most skilled processor cannot make a good final product from very poor raw material. Quality payment schemes therefore are designed to encourage the production of a raw material that will suit market requirements.

The major cost factor in products such as butter and cheese is not labour or capital investment, it is the cost of the raw material since liquid <u>milk accounts for about 85 per cent of the cost of the final product.</u> From this figure one can see the dramatic effects that milk quality can have on the cost of the final products. For milk going to the liquid market the com-positional quality does not have an effect on the yield of the final product unless the industry standardises the fat content. However, in recent years progressively more milk has been used for the manufacture of products than for liquid consumption, and the effect of compositional quality of the raw material in terms of market returns has therefore become much more significant.

If we take two large milk producers, one producing milk of poor compositional quality, the other milk of high compositional quality, the value of each litre of milk in terms of product produced is quite different. Producer A, with 3.0 per cent fat, needs to supply 30,355 litres of milk to produce one tonne of butter, whereas producer B, with 3.9 per cent fat, needs to supply only 23,350 litres of milk to produce one tonne of butter. Producer B therefore enables the manufacturer to sell 23.1 per cent more butter than producer A for the same volume of milk supplied. If each consigns the same amount of milk, logically the price paid per litre of milk to producer B should be correspondingly higher than that paid to producer A. There are similar arguments with respect to solids-not-fat, in that if producer A has a solids-not-fat of 8.4 per cent he needs to supply 11,523 litres of milk to produce one tonne of powder. Producer B, however, with 8.8 per cent solids-not-fat, requires only 11,000 litres of milk to produce one tonne of powder. Producer B, therefore, enables the processor to make 4.5 per cent more powder per litre of milk than his colleague, producer A, and he should therefore be paid a higher price per litre of milk.

These figures demonstrate clearly that quality payment schemes should be based on compositional quality in order to encourage the production of a better quality milk and in order to ensure that good quality milk producers who contribute most in terms of financial returns from the market-place are paid most for their milk.

Heat-stable lipolytic and proteolytic enzymes can create unpleasant taints in milk, making the product unacceptable to the consumer, and may therefore bring reductions in the selling price of the products. Specifications for milk products are also very much more demanding than in the past and demand a good quality raw material.

The presence of antibiotics in milk is unacceptable from a health point of view, because some people are sensitive to antibiotics and may suffer an allergic reaction if they are present in foodstuffs. Others may build up an immunity over a time if they are given regular small doses of antibiotics. In addition to these health aspects, there is the serious problem for the manufacturer of cultured products such as yoghurt and cheese: antibiotics are used in the udder to control bacteria causing mastitis, and antibiotics, if present in the liquid milk delivered to a cultured product factory, will perform the same function by killing the bacteria that the manufacturer is using in the production of cheese or yoghurt.

Other contaminants, such as blood or dirt, are not only aesthetically undesirable but may have bacteria associated with them which can create unpleasant taints in the final products.

These facts demonstrate that yield and quality of products can be seriously affected by poor quality liquid milk. These therefore are the reasons why there are quality payment schemes in the United Kingdom and in many other countries in order to improve the quality of the raw material and to ensure that the producers of good quality milk are paid more than producers of poor quality milk.

*Central testing*

The milk testing schemes currently in operation are under-taken by about 250 buyers' laboratories which act as agents for the Milk Marketing Board, undertaking sampling and testing for quality payment under the surveillance of the Liaison Chemist Service. While this procedure has been satisfactory in the past, there are a number of developments that make it desirable to follow the lead set by many continental countries and to centralise quality payment testing. The major reason for central testing is that it will bring a flexibility that is lacking in the scheme operated by 250 buyers. With the present system we are restricted to the capability of the smallest laboratory in terms of the number of tests that can be done each month and in terms of the methods adopted for payment purposes. Examples of the flexibility that will be achieved are:

1.  The frequency of testing for compositional quality will be increased from once per month to once per week.
2.  It would be possible, if it were thought desirable, to pay on a contemporary basis; that is,

for the average of test results in each month and not, as is the present practice, on an average for the year.

3.  If it were thought desirable, there could be tests for fat and protein instead of fat and solids-not-fat.

4.  A total bacterial count will be used as a measure of hygienic quality rather than the current resazurin dye reduction test, which is inadequate for cooled, bulk-collected milk. The EEC already have a proposed health and hygiene directive, which requires that the total bacterial count should be used as a parameter for payment on ex-farm milk, and this system certainly could not be operated in the 250 buyers' laboratories. The total bacterial count is much more meaningful for the buyer and to the producer in that it gives a clear picture of quality and trends rather than the simple pass or fail of the dye reduction test.

There are other benefits to central testing in that it will be possible to harness the recent technological developments in methods of data handling, sample identification, and testing. In harnessing this technology, one can gain the economy of scale with, I believe, a significant cost saving (on a cost per sample basis) to the industry compared with the present sampling and testing systems. Further benefits from central testing will be a greater uniformity and reliability of methods and an enhanced technical image for the industry.

There will obviously be some disadvantages also in the move to central testing, because until buyers gain confidence in the new system it is expected that there will be a duplication of testing.

The move from simple 'pass' or 'fail' tests to more definitive tests such as the total bacterial count will enable the Board to give advice to producers who either show a significant decline in hygienic quality or have a hygienic quality close to a penalty border line. It is hoped that central testing will enable much more advice to be passed back to the farmer, so that action may be taken wherever possible before failures occur….

Initially, self-adhesive bar code labels will be issued to each producer, which will be unique to his farm…. The tanker driver will take two samples at each farm into clean, sterile, 30 ml sample containers, and each will be identified at the farm using the bar code label and placed in specially designed cold boxes which will maintain the samples at below 4°C, and probably below 2°C, for the whole period between sampling and testing. The boxes will be delivered to either a buyer's premises or to a transport depot from where they will be collected by Central Testing Laboratory vans and delivered to one of the Central Testing Laboratory sites. The six laboratory sites are: Harrogate, Newcastle under Lyme, Llanelli, Plymouth, Thames Ditton, and Watercombe, near Yeovil. At four of these sites the laboratory will be attached to the MMB Regional Office, and at Watercombe the laboratory will be within two miles of the Regional Office. At Thames Ditton the laboratory will be annexed to the MMB Technical Division. The sites were chosen by the Milk Marketing Board's Operational Research Department to minimise transport and running costs and to enable samples to be received within 36 hours of sampling.

When samples are received at the Central Testing Laboratory they will be racked up and fed to the automatic testing instruments. One sample, after being warmed to 40°C, will be presented to the automatic Milkoscan. The Milkoscan will automatically identify the producer number from the bar code label, which has been put on to the bottle at the farm, and will read out the fat and solids-not-fat or fat and protein result as appropriate, and all of this information will be automatically captured on a typewritten sheet and on a tape or disk in computer language for transmission to the main frame computer at Thames Ditton. A small local computer will be used to control the testing instrument and to asterisk results that are abnormally high or abnormally low or to stop the instrument if it records inaccurate results for control milk samples which will be tested at regular intervals to cheek the instrument's accuracy.

A second sample will be presented to the Petrifoss, which is an automatic means of measuring the number of bacteria in a supply by taking a sub sample of milk and plating it. The plates will then be incubated for a period of three days to allow the bacteria to grow, after which the bacterial colonies will be counted. The sample, having been held cold between the farm and the testing centre, will have the same bacterial count as the milk in the bulk vat at the farm.

Again, the sample identity will be read automatically from the sample bottle at the time of test, thus preventing any sample misidentification. The automatic plating of samples will also minimise any possible errors. After three days of incubation the numbers of bacteria that have formed colonies will be automatically counted by a Biomatic, which rapidly scans the plate electronically and gives an accurate reading of the number of colonies present on each plate. This information will be recorded on the same tape as the sample identification, on a typewritten copy for use in the laboratory and in computer language on a disk or tape for forward feeding to the mainframe computer at Thames Ditton.

For the next generation of test instruments it is possible that technology will have advanced to give bacterial counts without the three-day incubation period. The NIRD development, 'the DEFT test', appears to be a strong contender for this purpose.

The antibiotic tests will be carried out by use of the Inofoss a mass inoculator which presents a small quantity of milk under test to a plate containing 96 small wells. Each well will receive a separate sample of milk. The well contains gelatine, bacteria, nutrient for the bacteria to live on, and a dye, so that when the milk is added and the mixture is incubated, the bacteria will grow and the dye colour will be changed from blue to yellow. If an antibiotic or other inhibitory substance is present this will inhibit the growth of bacteria, and the dye will remain blue instead of going yellow. Any positive samples found by this test method will be withdrawn and re-tested, attempts being made, if possible, to identify the antibiotic that has caused the failure. Positive results will be recorded and sent to the mainframe computer. Central testing laboratories will also have facilities for extraneous water testing and sediment testing, although these will riot be done as a routine. Each laboratory will be equipped with a visual display unit, which is a communication link to the mainframe computer and will be used to enable the laboratory manager to compare test results with the past history for any producer. It will be possible also to program the computer to compare test results with past test results and to call for action should results be out of line.

It is intended to notify producers of their test results once per month with their pay cheque, and it is hoped that the results of all tests can be given in this way. It would not be practical to notify producers' test results on a weekly basis because of the high cost. (It is estimated that notification of weekly tests to each producer would cost about £536,000 per year.) It is, however, the intention to notify producers who have failed a test or are close to a borderline or have had a significant change in quality as soon as the results are known. It may be possible to offer a more frequent notification service to producers at cost. A similar system of notification of test results to buyers is envisaged.

Presently, close collaboration with the dairy trade is being maintained through a working party of the Scientific Consultative Panel of the Joint Committee. This working group is helping to ensure that central testing develops satisfactorily to the benefit of producers and processors.

*Conclusion*

It is appreciated that the development of central testing is an enormous task which can be achieved only by close cooperation between the producers, the Board, and the dairy trade. Working parties with the dairy trade are already in operation, and the first Central Testing Laboratory was built and fully equipped at Watercombe, near Yeovil, at the end of last year. It is hoped that the other five laboratories will be built by the end of 1981 so that the practical experience gained at Watercombe will help to set the pattern for the other laboratories. Quality payment from central testing will not be introduced until all laboratories have been operating satisfactorily for at least three months, and it is anticipated that it will be October 1982 before the responsibility for payment is transferred from the 250 buyers' laboratories to the six central testing laboratories.

Central testing is concerned with quality payment aspects, and the buyer will still need to examine milk for quality acceptance tests, Liaison Chemists continuing to operate to safeguard producers' interests and to maintain a uniform standard in trade laboratories.

A significant proportion of the capital cost of central testing will be offset by EEC Co-

Responsibility Funds, and it is believed that the running costs on a cost per sample basis will be lower than those applying to the present quality payment system. I strongly believe that this move will permit more frequent and more useful tests to be made and so provide the milk producer and the buyer with the information they need progressively to improve milk quality. This will provide the dairy trade with the quality of milk demanded by our highly competitive and challenging market. By this means I hope that we shall be able to expand our sales at home and abroad to the benefit of the United Kingdom dairy industry.

## THE ROLE OF THE MILK MARKETING BOARD IN MARKETING
### M. E. BESSEY
(JSDT Vol. 35 No. 1 Jan 1982)

*Milk Marketing Board of England and Wales*
Throughout the whole of the EEC, if not the whole of the western world, milk producers have seen fit to invest substantial effort and resources in farmer-controlled organisations whose function is to market the milk produced by their members. Nowhere is this more true than in the Republic of Ireland, but the same pattern is reflected strongly, for example, in Denmark, Germany, and the Netherlands and, further afield, in the USA and New Zealand.

It is against this background of producer-controlled marketing (and processing) of milk playing such a dominant role in so many countries that I would set my brief examination of the role of the Milk Marketing Board. Because the constitution and role of that Board differs somewhat from that of other producer co-operatives in the dairy sector, questions and misunderstandings as to the nature of the Board have arisen from time to time. What I would hope to do, however, is to address myself indirectly to three questions:

1. Is the Board a true producers' marketing co-operative?
2. How, if at all, does its role differ from that of other co-operatives?
3. What is the significance of any differences?

In discussing the constitution and role of the Board, I shall, of course, be referring to the body operating in England and Wales, which is, perhaps with a touch of arrogance, legally constituted as *The* Milk Marketing Board. Similarly constituted bodies exist for Northern Ireland, for Aberdeen and District, for the North of Scotland, and for the rest of Scotland outside these areas. Much of what I will say would apply also to those bodies but I would obviously claim less detailed knowledge of their particular operations.

Firstly then I offer a few thoughts as to why the sort of producer co-operatives found in the Republic of Ireland and elsewhere did not develop in England and Wales. I think this is important because it was, in a sense, the vacuum that existed that led to the creation of the Board.

Having suggested, in effect, that the Milk Marketing Board sprang into being to fill a vacuum, I think it reasonable to ask why such a vacuum existed; why, indeed, England and Wales did not follow the course of development of producer cooperatives in the dairy industry that clearly took place in so many other countries.

My own belief is that the answer lies, at least partly, in the nature of the market for milk and milk products in the United Kingdom which even 100 or more years ago was radically different from that operating in other European countries in particular. I can well imagine, for example, the small peasant dairy farmer in vast areas of France and Germany, in the first half of the 19th century, converting the milk on his farm into butter or cheese for family use or for sale through local markets to neighbouring villagers and townspeople. These farm surpluses competed with those from local farmers - not of course with imported produce - and products rarely moved far from the place of production.

Later in the 19th century came the radical innovation of the mechanical separator. I can well imagine that this emancipatory piece of equipment was too expensive for each individual farm to buy and was the foundation of the first village co-operatives.

Having combined to produce butter together, the skim of course being returned to the farmers, what could be more natural than that they should combine together to sell their butter. Again in such countries as France and Germany, this butter - cheese conceivably being dealt with similarly - was still being sold in local markets. Here the farmers would soon learn the benefits of collective marketing rather than unbridled competition, remembering again that the competition would have been only with each other and not with product shipped across the length and breadth of the country and certainly not with product imported from overseas.

Another strand of development, probably in countries such as Denmark and the Netherlands, was the need to market volumes greater than the local markets could absorb. What again could be more natural than for farmer co-operatives to place their surpluses on more distant markets, even those lying overseas.

But apparently this evolution did not take place in England and Wales.

Due to the industrialisation of Great Britain, with the rapid development of a large urban population, probably three important marketing factors emerged. Firstly the urban market became dominant much earlier than in many other countries, that is, the markets were distant from the point of production. Secondly the rising demand for liquid milk raised the importance of this outlet for home-produced milk at an early date. Thirdly the rise in population (with the demand for liquid milk) created a strong import demand for dairy products. Probably, therefore, a very different market framework developed during the 19th century in England and Wales.

The significance of all this, in my hypothesis, is that producer co-operation in England and Wales, which certainly existed in the latter 19th and early 20th centuries, proved to be a sickly child. The market structure, with the early importance of liquid milk sales and the growing imports of New Zealand, Canadian, Danish, Dutch, and even Irish butter and cheese from the mid 19th century onwards, did not provide English and Welsh farmers with the right climate for co-operation.

What has all this to do with the Milk Marketing Board? Simply this, that by the early part of this century the market for milk in England and Wales, whether for liquid sale or for manufacture, fell almost entirely into the hands of private companies or, in parallel, the growing consumer co-operative movement. The producers' influence, despite many efforts to the contrary, ceased at the farm gate. As I have suggested, this was in complete contrast to the structural development in virtually every other developed dairy industry.

It was from this lack of successful producer co-operation that the Milk Marketing Board was born....

This then was the origin in 1933 of the Board: a body designed to fill the absence of producer co-operatives and to balance the power of the private dairy companies - a body owned by, financed by, and controlled by dairy farmers in England and Wales. It is by any reasonable test a producer co-operative but, if you will excuse the Irish flavour of the description, a *compulsory co-operative*.

From its birth in 1933 until the present time, the Milk Marketing Board has clearly faced many pressures and significant changes in its working environment. As a body it has had to adapt and, like any virile creation, it has developed and expanded. To appreciate its present role and policies, I think some brief appreciation of its development and adaptation may be beneficial.

From its inception the Board was welcomed by many producers but inevitably resisted by a few. Similarly, the buyers of milk viewed the new animal with mixed feelings, particularly as the main initial role of the Board was to seek outlets for the milk offered to it by its producer members and obviously to get the best price possible for that milk. Linked with this basic marketing task, there was the role of collecting the money from the buyers, pooling the proceeds, and sending the farmer his monthly milk cheque. In those early days, therefore, the Board acted as no more than a milk bargaining or selling agency.

The concept of differential pricing for milk, according to its ultimate use, was introduced at the outset, since the UK market was already well accustomed to the fact that liquid milk could earn a higher market return than manufacturing milk....

With the requirement placed upon it to find a market for all milk offered, the Board was faced at an early date with a bankrupt dairyman and no outlet for certain producers' supplies in an area in north-west England. With commendable foresight the Board stepped in to handle the milk itself and thus started its first ever creamery - now one of the major Cheddar factories in the world at Aspatria. In another tentative diversification, the Board started, in conjunction with local dairymen, the promotion of school milk schemes to help to expand the market for liquid milk.

And then came the Second World War and, within a few months, almost total regulation of the food industry by Government.

While constitutionally the Board remained unchanged as a producers' organisation, in practice it became the agent through which the Government regulated the industry. Producer prices were guaranteed, these guarantees being implemented through the Board's Milk Fund, the buying and selling prices, and hence the margins, of liquid milk and of manufactured products were regulated, again partly through the Board; and the Board became the Government's agent in allocating the use and controlling the movement of all milk supplies. In the collection of milk from farms, which previously had been left for producer and buyer to arrange, the Board became responsible for organising and rationalising all movements.

Through the 14-year period of Government control up to 1954 the Board therefore broadened its base and range of activity to the point of substantial involvement in nearly every facet of the industry. This was, however, largely as the agent of Government or at official behest in some form. While its links with Government have changed dramatically since that time, and in no meaningful sense has the Board ever since been an agent of Government, the uncertainties that still prevail in some quarters over the Board's true breeding as a producers' organisation are clearly a hangover from these wartime arrangements. But these, let us be clear, came to an end in 1954.

From the year 1954 the Board's true independence as a producer organisation was restored. Two features of the Government involvement, however, remained.

Firstly the milk producers' returns were still guaranteed, at least in part, by Government. When the guarantee needed to be implemented by injecting public funds it was still a matter of administrative convenience to do this via the Board's producer pricing system.

Secondly the Government retained control over the maximum retail price of liquid milk and of the Board's selling price for that market, and hence of the margin available to liquid milk distributors.

In every other significant sense the industry had moved out of Government control....

Back on the farm the range and sophistication of the services offered to its farmers expanded as the Board sought to assist the efficient development of its members' operations. Similarly, in the market place the Board, both by itself and in partnership with the dairy trade, devoted large sums of money and manpower to promoting, advertising, and developing the markets for liquid milk and products in the United Kingdom. Wherever the need was evident, the Board stepped in - anywhere between the farm and the final consumer. Always, however, these activities were undertaken on behalf of the dairy farmer who, of course, financed and benefited from (he hoped) the Board's activities. Always, as well, this broadening range of activity was as an adjunct to the still remaining basic and original purpose of the Board which continued as that of marketing agent for the farmer's milk.

And so we come to the latest chapter in the Board's history which has only recently begun. The seeds of change for the Board were sown when the United Kingdom joined the European Economic Community. Although it took five years (of transitional arrangements) before full membership was realised, the writing was on the wall that the environment would significantly alter for UK milk producers and for the Board.

There were four eventual areas of challenge and change.

(1)　From 31 December 1977 the producers' Guaranteed Price system came to an end and producers became totally dependent on market returns.

(2)　From a market environment in which manufacturing milk returns were low, being geared largely to world price levels for dairy products, the position changed over five or six years to one in which manufacturing milk returns rose to common EEC levels (subject to monetary adjustments), which were far higher than ever known before.

(3)    It was clear that the Milk Marketing Board system was going to be challenged in some essential respects, and there was uncertainty over the future powers of the Board.

(4)    The UK industry moved into an environment where the main competitive challenge was from large producer co-operatives situated elsewhere throughout the Community.

The Board, over a period, has responded to these challenges in two main ways:

(1)    It has successfully defended, with the overwhelming support of its producer members and of the British Government, its essential powers. Its legal position is now fully covered by EEC Regulations.

(2)    Recognising the growing importance of the dairy product markets to its producers, and as an insurance against the one-time possibility of some loss of its powers, the Board took the bold decision in 1979 to buy the majority of Unigate's manufacturing interests in England and Wales.

Thus, at one and the same time, the Board has preserved its traditional role and made a major step in bringing itself more into line with other dairy co-operatives by manufacturing a very substantial proportion of its members' milk supplies.

In recognition of its new position and as the latest step in adapting to its environment, the Board has within the last few weeks totally restructured itself into two quite distinct parts, both operating under its 18-man farmer Board. Within one structure the traditional MMB functions, such as producer services, milk allocations, milk pricing, and generic sales promotion, are to be found. At arm's length from this structure is the whole of the separately run commercial business of Dairy Crest, embracing the transport business, the liquid dairy businesses, and the very substantial creamery business.

In this historical review I have said quite a lot about the marketing role of the Board but perhaps I could now try to draw this all together with a brief review of the main current marketing activities. Let me do this by tracing the progress of milk from the farm to the consumer.

With its responsibility to take all milk offered to it (subject to quality considerations), the Board also has responsibility to arrange its collection. This it does, exclusively by tanker collection, partly by 'employing' its own Dairy Crest transport business and partly by paying other buyers of the milk or independent contractors to carry out the haulage. Where it employs Dairy Crest transport, this is on an arms-length basis on negotiated rates with exactly the same terms as any other contractor. Incidentally, Dairy Crest now carries just over half the milk.

As the producers' marketing agent, the Board then allocates milk supplies to the most remunerative outlets, that is, essentially on the basis of price and location.

The Board sells milk at different prices according to the end-use or value of the milk.

Finally, as a traditional Milk Marketing Board activity, and acting on behalf of all its producers, the Board devotes substantial staff resources and money on general sales promotion and development, often in association with the dairy trade. Naturally this activity is slanted towards the highest value outlets, such as liquid milk and cream, but all markets are embraced to some degree.

*The Dairy Crest commercial business*

Last, but by no means least, there is the direct involvement of the Board in the market-place through its own commercial businesses. I have referred to the arms-length nature of the relationship, milk being transported for the Board and milk being bought to operate the liquid milk dairies and creameries. The dairy business, operated through six wholly owned and two partly owned dairies in four areas of the country, has about 5 per cent of the liquid market in England and Wales. The creamery business operates 33 creameries and handles about half the milk manufactured in England and Wales, mainly into butter (a 75 per cent share) and cheese (a 50 per cent share). The full range of activities is, however, very wide, embracing also a full range of milk powders, whey powders, lactose, cream, catering products, etc. Altogether some

10,000 people are employed in this business, and overall it has a turnover in excess of £600 million per annum....

I must, however, conclude by comparing these activities with those of any other producer-owned dairy co-operative.

(1) Firstly, the Board accepts the milk offered to it by its members, and it then allocates that milk, by sale to itself or to other parties, in order to optimise returns.
(2) In pricing or valuing its milk, the Board is able, like any other co-operative, to get better returns from some markets than from others. It has, of course, no control over the market-place since this is freely open to overseas suppliers with the exception of liquid milk, where the UK Government maintains its own regulations on health and hygiene aspects in the absence of Community legislation.
(3) Like any enterprise, private or co-operative, the Board devotes part of its revenue to sales development of one part or another
(4) Finally, the Board takes its total revenue and shares it out equitably among all its members, a system common to all co-operative enterprises.

In conclusion, let me return to the three questions I posed at the outset. I hope that what I have said as to the development and role of the Milk Marketing Board has a totally familiar ring to any who are knowledgeable on the question of producer co-operatives. I think there can be no debate that the Board is a truly produced, owned and controlled dairy co-operative. Its role and functions emulate those to be found in any large co-operative throughout the dairy world. It may be larger than some but it certainly has its equivalents in size among the large French and Dutch co-operatives, for example. The statutes of the Board may differ in so far as all producers in England and Wales are members of the co-operative but this strength, as I have said, is tempered by the power vested in producers to wind up the Board at any time.

It seems clear, therefore, that there are no differences in principle between the Board and other dairy co-operatives. Any differences that may exist stem solely from relative size and from breadth of activities.

## THE EUROPEAN ECONOMIC COMMUNITY AND UNITED KINGDOM DAIRYING, WITH SPECIAL REFERENCE TO NORTHERN IRELAND – A PERSONAL VIEW
(JSDT Vol. 36 No. 3 July 1983)
D. L. ARMSTRONG - Milk Marketing Board, Brussels, Belgium

"Much has happened in Northern Ireland and to me since I was General Manager of the Northern Ireland Milk Marketing Board. I have now come from Brussels where I represent the Federation of Agricultural Cooperatives (UK) Ltd, including The Federation of Milk Marketing Boards. To ensure that I get paid, I am also the Director of European Affairs of the Milk Marketing Board of England & Wales at Thames Ditton.

I have given myself a wide-ranging brief in preparing this paper. Since the United Kingdom joined the European Community in 1973 the subject has become more and not less controversial. The debate on membership still continues, and the Labour Party, hoping to deal an ace from an otherwise bankrupt pack, has thrust membership of the EC into the limelight once more.

Where do we start? As first-hand experience of other ages is hard to come by, it is risky to say we live in the most cynical and hypocritical age ever. I believe we do, because our lives are dominated and degraded by the experts in publicity, whose main product, packaged in variety, is a vicious mixture of duplicity and hypocrisy. Hypocrisy and cynical politics go together, and the lofty language used at times by the exponents of both only emphasises to the critical and cynical ear the paucity of content and the amoral standards of the time.

In 1957 the European Community was founded on an ideal, the concept of European unity, but it was also practical in seeking to eliminate the causes of the internecine conflicts and in quite literally opening up the frontiers not only to trade but to cultures. Sadly, that ideal, that concept of unity, that opening of frontiers is now bogged down by economic recession and political stagnation. There are those who recall the ideal and seek to re-launch the Community, but their voices are heard only faintly. So the first lesson for the modern man in Brussels is that politics of the low or Machiavellian sort are an essential part of the job. We have to mix it and, believe me; we are dealing with real experts at the game.

Move and counter-move in the battle between the Italians and the French for the livers of the European wine drinker demonstrate splendidly the cynical realpolitik. According to EEC rules, the French are required to accept Italian wine, which is cheaper and better for certain purposes. French wine growers cannot compete, so on various pretexts, such as waiting for documentation, the wine is held up in customs stores or, worse still, the vehicles carrying it are wrecked. Would the English milk producers react in the same way to imports of milk from France? I doubt it, and there is the dilemma of the age. The advantage appears to lie with the lawbreakers and not with the principles.

You may feel I overstress the point, but I believe that the essentially cynical and Machiavellian nature of the Community and the hypocrisy associated with it must be recognised if we are to gain material benefit from it. In fact how different is it from politics at Westminster? Is there a man of principle in the House?

Personally, I had grave doubts about the wisdom of joining the EEC many years after the original six had drawn up the rules, after very hard bargaining. I believe also that the British Government could have made a better deal. The prize offered to France was the British food market, which would have met French aspirations to be the major exporter of agricultural products. In return British industrial exports were to expand substantially. Despite muted encouragement British agriculture expanded to such an extent that, combined with continued imports from New Zealand, French aspirations were frustrated, while the entrenched feudal attitudes of the British trade unions, combined with the lack of will of much of British industrial management, frustrated the export drive. In mitigation, the unprecedented increase in oil prices fuelled inflation, already gathering pace, and increased the cost burden of British industry (though all western countries were affected in the same way). At the start, much greater emphasis should have been placed by the negotiators on the expansion of British farm production, and matters such as the retention of marketing boards without qualification should and could have been established.

The failure to do so was a serious error from which we are still suffering. Withdrawal from the EEC is not now a practical proposition. The EEC would be weaker without the UK but would be much easier to manage. There are still ties binding continental Europe, which do not apply to the UK with the same force. Gradually, however, by recognition of the true nature of the Community and acting accordingly we can get more out of it. National sovereignty does not after all mean so much to a toothless lion whose roaring days are over.

There is, of course, always the possibility that the Community may collapse because of the centrifugal forces generated by the member states in their pursuit of national aims, but the UK would not benefit if that happened.

The institutions of the European Community consist of the Parliament, now directly elected; the European Court; the Council of Ministers; and the Commission, which can be regarded as the Civil Service of the Community. Parliament debates, the Councils horse deal, and the Commission tries hard to avoid upsetting the member states, particularly the French, the Germans and the British. Nevertheless these institutions influence our lives and agriculture (our special interest) to an ever-increasing degree. To operate successfully, 'our man' in Brussels has to keep the Euro MPs informed, to work on the Ministers and civil servants of the member states, and to try to push the Commission along in the direction that suits his members best, and, if what suits the members also helps Europe, then it is both an accident and a bonus. Self-interest is still the prime motivator.

What has been the impact of joining the EEC on British agriculture generally and on the milk industry in particular? This is a subject that still awaits a definitive study, though the House of Commons Agriculture Committee has given it attention. Professor & J. Sheehy, of University College, Dublin, has written a paper on The Impact of EEC Membership on Irish Agriculture. He summarises his conclusions thus:

"The accession of the Irish Republic to the EEC in 1973 has provided an interesting case study for agricultural economists. The change from dependence on the low-priced UK market to participation in the EEC market involved an increase of 45% in real prices received by farmers between 1971 and 1978. The consequences of this for farmers' incomes, for the balance of payment and for income distribution in the country generally were enormous. However, for economists, perhaps the most interesting issues relate to the consequences for resource use. In general there has been little noticeable impact in this area: capital formation has increased but the number of workers has continued its pre-EEC rate of decline. *The overall growth rate has not accelerated in response to the massive price increase. Furthermore, EEC structural policies have had little, if any impact. The EEC effect on Irish agriculture has thus been mainly a monetary one."* (Sheehy, 1980).

There has been a spectacular rise in producers' prices, which was in recent years affected by an equally spectacular rise in inflation - which led to the irony of producers seeking interest subsidies to finance the inflation which the increase in prices in part brought about.

Other trends in Britain have been:

(a)    a substantial increase in milk production;
(b)    a substantial increase in cereals output, and a growth in the imports of cereals substitutes because of the high prices set for cereals;
(c)    a substantial reduction in the number of milk producers, accompanied by an increase in the size of herds and increased yields;
(d)    unequal effects in different parts of the UK. Northern Ireland, in particular, has been adversely affected in recent years;
(e)    substantial increases in consumer prices for basic foodstuffs, but especially for liquid milk and dairy products; a dramatic fall in butter consumption, mainly because of the huge increase in butter prices, both absolutely and in relation to margarine;
(f)    stagnation in the cheese market;
(g)    depression of fishing, poultry and horticulture;
(h)    creation of newsworthy stockpiles and equally newsworthy disposals to the Soviet Union and elsewhere;
(i)    an elaborate machinery of regulations to maintain the facade of a common market;
(h)    the replacement of outside supplies (with the partial exception of New Zealand) by European supplies. Substantial growth of imports of food from Europe.

It has often been said that the Common Agricultural Policy (CAP) is neither common nor a policy. Nevertheless it is the only general set of rules applicable throughout the Community, and therefore attacks on it meet with the most determined opposition, especially from the farming Communities of the original six.

The essence of the problem is money, as is the case with most problems. Community taxpayers as represented by their Governments are reluctant to provide more cash to finance production of commodities which they do not want at the prices requested. The liquids, milk and wine, and cereals are the key areas. It is interesting that bread, wine and olives, the staple commodities of the Mediterranean area in ancient times, are still as big a prize today.

There is a struggle between milk producers and wine producers for the limited funds available. It is a struggle between northern Europe, with the emphasis on milk, and southern Europe, with the emphasis on wine. When Spain and Portugal join the Community, the pressures of the wine producers may become irresistible: the cork is quite likely to be blown from the bottle.

In the curious language of the EC, the Commission has produced its response to the Mandate, that is, an instruction from the Council of Ministers to the Commission to make proposals for the reforming of the CAP. The document bears all the scars of compromise but is nevertheless a programme which the Commission will try to get implemented. The main points of the Mandate Document are:

(a) A proposal to adjust the UK contribution to the EC budget to compensate for the lower level of financial benefit from the agricultural guarantees. It is proposed that member states which receive most from the CAP should in effect share the proceeds with the UK.

(b) EC prices should be brought more into line with international prices, the inference being that EC prices would be reduced.

(c) The development of food exports which appears to mean the export of surplus production to non-EEC countries.

(d) The introduction of quota for products in surplus such as milk, and the linking of price guarantees with levels of production.

(e) Reconciliation of lower price levels with support of farm incomes, with the introduction of income subsidies for disadvantaged farmers.

(f) More control over national aids, both to agriculture and to industry.

(g) Regional funds to be directed more to the poorest regions, and more support to be given to Mediterranean products.

(h) Use of the Social Fund for job creation.

(i) Expansion of the European Monetary System to include the UK.

(j) Elimination of barriers to trade.

(k) Support for research programmes in energy, food and environmental matters.

(l) Tighter quality control to reduce the supply of second grade products.

The Commission's document ranges over all aspects of Community policy, but inevitably the CAP occupies the centre of the stage. The Commission makes the ritual re-statement of the principles underlying the CAP. These are: market unity, Community preference and financial solidarity.

The Commission would expect that decisions on their proposals would be taken before the next price review, which means that the Council of Ministers, under UK Presidency, as it happens, would soon have to come to conclusions. As a demonstration of determination to achieve something, a special task force of the Permanent Representatives in Brussels has been set up at ambassadorial level.

Expectations of action are low. The UK Government has given the report a muted reception. Milk, wine and cereals are the key commodities. Milk is one of the greatest interest to this audience as producers if not consumers. Milk absorbs a substantial proportion of the Community budget. As a result of higher prices and yields (the increase in yields being a consequence of better techniques, better cows and better management) milk production in the Community has expanded and has given rise to the highly publicised butter and skimmed-milk powder mountains, which have for the moment subsided, partly because of fortuitous circumstances and partly because of exports. The Commission regards milk as a product in excess production and, not for the first time, proposes co-responsibility measures to control the flow, if not close the tap. They have already devised an instrument for the purpose, the co-responsibility levy, which is a rather cumbersome way of partly cancelling the effect of higher prices. The amount of money involved in the levy is substantial, being £35 million for the UK in the past year, and estimated currently at £56 million. Northern Ireland producers will contribute £4.2 million to the fund. Up to the end of 1980, the UK had got back in various forms about £27.5 million.

The Commission now wish to extend the principles of co-responsibility by linking the levy to a production target, rather like the old idea of the standard quantity. A variation is the so called super levy, which would operate at a penal rate on those producers expanding production above the quantum, and could either be used in conjunction with the co-responsibility levy or replace it completely.

The milk position has been made more hazardous certainly in the UK by the dramatic fall in butter consumption.

*The position of Northern Ireland*
Northern Ireland is in a specially difficult position. Milk is of key importance to the economy, and not just to agriculture, because it provides the rural community with a steady and reliable

cash income. The special measures taken in the last month will go some way to redress the adverse balance. The seriousness of the position is shown by the decline in income from £69 million in 1977 to £9 million in 1980.

Northern Ireland is one of the areas that should benefit from the regional funds. If the proposition to align producer prices for milk with world traders' prices for butter and SMP were to be pursued logically, there would be a substantial reduction in producer prices. This would meet with so much hostility that it is very unlikely to happen.

*The future*

Dramatic change in the CAP is most unlikely, but at the same time the pressure for better value for money from the CAP from urban taxpayers will grow relentlessly. All the time political power is moving from rural to urban areas.

Cereals policy is also of concern to milk producers. The growth of imports of cereal substitutes and particularly of manioc from Thailand is regarded by the French and other cereal producers as a twofold threat. Not only is there a reduced outlet for European cereals, but the Dutch for example, are enabled to produce milk more cheaply than their competitors. The value of Rotterdam as an efficient port is relevant in this context, as is also the Dutch talent for adapting feed practices to the cheapest sources of feed. There is a demand for restrictions on imports of cereal substitutes, mainly manioc and maize gluten, but milk producers must welcome the reduction in feed costs. This is an example of a situation in which it is very difficult for farmers to act in unison. An attempt is being made to develop an umbrella policy covering cereals production, fats and proteins, but little progress has been made.

Critics of the CAP would argue that the need for such a device [quotas] arises because the Community will not grasp the nettle of adjusting prices to demand. As prices to producers continue at levels which produce more milk than the market needs on a European basis, the Commission will be forced to use devices such as co-responsibility levies and production controls in order to keep the cost of support within some bounds.

The export outlet is a substitute for intervention, and EEC exports depend on subsidies to make them competitive with New Zealand and other suppliers. The Commission has an extremely difficult task in trying to ensure that refunds are pitched at levels which enable European traders to compete without windfall profits accruing. There are proposals for long-term agreements with particular countries. On the face of it this appears attractive but again would be to the advantage of those member states where the dividing line between state and private finance is blurred. France is a substantial exporter of agricultural produce (the green oil) and has a very strong interest in any measures that will encourage French exports. It is doubtful if this development would be in the interest of the UK though there is merit in seeking to establish a stable financial framework for Community exports. The masterly New Zealand deal with the United States of America for 100,000 tonnes of butter may tempt the Commission into thoughts of direct trading, but apart from opposition from the Community's traders, it is extremely doubtful if the Commission has the commercial 'know-how' to mount such operations.

Undoubtedly, money, that is, the Community budget, is a major difficulty in dealing with farm policy. The resources of the Community are presently limited to income from levies on imports and up to 1% of VAT. The Parliament and the Commission would like more money to spend on social and regional policies, but there is no indication that there is any prospect of member states agreeing to raise the level of their contributions. The only other way is to reduce expenditure on the CAP, which can only be achieved significantly by lowering prices to producers or restricting the volume of products to which the guaranteed prices apply....

One can fully support the plight of farmers, hit by rising costs and frustrated expectations (particularly milk producers in Northern Ireland), but the truth is that agricultural incomes depend too much on political decisions and state subventions, including consumer subsidies such as the butter subsidy in the UK. Most urgent is the need to tackle the decline in the butter market. The price relationship with margarine is likely to continue in favour of the latter, and we must therefore redouble efforts on the promotion and presentation of the products. The Swedish example of a product made mainly from butterfat but with the physical characteristics of margarine may be the road to follow. Meantime the EEC butter subsidy must be maintained.

Cheese is an area where there is scope for growth, but here again the temptation to squander resources on the fanciful needs to be resisted.

*Conclusions*

We are in the EEC and, despite talk about withdrawal, we will stay in it. Therefore we must make the best of it. We should stop thinking about what might have been and concentrate on playing the game the French play, that is, miss no opportunity to gain undue influence, exploit the rules ruthlessly and put the UK interest first.

We need to develop a plausible and overall philosophy to embrace our particular needs and to push it at every opportunity. It is important that selfish acts should be given the cloak of legitimacy. The ban on poultry imports to the UK because of the risk of spreading Newcastle disease is an example.

We should not get tethered to any particular system of production, manufacture or distribution. If daily delivery of bottled milk becomes so costly that the benefits are outweighed, producers should not hesitate to seek cheaper means of getting the milk to consumers, but it is not sufficiently stressed that the glass bottle is ecologically a much more desirable and acceptable container, as well as less costly than any of the other alternatives.

Milk is vulnerable, and measures to expand consumption (and combat imports) are essential if milk production is to continue to expand.....

Finally, producers must present a united front. In the Northern Ireland context, this means that the cooperative movement, which includes the Milk Marketing Board, must work together. John Morley, of the Plunkett Foundation, has done valuable work in demonstrating that the Milk Boards meet the essential criteria of a cooperative. He concludes that, in terms of cooperative principles and the provisions of UK and EEC legislation, there are no difficulties in recognising the marketing boards as part of the cooperative family. The Milk Boards are members of the Federation of Agricultural Cooperatives which I represent in Brussels at meetings of COGECA, which is the professional cooperative body for the EEC.

## TIME, TECHNOLOGY AND THE DAIRY INDUSTRY
### T. R. ASHTON
(JSDT Vol. 36 No. 3 July 1983)

This year is the Golden Jubilee for the dairy industry. It was 50 years ago that I came into the industry, too - in a period of depression which in many ways was similar to the times in which we now live, although now we speak of a recession, which is a better word as it implies hope of recovery.

In my time there have been marked changes in our scientific and technological approach to the problems of milk production, handling and treatment. I think it would be true also to say that these changes have outstripped marketing expertise and imagination to the extent that we are faced with a declining liquid market and a products market with but few innovations, strong competition and a lack of inventiveness and initiative to keep pace with the march of events.

*The status of the technologist*

In my early days technology was largely in the hands of management whose basic qualification was experience, the technician, despite some formal training, being confined to the laboratory determining fats, gravities and acidities and sometimes performing other mysterious tests - counts, coliforms, reductase tests, etc, the results of which were not only of dubious accuracy but difficult to interpret and translate into practice. The war and post-war period, however, caused some rethinking and reshuffling of the cards into a somewhat more orderly arrangement resulting in the laboratory technician acquiring the status of technologist, often with an extension of duties to include quality control, as it was then and frequently still is call-ed, and perhaps with some involvement in plant operation and performance.

With management becoming more and more concerned with labour, distribution and marketing problems and later the rebuilding, replanning and replacement of old and worn-out plant and machinery with more modern equipment, the niche of the technologist became established and he came to have a new role - to act in an advisory and later in a more active capacity on matters such as choice and size or capacity of plant, plant layout, installation and operation, CIP systems, mechanisation and automation of handling, products development, effluent disposal, etc, and to monitor plant performance and end-product quality. It was indeed a time when the dividend and scrip issue was the yardstick of achievement. Today it is largely a question of staying in business. For the technologist it was a time of harassment and difficulty.

To me, technology is a logic which can be acquired only from the combination of knowledge and experience; it can be taught only in rudimentary principle. I say this because every problem or situation has different variables. Technology is a challenge to resource. Properly applied it is rewarding of effort and thrilling in results.

In this section of my talk I want to review some of the factors, circumstances and events that I consider have effected change and changed the face of technology in the dairy industry since the war. In doing so, I hope that I might be able to extend the horizon of Professor Crossley's address 'A Half Century of Dairying; From Tradition to Technology' which he presented to the Society in 1976 on the occasion of his award of the Gold Medal.

Should my comments be provocative or should I be 'prickly', as I am told I can be, my only plea or excuse lies in my belief in Kipling's words, 'Experience is a fine teacher, but a fool follows his own bent.'

I have always thought that Professor Crossley had a feeling of nostalgia, if not regret, when dairy science was put under the umbrella of food science. To me and, I think, others, the science of dairying had become so well-defined and well-established in our teaching institutions that it seemed a pity to abandon it in favour of food science, which, though wider in scope, was less well-defined and somewhat abstract. I think food science and technology has much to gain from the fundamental study of dairying practices which I hope will continue to be given specific attention in the training curricula of universities, colleges and other educational institutions.

It goes without saying how much the dairy industry is indebted to the post-war committee on education (DITEC), which placed emphasis on technology, and in this connection to the efforts of, among others, the late Mr Ridley Rowling and Professor Crossley.

By reputation and record the National Institute for Research in Dairying (NIRD) and the Hannah Research Institute (HRI) have had no equals in the world of dairy science but not until lately, and then only modestly, have they had much concern for the problems of utilisation and those in the field of technology on which the prosperity of the industry largely depends. In the past, and even today, this outlook has, in my opinion, been a severe handicap to the growth and development of the liquid milk and milk products sides.

The value of direct collaboration, through committees, between research leaders and representatives of industry and shaped by the two national research institutions is difficult to assess. The committee established at NIRD in the 1960s was too short-lived to be of much benefit to either side; that at the HRI started later and, still in being, has proved to be useful in identifying and selecting certain subjects for study which might be slanted to practical application, i.e., technological aspects. More basic technological research is planned at the HRI when suitable housing and equipment becomes available, which should be soon.

Now defunct for many years, the Milk and Milk Products Technical Advisory Committee of the Ministry of Agriculture, Fisheries and Food (MAMPTAC) was a useful forum for the exchange of views and the publication of technical and technological information on dairying subjects. With direct access to Government, this committee, representing academic interests, advisory services, trade and industrial opinion, helped to foster enquiries into many technological aspects of dairying.

There followed the appointment of the Milk and Milk Products Committee of the Joint Consultative Organisation of MAFF under the Chairmanship of Bob Crawford, the task of which was to survey the whole scene of dairying and the brief to point out those areas in which research effort was lacking and was urgently needed. Owing, however, to the national purse on agricultural research having many demands made on it, the allocation of funds for purely dairy research came low on the list of priorities.

The technical and technological content of Acts and Regulations promulgated by Government, chiefly by MAFF and the Department of Health and Social Security, has always been of consequence to dairy practices. Of particular significance lately and necessitating attention has been legislation on environmental pollution, covering noise, the atmosphere and industrial effluent.

## International

While there must be full and appreciative recognition of the influence of the International Dairy Federation (IDF), suffice it to say that in providing a meeting ground for discussions on technological and other matters the publication of reports has often been long drawn out, and sometimes when issued the reports have been of limited value because of delay.

Participation in IDF affairs and studies is vested in the United Kingdom Dairy Association...

Collaboration between governments is the function of the Food and Agriculture Organisation of the United Nations and the World Health Organisation of the United Nations, the former acting as a co-ordinating body for food standards and the Codex Alimentarius, the latter as the co-ordinator of health and hygiene regulations, both with a view to incorporation in the national legislation of member countries.

More specific to Europe are the Common Market Regulations of the European Economic Community which, once agreed, have the force of law in member countries.

From the technological standpoint national and international legislation has far-reaching effects on science and technology. So, too, can foreign legislation be of importance because of the imposition of quality standards sometimes more exacting than our own and to which dairy products must conform at the point of delivery.

## The dairy, industry

This indeed amounts to self-help - in earlier times there was almost an indifference to need, a situation which improved to one of tolerant acceptance, then to appreciation and today to positive interest and action.

Introduced as an advisory body to the Joint Committee of England and Wales, the Scientific Consultative Panel, aided by a milk quality sub-committee, was given the primary responsibility of instigating investigational work in connection with quality payment schemes to producers. The partnership between the England and Wales Milk Marketing Board and the trade under the authority of the Joint Committee, which includes MAFF personnel, has proved to be a good working arrangement and eventually led to the sponsorship of a few investigational projects which were beyond the scope of members' industrial expertise. The work, financed by the Joint Committee and carried out mainly at NIRD, was found to have rewarding results.

With the idea of the sponsorship of research becoming a feasible proposition, due in large part to the outcome of the deliberations of the Joint Consultative Organisation Committee, the logic of the contact group was developed. It was in fact taken up from experiences in Holland following a visit by Professor Rook. From these beginnings the Dairy Trade Federation has now formed a Research Policy Committee bringing together all UK technical and technological interests, the aim of which is to 'define research priorities and arrange sponsorship of research work based on recommendations of its five Contact Groups.'

The watch-dog committees of the Creamery Proprietors' Association and the Dairy Trade Federation - the Technical Advisory Committee and the Scientific Liaison Committee - both have concern for legal and technical matters relating to milk and milk products in their respective spheres of interest, and will undoubtedly exercise effect on research policy and the prosecution of technical and technological research work.

The Technical Committee of the Federation of Milk Marketing Boards will also have an influence on the choice and execution of research projects.

*Other associations/organisations*

These include such organisations as the Packaging and Allied Trades Research Association and the Leatherhead Food Research Association who have direct or indirect contributions to make to dairy science and/or dairy technology. As a coordinating body, the Food and Drinks Industry Council must have benefits to gain for its members and knowledge to give to the dairy industry.

*Societies*

Much credit for the present healthy outlook on dairy technology is due to the lead given by our own Society, also to our kindred societies in Australia and New Zealand, all of whom have common aims and ideals and are members of a new federation pioneered by Vic Cottle. The enthusiasm shared by these and other societies of like kind is evinced by their work and the world-wide interest which they evoke in dairy science and technology at the present time.

*Technical press*

The technical press, which includes trade journals of the dairy industry, has much improved coverage of technical and technological information in recent years. This is especially true of Society journals which now contain reports of original work in addition to papers given at meetings and accounts of proceedings. By comparison it is my impression that the two main research journals, namely, the *Journal of Dairy Research* and the *Journal of Dairy Science,* now have relatively few papers on technological subjects, with numbers still decreasing....

Competition in the marketplace tends to occupy fully the attention of trade laboratories as well as, of course, process control activities as they are called. The development of new products or improvement of existing products is a healthy sign, demanding expertise as it does. As already indicated, specialised research is likely to be sponsored by industry as a whole, some sponsorship being carried out on a private basis.

Before I go on to my conclusion I would like to raise but not answer a few questions:

1.   Why has it taken so long to come to the decision that technological research must be sponsored by industry?
2.   Why, in spite of repeated entreaties from industry, have the research institutions shown a disinterest in technological research? Is it because they were and possibly still are distrustful of links with industry because it savours of financial gain or exploitation?
3.   Why did misunderstanding or lack of sympathy, or could it be procrastination, on the part of MAFF or the research institutions, unnecessarily delay technological development so necessary to the prosperity of the industry and the country?
4.   What of the future and relationships between research and industry? Will there be a continuation of tradition or radical changes to encourage progress?

And now I must bring my talk to a conclusion and, in doing so, mention a few last points:

## POSSIBLE FUTURE RESEARCH

1.   I hope studies already begun will be continued. I have in mind:
      (a)   work on spores, their source, enzyme systems and effects on the keeping quality and flavour of milk and milk products;
      (b)    work on direct epifluorescent filtration techniques as a means of quickly assessing the bacteriological content of milk and possibly milk products and, as an extension of this work, to look into the problems of rapid methods of assessment of enzyme activity and enzyme thermo resistance.
2.   I would like to see new programmes of work started on membrane processes, detergency and plant cleaning, and energy usage and economy.

In this new age of the micro-chip and computer I feel that all students of dairying at colleges and universities should have tuition in computer orientation and the application of computer science to dairying in its wider aspects. I think too that there is an urgent need for research workers to be given training in computer science to help them in their work.

And now let me say once again thank you for awarding, me the Society of Dairy Technology's Gold Medal, which I will cherish to the end of my days. Let us hope that the new changes in attitude to technology will have as reward expansion and economic improvement in the industry in the future and for the common good.

## THE IMPORTANCE OF AGRICULTURE IN NORTHERN IRELAND
(JSDT Vol. 36 No. 3 July 1983)

J. A. YOUNG

Department of Agriculture for Northern Ireland, Dundonald House, Belfast

(CF. Young Agriculture in N.I., JSDT Vol. 30 No. 4 Oct 1977)

"I propose to discuss the importance of agriculture in Northern Ireland in terms of five different factors or yardsticks, namely:

as a provider of employment and contributor to gross domestic product (GDP);

as a user of Northern Ireland's main national resources - the soil;

as a provider of raw material for processing and further employment;

in comparison in output and performance with agriculture in Great Britain and in the Irish Republic;

as a stabilising influence in the economy.

I will then refer briefly to the make-up of Northern Ireland's agricultural output, to trends in farm income, to the arrangements for supporting the farm income and to the importance of quality in improving returns from the market.

Despite the steady reduction in employment on farms over the years, farming accounts for almost 10% of the working population. The total agricultural labour force including the farmers, family workers, full-time hired workers and a small proportion of seasonal or casual workers is approximately 60,000, that is, about eight times as great as the number in shipbuilding and more than double the number remaining in the textile industry.

Farming contributes approximately 5% of the total GDP in Northern Ireland.....

Over the years there has been a steady movement of population from the land to urban areas and to other industries. Putting it another way, the volume of production from Northern Ireland farms is now more than three times that of 50 years ago, whereas the labour force producing that extra production is less than one-third of 50 years ago. The result is that output per man has increased about ten-fold during the past 50 years. During the same period the average size of farm has trebled, although it is still less than half that in Great Britain.

Northern Ireland has few natural resources apart from its soil. It has always been the aim of both scientists and farmers not only to increase output per acre but also to conserve and improve soil resources. Both aims have been pursued through a variety of measures, including the breeding of better varieties of plants, greater use of fertilisers and programmes of drainage and land reclamation. Grants are available to encourage farmers to apply lime and, in the less favoured areas, fertilisers to the land as well as for field drainage and reclamation. The Department of Agriculture has a continuing programme of arterial drainage to ensure that, as far as is practicable, there will be a network of free-flowing watercourses in Northern Ireland capable of taking surplus water from field drains.

Practically all farm produce is now subject to further processing and/or packaging, and the fact that the total volume of production is now about three times what it was 50 years ago means that there is that much more material available for the local food processing and packaging industries. The slaughter of fat cattle and sheep and processing of the carcasses is still at a relatively early stage of development. As recently as the early 1960s, about two-thirds of our cattle and sheep left Northern Ireland on the hoof. Now our plants are slaughtering over 80% of all our fat cattle, and hopefully with the recent change in the support arrangements for fat sheep and lambs the proportion of those slaughtered in Northern Ireland will quickly increase.

Currently, the number of employees in food processing and packaging in Northern Ireland is about 12.000. In addition, of course, there are nearly 5,000 employed in industries that supply farmers' needs, for example, feeding-stuffs and fertilisers.

In brief, the volume of output from Northern Ireland agriculture is about 7% of that in Great Britain and nearly one third of that in the Irish Republic. Employment on farms in Northern Ireland, as already mentioned, is about 10% of the working population. That compares with about $2^{1/2}$% in Great Britain and over 20% in the Irish Republic.

Although farming and food processing and the associated supply industries can never be protected from recessions, it is nevertheless a fact that they are more stable during recession than most other industries.....

From the 1930s until United Kingdom entry to the European Community (EC) the United Kingdom economic support arrangements for the agricultural industry applied in Northern Ireland n the same way as in Great Britain.

In practice, we have found in Northern Ireland that reliance on the EC support arrangements would not in all cases have been satisfactory. This is partly because of our relative remoteness, partly because of our dependence on imported feeding-stuffs and partly because of the fact that we are much more reliant on livestock enterprises than, say, farmers in Great Britain who are in a position to grow substantial quantities of cereals, horticultural products and sugarbeet. As well, the fact that we are the only part of the United Kingdom with a land frontier with another Member State has created problems arising largely out of the fact that the UK and the Irish Republic Governments chose to pursue different Green Pound policies, and, more recently, the Irish Republic chose to link its currency with the European Monetary System rather than with Sterling.

You will also be aware that in the EC as a whole the main support for the price of milk is intervention buying of butter and skimmed milk powder. However, the UK Government has taken the view that that would provide inadequate support in the UK and, by fixing relatively high maximum prices for liquid milk, it has therefore enabled the Milk Marketing Boards to obtain a higher return from the liquid milk market. But because we in Northern Ireland sell only 18% or our milk for liquid consumption compared with 50% in Great Britain we gain less from this arrangement, and therefore average returns to the Milk Marketing Board in Northern Ireland have for some years been lower than those in Great Britain. In 1978 and 1979 Government, with EC approval, made up most of the difference in Northern Ireland by paying over £22 million to the Milk Marketing Board. This was not possible in 1980 because the EC approval of the payments came to an end, and in any case funds were not available following the UK Government's decision to curtail public expenditure as an aid to the control of inflation.

The pig and poultry industries also had difficulties because under the EC support arrangements for cereals they no longer had access to the relatively cheap grain from North America on which they had been built up....

Largely because of the need to curtail public expenditure, no special financial aid was made available to Northern Ireland agriculture in the financial year 1980/81 except the sea transport subsidy on eggs going to Great Britain. In 1981/82 approximately £10 million was paid in special aid, mainly in respect of beef, milk, pigs, eggs and broilers. For the year 1982/83 that amount was increased to £16 million, and in addition Northern Ireland agriculture will be in receipt of approximately another £4 million over and above Great Britain farmers through the higher level of aid for feeding-stuffs projects and the agricultural development programme in the Less Favoured Areas. Of the total of £20 million expected to be paid out in grants in 1982/83, approximately £5 million will be recouped from EC funds. In addition, the EC is financing the calf subsidy, which was introduced on 20 May 1982 for the 1982/83 marketing year. The rate is approximately £20 per calf.

You will appreciate that the provision of the various special aids has involved a great deal of discussion and negotiation by both Ministers and officials in London and in Brussels. As a result of the recession, farm incomes fell substantially from the relatively high levels of 1977 and 1978 to their lowest level for many years in 1980….

*The importance of quality*

Despite the Government's help in topping up the basic EC support measures in Northern Ireland and its success in persuading the EC Commission and Council of Ministers to agree to these additional measures and even to contribute to their cost, the top priority will continue to be the maximising of returns from the market. And the market for two-thirds of our produce is outside Northern Ireland.

The Department's education, research and advisory services will continue to help in every way possible. Our staff at Loughry and Newforge Lane are equipped to help food processors with their problems. I am glad to say that over the past five years they have helped over 100 processors to solve about 400 different problems. But of course the main thrust in meeting market demand and identifying new markets must come from the commercial processors who in turn need the co-operation of producers of the raw material.

Finally - the Department does not have a crystal ball to look into the future. But it seems certain that consumers will continue to become more choosey and price-conscious while competition from producers and processors in other countries will increase. Nevertheless I can see no reason why our producers and processors with the help of the Department and the marketing organisations should not continue to compete successfully provided they make the effort to anticipate and supply the products in demand.

**Protection of the environment: Friends of the Earth's view of the role of the dairy industry (Paper given at symposium 'The Dairy Industry – Clean and Green.' 22 October 1991, London) LIANA STUPPLES, Friends of the Earth Ltd**

*"Clean and Green?*

The topic of today's symposium is 'The Dairy Industry - Clean and Green', although I prefer to pose it as a question rather than state it as a point of fact…. As the public, the regulators of industry and environmental groups all become more sophisticated in their appreciation of environmental issues, 'green' claims come under closer scrutiny…. Furthermore, the impetus is for this 'environmental auditing' to become an activity which is not superficial or separate from the normal running of the industry but one which is part of, and crucial to, successful business. The dairy industry affects the environment in a number of ways from the production and processing of milk through to the use and disposal of packaging it uses.

What role does the dairy industry have to play in the protection of the environment? Although I will emphasise the importance of considering the full range of environmental impacts, I will focus on water issues, for this is the field in which I specialise.

The lovable and worthy returnable glass bottle has provided a focus for considering the 'greenness' of the dairy industry. Unfortunately we appear to be experiencing its sad decline. Friends of the Earth has had a number of positive relationships with the dairy industry over the years including this autumn's joint promotion with the National Dairy Council. Through this campaign we wish to promote the reuse of returnable milk bottles. Why should we be concerned about the disappearance of returnable packaging? I do not necessarily just mean glass bottles; other types of packaging are also potentially returnable. What is the packaging option which will have the least environmental impact and which will help the dairy industry to minimise its impact on the environment?

The environmental impact of refining the raw materials and manufacturing the product also has to be considered; for example, the environmental impact of styrene includes the use of benzene in its production (van Duin and de Graaf, 1987). The use of energy throughout the life cycle of the product is also important, because energy efficiency is the most effective way of reducing the environ-mental impact of energy generation.

Environmental impacts may also arise from the use of the product. Finally, the impact of the product once its useful life is over needs to be assessed.

Compared to one-trip packaging, reusable packaging has an impressive environmental record.

For each common pack size, the refillable container at ten trips uses considerably less energy than the best equivalent non-returnable option, even with a 10% recycling rate.

Other studies have also found that returnable packaging has less environmental impact than one-trip packaging. The dairy industry must continue to promote milk in returnable bottles.

Although the UK's returnable bottle milk delivery system is the envy of environmenta-lists around the world, we cannot leave our assessment of how clean and green the dairy industry is to the consideration of the packag-ing alone.

The debate about packaging has received a lot of attention and the link between the production of paper, for example, and the impacts of forestry is well established in the public's mind. However, the greater pollution hazard from the dairy industry may lie with the product itself, although the wider public may not be aware of this - yet. Milk and the systems which produce it are potentially very polluting.

The intensification of agriculture over the last 20 years has had an immense impact on the quality of some of our most sensitive rivers. The principal one is the amount of waste produced. Large volumes of cattle slurry and silage liquor are created and these have to be stored before they can be disposed of.

The relative impacts of farm wastes on the environment is high. The effects of slurry and silage liquor entering a water course can be devastating, deoxygenating water and wiping out fish in the affected stretches.

A dairy farm with twice daily cleaning of the milking parlour, collection yard and dairy can produce up to 86 litres per head per day (1/hd/d) of animal waste (Beck, 1989). A dairy herd of 53 cows has a potential water pollution load equivalent to that of a community of 465 people (Ministry of Agriculture, Fisheries and Food, undated A). Animal waste pollution can lead to ammonia levels in rivers which are toxic to fish, which interfere with potable water supplies, and which can create blankets of decomposing solids which contribute to the loss of oxygen.

Silage making produces large quantities of liquids that seep out of the silage. The potential pollution load of 19,000 litres per day of effluent is equivalent to that of a community of 10,800 inhabitants (South West Water Authority, 1986). Silage effluent is very acidic and can be 200 times more polluting than human sewage (MAFF, undated A).

Farms are potentially very polluting places and unfortunately they are also very leaky places. Agriculture may now be causing more damage to the water environment than the whole of the chemical industry. For the rivers which were saved from industrial pollution are now being contaminated by the new industry: agriculture.

Farm pollution is not a problem which is only relevant to rural populations. People in the cities cannot afford to ignore it. In 1989, in the run up to privatisation, the water industry had to face the possibility of major outbreaks of water-borne disease. An outbreak of a diarrhoeal infection called 'cryptosporidiosis' was linked to the water supply system in the Swindon/Oxford area. Several hundred people had confirmed cases of the illness and it was estimated that about 5000 people were affected.

Cattle slurry (and possibly other livestock slurry) probably provides the major potential for the introduction of *Criptosporidium* oocysts into surface water sources. Young calves routinely get cryptosporidiosis infec-tions and slurry spread on fields or discharged to a waterway is an obvious source of oocysts. In Ayrshire, in 1988, contamination of a water supply through a cracked pipe led to an outbreak of cryptosporidiosis (Smith and Rose, 1990; Smith *et al*, 1988). The ground adjacent to the broken pipe had been sprayed with cattle slurry.

The Department of Health set up an inquiry into the more general implications of *Cryptosporidium* for water suppliers and published a report in 1990. This report confirms that agriculture is the main source of *Cryptosporidium*, although sewage treatment works may also contribute to the problem. *Cryptosporidium* is an example of why the city dweller cannot afford to be complacent about the impact of modern agriculture.

The processing and bottling of milk, at any scale, also creates potential environmental problems. The latest report on water pollution from farm waste notes a disturbing new trend in farm pollution. The North West region of the National Rivers Authority reports: 'Two new problems have emerged this year. The first relates to farm dairies, where milk is brought in for bottling purposes [which] gives rise to increased volumes of wash-water and the second involves effluents from dairy practices' (National Rivers Authority, 1990, p 14).

The Welsh region noted that 'worryingly serious effluent problems are occurring where farmers have diversified into cheese making and milk bottling, the resulting effluents seriously overloading existing effluent management systems.' (National Rivers Authority, 1990, p 16).

The dairy industry must not be complacent regarding pollution damage.

*Solving the problems: External regulation*

How are we to combat these serious problems? Environmental legislation is getting tougher but the dairy industry also has a responsibility to regulate itself. Under the Water Act 1989 it is an offence to permit any polluting matter to enter water without a 'consent' from the National Rivers Authority. This enables the National Rivers Authority to prosecute polluters after the pollution has occurred. Ideally, controls should prevent the pollution occurring.

The Government has recently introduced The Control of Pollution (Silage, Slurry and Agricultural Fuel Oil) Regulations 1991, which prohibit the making and storing of silage in a field without a constructed base and effluent containment system. Similar requirements are made for slurry stores and effluent collection facilities. The rules apply to all 'new' constructions and to existing installations which are substantially enlarged or reconstructed.

*Solving the problems: Regulation from within*

The solution of the farm pollution problem is not the sole responsibility of individual farmers. It is the responsibility of the whole industry to get its environmental house in order, and large dairies and bulking plants cannot stand back and expect the farmers to comply with regulations which they themselves are ignoring. Should not the dairy industry be examining the systems that produce its raw materials? After all, the dairy industry and the Society of Dairy Technology are not insignificant bodies which presumably could influence both suppliers and processors. Any pressure applied now would be timely.

The impact of agriculture can be expressed in fundamental terms that are very easy for the public to understand. In 1990 there were 2,853.000 dairy cows in the UK and they each produced 5010 litres of milk (Milk Marketing Board, 1990). Based on the quantities of slurry and silage liquor produced by an average herd, it is possible to approximate that in order to produce a pint of milk, six pints of slurry and 26 pints of silage liquor are produced. Adequate treatment of these wastes is a big responsibility.

When the link is established in the public mind between the impact of farm pollution and their health, and between the impact of the dairy industry as a whole and the health of the local river, the connections between milk and good health will no longer be accepted.

Protection of the environment is everybody's responsibility. If we don't all act then we all lose. Perhaps the dairy industry has more to lose than most, and it has a crucial role to play in protecting the environment.

# APPENDICES

CONDENSED REPORT OF THE WORK OF THE SOCIETY OF DAIRY
TECHNOLOGY COVERING THE YEAR ENDED JUNE 30th, 1946

The problems, makeshifts and uncertainties made manifest in the transition period from War to Peace have affected the Society as it has affected all connected with the Dairy Industry, but the work of the Council and the Committees has gone steadily forward.

## Membership.

The Society's membership has increased considerably - from 860 to 1143 during the year. Of this total, 561 are mainly connected with processing, distributing and manufacturing of milk; 155 are chiefly concerned with dairy engineering etc; 338 are mainly connected with research, education, advisory work, public health services, production of milk etc., and there are 89 Associate Members.

Full or associate membership of the Society is open to all interested, and it is hoped that firms connected with the Industry will use every endeavour to encourage members of their staffs to apply for membership of the Society, full particulars of which can be obtained from the Secretary, 19, Bloomsbury Square, London, W.C.1.

## General Meetings.

The first meeting of the year was held at Cardiff, the subject being "The effects of wartime conditions on milk production".

The second meeting in May, in London, was a combined meeting held in conjunction with the Society of Instrument Technology. The attendance was excellent and those attending heard a great deal of interesting and controversial matter relative to temperature measurement and pasteurising control.

In July, at the kind invitation and assistance of the National Institute for Research in Dairying, the Society met at Reading. This meeting, agreed by many to be one of the best ever held by the Society, took the form of a tour of the Institute, the Artificial Insemination Centre and the various laboratories during the morning with a paper "The problem of low-solids-not-fat in milk" read during the afternoon.

The fourth meeting to be held on November 12th and 13th next will be preceded by a Dinner and Dance held during the evening of Monday November 11th, 1946.

The Council of the Society are to give consideration to a suggestion that eminent members of the Dairy Industry from other countries be invited to address members of the Society on relevant subjects.

## Sectional Activities.

A great many very successful meetings have been held during the year under review by each of the Sections - Midland, Welsh, Scottish, North Western and Western.

The Welsh Section's initiative in arranging a visit to Southern Ireland was well rewarded by the great success attending their efforts.

## New National Dairy Technological Institute.

Some progress has been made towards the foundation of a National Dairy Technological Institute. The proposal has been drawn up and circulated to many influential members of the Dairy Industry and has won the blessing of the Dairy Engineers Association, the Dairy Appliances, Manufacturers and Distributors Association, the British Refrigeration Association and the Dairy, Ice Cream and Equipment Association. A great deal of money will be needed and those concerned with the project feel that the next step is to discover exactly what support the Trade would give before again appealing for Government assistance. Discussions are proceeding as to ways and means of approaching members of the Industry in order to obtain definite promises of initial support.

CHARITY NUMBER 213901

CHARITY COMMISSION
14 RYDER STREET
ST. JAMES'S
LONDON SW1

**CHARITIES ACT, 1960**
**REGISTRATION OF CHARITIES**

Dear Sir

**Name of Charity: SOCIETY OF DAIRY TECHNOLOGY**
**Place: NATIONAL**

The above-mentioned charity has been entered in the Central Register of Charities and a copy of the index slip relating to it is enclosed for your information. In order to maintain the Register with up-to-date particulars it is requested that any changes in the information contained in the index slip may be communicated to the Commissioners as they occur. The number at the head of this letter should be quoted in this, and in all other, correspondence with the Commission.

The submission of accounts by the charity has been considered by the Commissioners in relation to the requirements of Section 8 (I) of the Act. Their decision is contained in the enclosed memorandum and it is requested that, in those cases in which the submission of accounts is required, the instructions in the memorandum may be closely complied with.

I return herewith any deed, or copy deed, forwarded with the application for registration and of which the Commissioners were not requested to take custody. Details, if applicable, are shown below.

Yours faithfully

p.p. W.E.A. Lewis
Secretary
RE. 16

It is regretted that the specimen Account
Form is not yet available. Prints of the
appropriate form will however, be sent
if you care to apply in three months time.

Enclosures: Index slip
Memorandum on accounts

M. Sonn Esq
17 Devonshire Street
W1

# INCORPORATION OF THE SOCIETY

As reported in the SDT News up date of October 2000, the Society was incorporated as a company limited by guarantee and registered as such on the 5 April 2000. The company is also, of course, a registered charity. The company number is 3965383 and the charity number is 1081615. About the same time the unincorporated society was removed from the registry. This was confirmed in a letter dated 5 December 2001 from the Charity Commission.

## CERTIFICATE OF INCORPORATION
## OF A PRIVATE LIMITED COMPANY

Company No. 3965383

The Registrar of Companies for England and Wales hereby certifies that

THE SOCIETY OF DAIRY TECHNOLOGY

is this day incorporated under the Companies Act 1985 as a private

company and that the company is limited.

Given at Companies House, Cardiff, the 5th April 2000

THE OFFICIAL SEAL OF THE
REGISTRAR OF COMPANIES

*C O M P A N I E S   H O U S E*

HC007A

217

*Diseases*

Farm animals like humans are prone to several diseases. The outbreak of BSE the true cause of which is still the subject of speculation, was a tragedy for British cattle farmers. The ravages of a dreadful disease were made worse by the statement made in Parliament by Mr. Dorrell that there was a possible link between BSE and JCD. Throughout the crisis the media persisted in describing the disease erroneously as "Mad Cow Disease" and references to it were usually accompanied by a photograph of a staggering cow. Panic set in. Almost overnight beef became unsaleable. In the end common sense did prevail but not before enormous damage had been done to the market for British beef at home and abroad. The forecasts of thousands of new cases of JCD proved groundless whether of the old or the alleged new variant. The BSE affair was a dramatic example of manipulation of the public's concern about health.

*Foot and Mouth*

There have been two recent very serious outbreaks of Foot and Mouth in the UK. The first in 1968 when the General Meeting of members in January of that year had to be cancelled (J.S.D.T. Vol. 21. No 1. 1968) and the second which has just come to an end. The second was made worse by the most shocking mismanagement. Foot and Mouth has been a threat for a century of more. In the history of the RABDF 1876 -1976 it is recorded:

"Indeed a resolution at the very first formal meeting drew attention to the devastating effect of Foot and Mouth Disease. 'The existence of cattle disease in the neighbourhood of the Metropolis' it stated, was evidence that the regulations of the Privy Council with respect to the importation of live stock are utterly inadequate for the purpose. They would recommend that all fat cattle should be slaughtered at the port of debarkation and that store stock and breeding cattle should be kept in quarantine for a fortnight, by which time the possibility of the propagation of the disease would have passed away" (RABDF, 1876-1976.)

Vaccination may offer a solution but vigilance and rigorous control of imports is the best protection.

Bovine T.B. which had been almost eradicated in many areas of the UK is now spreading, having a very serious impact on the welfare of milk producers especially in the West Country where badgers are a serious problem. To date, largely because of pressure from animal welfare activists, no effective strategy has been introduced to curb the spread of this dreadful disease. At the same time there has been a resurgence of cases of T.B. in the human population.

*The Super Levy Quotas*

The well publicised mountains of butter and skim milk powder and the public hostility which was whipped up by the media led to demands to find alternative payment systems for milk which would discourage milk production. The Commission's solution was to impose quotas on milk production (the super levy) at farm level. The UK resisted the introduction of quotas realising that the impact on British dairy farming would be adverse and unfair. The Irish were even more vehemently opposed and emphasised with passion and skill the serious consequences which would follow any attempt to curb milk production in the Republic. The Commission got their way because the British minister of Agriculture caved in, and equally important concessions were made to the Irish which blunted the impact of quotas in the Republic.

As the Milk Marketing Board Director of European affairs, I was present at the crucial meeting in Athens. I headed my report (published in the Milk Producer, January 1984) "Athens belongs to the Irish" and it was true.

As was inevitable with compromise solutions the quota system (still in operation) was inequitable and unfair and caused market distortion. Quotas were introduced in 1984 and were applied with significant variation in rigour in the various member states. Italy which is a major milk producer did not bother at all, and even today about 20 years later, there is doubt if the owners of the large dairy herds in Northern Italy have suffered any inconvenience because of milk quotas.

It was obvious from the start that quotas would become a marketable commodity - effectively a licence to produce milk. There was an attempt to tie quotas to the land, which failed. The question of the value of quotas remains a stumbling block to attempts to alter the system.

## TECHNICAL COMMITTEE

The following write-up on the Technical Committee of the Society of Society of Dairy Technology is by Mr A J D Romney, Chairman of the Committee from January 1991 to January 1992.

The origins of the Technical Committee are as old as the Society itself. The SDT was founded on 2 March 1943 and Council minutes of August 1943 report on a committee formed to consider the Standardisation of Dairy Plant and Equipment, its Terms of Reference being 'to explore the question of standardisation of equipment and supplies for the Dairy Industry'.

The view was that this committee would liase with BSI who would actually produce the standards. Reading between the lines of later minutes, it appears however, that BSI did not take quite the same view of the urgency of producing standards as did the SDT.

In September 1947, Council decided that 'The Standardisation' Committee was a fact-finding committee and not a Technical Committee, whose function is to make recommendations regarding subjects for standardisation rather than to draw up specifications.

The Committee was involved throughout the 1950's and 60's with both BSI and with the production of various publications for the Society, the earlier editions of the Pasteurizing Plant manual, In Place Cleaning of Plant and Equipment, the Cream Processing Manual and a short publication on Bottle Washing being of particular note.

Early in 1985 the Society was approached by the UKDA on the need for the production of a completely new edition of the CIP Manual, originally published in 1959, because of major problems being encountered with Road Tanker cleaning standards within the Industry. As a result of this request, Council decided to re-constitute the old 'Dairy Equipment and Standardisation committee' to get this work under way. The first meeting of this reconstituted committee now re-named the Dairy Equipment Advisory Committee (DEAC) met in July 1985 chaired by Dr. John Gordon who explained the background to the UKDA's request and that I had been nominated by them to co-ordinate and edit the work - a request which Council had been pleased to endorse. By October, the layout of the CIP Manual had been agreed, authors identified and drafting was under way.

It was agreed that DEAC would also revive the Society's BSI and ISD links which had become somewhat vague.

A few months later, John Gordon handed over chairmanship of DEAC to myself and throughout the second half of the 1980's, work continued both with the CIP Manual and the BSI and UKDA etc.

Suffice it to say that after much debate, it was decided that DEAC and the Education & Research Committee should combine to become 'The Technical Committee' and I was invited by Council to become its first Chairman. The first meeting of this Committee took place in January 1991. Its first task was to look at the Terms of Reference of the two original committees and propose a new one for Council's approval which are as follows:

1. To provide a forum for the dissemination of information and advice on Education and Training, Research, Dairy Equipment and Practice both at home and abroard.
2. To provide the link between the Society and those organisations eg. BSI, ISD etc involved in standardisation where such standards relate to dairy practice.
3. To bring to the attention of Council and to the Society's members such other activities as might be considered by the Committee to provide a useful function.
4. The Technical Committee is to work closely with the Editorial Board in the production of technical documents.

It was agreed that the Chairman of the Editorial board, or their delegated nominee, shall be a member of the Technical Committee.

Initial work is under way to produce a series of 'Technical Guidelines' for publication. Drafts of 'Hygiene in the Packaging Room' and 'Temperature Control throughout the Cold Chain' are under way.

Proposals for future 'Technical Guidelines' include:

> 'General Hygiene for Food Handlers'
> 'Standardisation of Milk'
> 'Post Pasteurisation Contamination'

The Technical Committee also continues to provide SDT representation on several BSI working parties, to UKDA etc while organisation of the Residential Courses has passed to Programme Planning.

The present Chairman of the Technical Committee is Mr Garry Royall and Committee Members are:

Messrs Gary Ackrill, John Bird, Peter Fergusson, Peter Fleming, John Gordon, Robert Hamilton, Brian Watt, Ivan Kettyle, Alan Stack, John Sumner, Ray Thompson, Andrew Wilbey.

(JSDT Vol. 45 No. 2 May 1992)

*Presidents (1943 – 2003)*

A list of presidents from the foundation of the Society to the present is given below. The first Lady President was Mrs. Wendy Barron who, as a Trustee, and a President played a notable role in the affairs of the Society.

Mr. Ian McAlpine introduced Mrs Barron at the AGM held on 10 Oct 1978. Mrs Barron had been chairman of the London, South and Eastern Sections in 1973-74 and a member of Council, as Hon Treasurer from 1974-78. There were subsequently two other lady Presidents, Miss S Moore (Mrs S Watson) 1985-86 and Miss E Richards 1988-89 both of whom made outstanding contributions to the well being of the Society.

| | | | |
|---|---|---|---|
| 1943-44 | Prof H D Kay* | 1944-45 | F Proctor* |
| 1945-46 | G Enock* | 1946-47 | Dr A T Marttick* |
| 1947-48 | J Matthews* | 1948-49 | E Capstick* |
| 1949-50 | J White* | 1950-51 | Dr J A B Smith* |
| 1951-52 | Dr A L Provan* | 1952-53 | E L Dobson* |
| 1953-54 | Prof E L Crossley* | 1954-55 | J C Taylor* |
| 1955-56 | H C Hillman | 1956-57 | H S Hall* |
| 1958-58 | J Lewis* | 1958-59 | F C White* |
| 1959-60 | J Ridley-Rowling* | 1960-61 | Dr R J Macwalter* |
| 1961-62 | Dr R Waite* | 1962-63 | W W Ritchie* |
| 1963-64 | C H Brissenden* | 1964-65 | J F Hunter* |
| 1965-66 | J Glover* | 1966-67 | Dr J Murray* |
| 1967-68 | Dr R M Crawford | 1969-69 | C S Miles* |
| 1960-70 | T Idwal Jones* | 1970-71 | P A Hoare |
| 1971-72 | Dr J G Davis* | 1972-73 | Dr E Green |
| 1973-74 | Dr J Rothwell* | 1974-75 | N L T Garrett |
| 1975-76 | Dr G Chambers | 1976-77 | V C H Cottle* |
| 1977-78 | S G Coton | 1978-79 | I A M MacAlpine |
| 1970-80 | Mrs S W Barron | 1980-81 | Dr H Burton |
| 1981-82 | F G Weldon | 1982-83 | W J Hipkins |
| 1983-84 | A C Jackson | 1984-85 | R A Dicker* |
| 1985-86 | Mrs S Moore (now Watson) | 1986-87 | G McL Grier |
| 1987-88 | T C Tamplin | 1988-89 | Miss E Richards |
| 1989-90 | Dr J G Gordon | 1990-91 | G R Royall* |
| 1991-92 | J M Bird | 1992-93 | P Flemming |
| 1993-94 | Dr A C O'Sullivan | 1994-96 | G C J Lee |
| 1996-98 | R McC Hamilton | 1998-00 | J S Westwell |
| 2000-02 | Dr D L Armstrong | 2003-05 | Mr M Walton |

*Deceased*

## The UK Dairy Industry

Mr <u>Tim Brigstocke</u> set out in a series of papers the immense problems now facing UK dairy farmers and the processors and manufacturers of milk. The papers were published in "Feed Compounder" in May and August 2001. It is not possible to include these papers but they are rewarding reading. He discusses the problems and changes facing dairy farmers, the collapse in farm prices for milk, the impact of the EU (including the perennial reform of the CAP and Foot and Mouth, the rationalisation and contraction of the UK industry, the ending of organised milk marketing, the future for cooperative action and many other issues including the overweening influence of super markets on prices of farm produce.

- I have cherry picked some main points:
- If the forecasts are correct there may be only 9000 active milk producers in England and Wales by 2007
- In 2002 the average UK milk price was 13% below the EU average.
- Farm gate prices for milk have fallen by almost a third in the last seven years.
- The number of large processors of milk has fallen from ten to three in seven years and the process continues.
- The farmer controlled share of milk processing is growing but painfully.
- The need for greater collaboration across the dairy food chain. DEFRA to achieve this has set up the Dairy Supply Chain forum chaired by Lord Whitty (RABDF has a seat on this forum).
- The encouragement of innovation in processing methods and new products. <u>Is there a role here for the SDT?</u>

*Study Tours*

During the period when the dairy industry in UK had a measure of prosperity and still a relatively large number of dairy businesses and the stability provided by the Milk Marketing Boards there was a demand for study tours.

Study Tours were organised centrally by the Society and also by sections. Some Society tours were long and extensive and others were short visits. There were also organised visits to the periodic dairy equipment exhibitions in Frankfurt and Paris. The most ambitious tour was that undertaken during the Presidency of Mrs. Barron (1979) when visits were made to China, Thailand, Hong Kong and Japan. There is an interesting report of these tours in the Journal by A G Barber (JSDT Vol 33 No. 2 April 1980). He paid tribute to the leadership of Peter Lee, the Secretary of the Society. His enthusiastic involvement in Study Tours (and conferences) is recorded in the Golden Jubilee Booklet, which also contains a selection of photographs of members on study tours. There are reports in the journals of most, if not all, of the study tours.

I believe the earliest study tour was one undertaken by the Welsh section in April 1946 when 46 members and friends visited Southern Ireland over a period of 4 days. A selection of other tours follows:

| | |
|---|---|
| 1959 | Short visit to USA and Canada. Report by L A McAlpine (JSDT Vol 15`Vol 2 1960 |
| 1963 | France, H J Anderson, Director Seamans, Gaywood Dairy Ltd, Kings Lynn. (JSDT Vol 17 No. 3) 1964). |
| 1976 | Denmark and Southern Sweden 1976 W G Duff (JSDT Vol 30, No. 3 July 1977) |
| 1977 | New Zealand and Australia. 51 members and guests. Presentation of plaques to Dr McGillivray by Mr Cottle. (JSDT Vol 30 No. 3 July 1977). |
| 1979 | Finland, E Richards. 24 members. (JSDT Vol 32 No. 4 Oct 1979) |
| 1983 | Canada, 27 Sept to 13 Oct 1983. Overbooked. (JSDT Vol 36 No. 3 July 1983) |
| 1989 | Bavaria, 1 - 7 Oct 1989. Rosa Pawsley (JSDT Vol 43 No. 3 August 1990) |
| 1988 | New Zealand and Australia 25 Sept - 14 Oct 1988. Report J Noble. (JSDT Vol 42, No 3 August 1989) |
| 1990 | Wisconsin/Chicago 10-28 Sept 1990. |

As Dr Chambers recalls, study tours did not always run smoothly. The trip to Denmark and Southern Sweden which he led as President and in which Ben Davis of Unigate, and Dick

*The plaque commemorating the 50th Anniversary of the foundation of the New Zealand Dairy Research Institute, presented to Dr. W. A. McGillivray, Director of the Institute, by Mr. V. C. H. Cottle, President of the Society of Dairy Technology on 15th March 1977.*

Godsill of Fry-Cadbury, Ireland participated, found their hotel was located in the "red light" district of Copenhagen (the party moved to a more appropriate hotel the next day) and the coach which was to take the party to board the hydrofoil to Malmo (Sweden) did not turn up and a fleet of taxis had to be organised at very short notice.

The much more ambitious tour to New Zealand and Australia in 1977 was trouble free. This was led by Vic Cottle, assisted by Peter Lee. Ron Dicker in his report identified a conference celebrating 50 years of the NZDRI as the "technical highlight of the tour". This assembled some 300 delegates from around the world to listen to and discuss almost 50 technical papers including papers presented by members of the SDT tour party. On the non-technical side the transport of a Stilton cheese to be presented to Mr Robert Muldoon on the occasion of the commemorative dinner caused a problem with the N. Zealand customs and was not, it turned out, to Mr Muldoon's taste! However, all ended well as the Prime Minister quaffed a few glasses of Baileys Irish cream liqueur which was to his taste.

### Honorary Members of the SDT Since 1980

| | |
|---|---|
| Dr E Green | August 1980 |
| Mr E McCormick (deceased) | February 1982 |
| Dr J Rothwell (deceased) | February 1984 |
| Mr P Hoare | February 1984 |
| Mrs S W Barron | February 1985 |
| D H Burton | February 1985 |
| Miss J Betts | February 1985 |
| Mr W J Hipkins | February 1985 |
| Mr F G Weldon | February 1985 |
| Dr R J M Crawford | February 1986 |
| Dr G Chambers | February 1987 |
| Mr H E L Harz (deceased) | February 1987 |
| Mr D Ralph | August 1987 |
| Miss B Fitton* | August 1987 |
| Mr P Lee** | February 1989 |
| Mr R Lawton (deceased) | February 1989 |
| Miss E Richards | February 1992 |
| Mrs S Watson | February 1992 |
| Dr P Young | February 1993 |
| Mr R Thompson (deceased) | C A 1994/96 |
| Professor D D Muir**** | May 1998 |
| Mrs R A Gale***** | August 1998 |
| Mr R Early | May 2002 |
| Mr J Lyons | September 2003 |

\*        *Past Managing Editor of the Journal*
\*\*      *Past SDT Executive Secretary.*
\*\*\*\*    *Also SDT Gold Medal Award (at 1996 AGM)*
\*\*\*\*\*  *Retiring SDT Executive Secretary*

*21st Anniversary Celebration, 26 October, 1964*

This celebration is recorded in the Journal (J.S.D.T. vol 18 No 1 1965)

A dinner and dance were held on Monday, 26th October 1964, at the Europa Hotel, London, W.1, to mark the 21st birthday of the Society.

This happy and informal function was attended by nearly 300 members and guests of the Society, the principal guest and speaker being, most appropriately, our Founder President, Professor H. D. Kay, C.B.E. The other guests of honour were Mrs. Kay, Mr. R. B. Giusti (President of the Dairy Engineers' Association) and Mrs. Giusti, Mr. J. White (Vice President, National Dairymen's Association) and Mrs. White, Mr. W. A. Nell (Honorary Member of the Society) and Dr. R. Seligman (First Gold Medallist of the Society).

The official hosts for the evening were Mr. J. F. Hunter, who began his presidential year at midnight during the dance, Mr. C. H. Brissenden, the 21st President of the Society, Mr. W. W. Ritchie, President of the Society, 1962-1963, and their ladies.

*The President, Mr. C. H. Brissenden, making the presentation to the Secretary, Mr. Michael Sonn.*

The toast of 'The Society' was proposed by Mr. Brissenden and Mr. Hunter proposed the health of the Society's guests, prompting an eloquent reply from Professor Kay.

An item which did not appear on the toast list was the presentation of a tea set to Mr. Michael Sonn, Secretary of the Society for the past 15 years, in appreciation of his loyal and cheerful service. It was evident that the presentation took Mr. Sonn completely by surprise and the thought which prompted it was heartily endorsed by the acclamation of all present which followed his words of thanks.

This was indeed a memorable occasion, attended by members and guests from all parts of the United Kingdom, it must have been particularly gratifying to Professor Kay, and the 14 of his 21 successors as President who were also present, to see such abundant evidence of health and vigour in the Society at its coming of age.

*Essay competition*

An essay competition was organised as part of the celebration of 21 years of Welsh Section. In 1967 the winner was "Half a Century" by Sion Tyllwyd, the nom de plume of Miss E Richards. After that the competition was confined to students! A short extract follows:

"We youngsters from the village were part of the farming community. The old shepherd could not get along without our help at lambing time. Washing and shearing days were to us as bank holidays are to the present generation. The rickyard would not have been filled for the needs of winter if the village lads hadn't lent a helping hand. Oh, the pride of sharpening the scythe or the hook or axe so that it was as sharp as that of the old master. His tools were so keen he could even remove the hair off his arms. Oh for the day when we could do just that! How we looked up to those 'old' men. The ploughing, the hedging, the ditching amongst other tasks, all of them were important in our lives. We learnt a great deal about them and even more from Mother Nature and her servants......"

*50th Anniversary Conference 20/21 April 1993*
*(JSDT Vol 45 No. 4 November 1992)*

The conference a successful and happy one and is recorded in a handsome booklet "50 years – A Commemorative Booklet to celebrate the Golden Jubilee of the Society of Dairy Technology." The photographs in this booklet are of special interest and value.

"The Society celebrates its 50th Anniversary in 1993 and the main event in the Society's social calendar is the Conference being held on 20/21 April 1993 at the Garden House Hotel, Cambridge.

The '1993' Committee have met regularly throughout this year to formulate a programme. Internationally renowned speakers will present papers on Milk Production, Milk Processing and Utilisation, Marketing, Technology and an Overview. Your Council is proud to award gold medals to Dr George Chambers CBE and Dr Borg Mortensen. It is hoped that our Founder Members will be able to attend, because it was through their determination, despite the exigencies of wartime, that the Society was formed to bridge the communication gap between scientists and industry and the Society still strives to do this today. We also hope that as many Past Presidents, Honorary and Retired Members, as possible will be in our midst.

The Conference Banquet promises to be a very prestigious occasion, with Dr George Chambers presenting the After Dinner Speech.

The Journal cover will take on a new look for 1993. A commemorative booklet is planned. A commemorative paperweight will also be available.

Sections are planning to hold at least one special function in their area to celebrate the 50th year of the Society.

*Links With Related Organisations*

With the decline in membership the Society initiated talks with organisations whose interest and objectives and field of activity correspond to a greater or lesser degree with those of the S.D.T.

THE INSTITUTE OF FOOD SCIENCE & TECHNOLOGY (IFST) AND THE ROYAL ASSOCIATION OF BRITISH DAIRY FARMERS (RABDF).

Discussions with the IFST in the 90's got to the stage of presentation of an outline scheme to the Council of the SDT. On advice, Council, while acknowledging the positive character of the discussions, decided to carry on independently. However, cordial relations were maintained with the IFST and consideration is being given to a reciprocal associate membership scheme. As normal with such propositions the problems tend to overshadow the advantages. As the contraction in the number of players in the milk and food industries in the UK continues there will be pressure from all sides to foster closer links.

Talks with the RABDF also took place in the same period and having identified ways in which the combined interests of both organisations could be satisfied the project collapsed. Talks resumed in 2000 and continued for several months intermittently. It was suggested that the SDT could use spare office space at the RABDF office in Leamington Spa to carry on its administration from there. However, the advantages to both organisations were not convincing enough for the Council of each organisation to endorse the proposal.

Following the first set of talks the SDT became one of the organisations affiliated to the RABDF and Dr. Armstrong. who was a long-standing member of the RABDF represented the SDT on the Council of the RABDF. He was also, for a time, a member of the General Purposes Committee of the RABDF. As part of a new organisation of the RABDF the concept of affiliated societies was abolished.

Mr. Tim Brigstocke, Chairman of the RABDF is a member of Council of the SDT which contributes to a constructive relationship between the two organisations.

As has been mentioned elsewhere the SDT is represented on the UKDA and through that body to the International Dairy Federation. The Society continues to seek actively the possibility of collaboration with the organisations in the agri food sector including D.I.A.L. which represents the manufacturers and processors.

*Confederation of Societies of Dairy Technology*

During the Society's study tour of New Zealand and Australia, in March 1977, an opportunity was taken to further an idea of Mr. V. C. H. Cottle, to institute a federation of Societies of Dairy Technology. A preliminary meeting was held between the then Presidents of the New Zealand, Australian and our own Society...

A subsequent meeting took place during the IDF Sessions in Paris, in June 1978, between the same societies and the Danish Society of Dairy Technology.

"The Society has recently received assurance from the other societies concerned that such a Confederation would be a very welcome one, and it is hoped that an opportunity may be taken during the IDF Sessions in Montreux, in September this year to discuss the matter further and bring the formation of a Confederation of Societies of Dairy Technology to fruition.

The draft aims and objections of the Confederation are:

1. To promote goodwill on an international basis between members of the dairy industry, and to further the well being of the dairy industry worldwide.
2. To assist members of sister societies, either as individuals or as groups, during visits to the countries of individual societies.
3. To exchange information on the programme of activities of each society. (Note: implicit in this aim that member of any federated Society would be welcome at the meetings of any other federated Society on the same basis as local members.)
4. To assist in providing international speakers from members visiting other countries by early notification of possible arrival of those members.
5. To create a channel for informal discussion on any problem of the dairy industry.
6. To endeavour, by scholarship or industrial exchange, to provide a means of education for members.
7. Where possible to establish joint meetings of member Societies.

On 7 September 1980 there was a meeting of the Confederation of Societies of Dairy Technology in the Grand Hotel, Bristol. (JSDT. Vol 34, No. 1 1981).

Present were: Mrs. S. W. Barron (UK and Ireland Society), Dr. A. H. Pederson (Danish Society), Dr. W. Sanderson (New Zealand Society), Mr. V. C., H. Cottle (Past President UK and Ireland Society) and Mr. P. H. F. Lee (Secretary, UK and Ireland Society).
An apology for absence was received from Mr. J. Ridley Rowling, OBE, a Past President of the UK and Ireland Society.
The main topic for discussion at the meeting was the possibility of holding an International Conference in 1982 or 1983, to inaugurate the Confederation.
All the delegates agreed in principle that this was an excellent idea... May or October were mentioned as the months most suitable to the New Zealand and Australian Societies for such a conference.
A second important topic discussed was the possibility of sponsorship of a UK student by the Ridley Rowling Trust under the terms of Item 7 of the aims and objectives of the Confederation. This had been offered by Mr. J. Ridley Rowling in a letter to the UK Society. In discussion on this proposal it was reported that both the Australian and the New Zealand Society had special funding for industrial training, exchanges, etc. The policy in New Zealand was to assist, say, Plant Operators as a whole, rather than one student for a lengthy period. Denmark already had five or six students yearly from the UK, and could easily assist in this matter. A list of contacts within the member Societies was appended to the minutes of the meeting.
On Society publications it was reported that a list and some publications from the Australian Society had been received by the UK Society, and it was intended to publish details of these in the January 1981 issue of the UK Society's Journal...
In discussing possible further members for the Confederation, both Norway and South Africa were suggested, although the organisation in Norway was not strictly a Society of Dairy Technology. It was agreed that, in time, investigations might be made to see whether South Africa should be invited to join.

*Memorabilia - The Society Gavel*
Information about the Society Gavel was reported in the Journal in 1981. (JSDT Vol 34 No. 1 January 1981).
In 1947 the Australian Society of Dairy Technology expressed a wish to develop a closer association with out own Society, and in token of this presented, through the late Mr J H Bryant,

of Bell Bryant Pty, one of their Founder Members who had been a member of our Society since 1943, a gavel made form a piece of cedar wood taken from the building which housed the first pasteurising plant in Australia.

The gavel was subsequently inscribed with the name and year of office of every President since our Society's foundation, and was for many years used by the President at General Meetings. It has now been adopted officially as the symbol of Presidential office.

Inevitably, with the passage of time, space on the gavel to continue the traditional inscriptions has run out, and at it's annual General Meeting on 29th October, 1980 the Society was therefore delighted and deeply grateful to receive from the retiring President, Mrs S W Barron, the gift of a stand, made from matching wood especially obtained from Australia, which will provide space for further inscriptions for many years to come.

Mrs Barron's thoughtful and generous gesture is commemorated on the plaque affixed to the stand.

### Ties and lapel badges

The Society also produced a tie and lapel badge. The tie had a motif which was the prize winning design submitted in a competition organised by the Society. Originally the ties were priced at £3.50 each but in 1984 the price of the tie was reduced to £2 and that of the lapel badge to £1.

### Society plaque (shield)

The Society also produced a plaque which had the same design as the tie which was presented on occasions to the hosts of overseas study tours.

Miss E Richards in her acceptance speech as President (JSDT Vol 42 No. 1 February 1989) remarks that Peter Lee complained about the weight of the Society plaques carried on Study Tours! Miss Richards added, "these have been distributed across many continents."

### 50th Anniversary commemorative paperweight

A limited number of glass paperweights hand engraved with the Society logo and in a presentation box priced at £14.50 were made to commemorate the Golden Jubilee of the Society in 1993 (JSDT Vol 46 No. 2 1993).

### Prominent Families in the Industry
### The Barhams of the Express Dairy Company. (JSDT vol 7 No. 2 April 1954).

Many of the diary companies which were familiar names in the UK were family founded. It would be rewarding and interesting to trace the influence of families such as the Davis, Gates, Horsleys but space does not permit. I have selected the story of the Barham's of The Express Dairy Company because it traces the development of the industry and especially milk delivery in the 19th Century.

When the full history of the remarkable development of the dairy industry of this country from mid-Victorian days onwards comes to be written, no contributions will deserve and receive greater prominence than those made by the late Sir George Barham and his son George Titus, based upon the foundation and development of today's Express Dairy Company.

### Sir George Barham

The life of George Barham (1836 – 1913) overlapped and supplemented by that of his son George Titus (1860 – 1937) covered practically the whole period of the rise and progress of British dairying from the medieval conditions of milk distribution and processing still prevalent in early Victorian days, to the gigantic and complex industrial organisation of today.

His active connection with the industry dated from 1858 when, at the age of twenty-two, breaking away from a different intended career, he started a milk retailing business in Dean Street, Fetter Lane, in the City of London. His success was immediate and his activities soon expanded into wholesale trading. He was quick to recognise the possibilities of extending the area from which his supplies of milk were drawn into the rural areas surrounding the metropolis

– a development which the rapid expansion of railway facilities helped to make practicable, and soon proved to be essential when the outbreak of cattle plague in 1865 closed the London cowsheds and threatened a collapse of the milk supply. This extension of his business into the wholesale field led to his establishment in 1864 of the Express Country Milk Company, with its headquarters in Museum Street, Bloomsbury, and branches in the strand, where his father had carried on business before him, while another branch was established at Blackheath. At the same time he entered the field of milk production with a well-equipped farm at Finchley… The herd of Guernseys established at this farm acquired a national reputation and provided foundation stock for herds in many parts of the world.

The marketing of 'country milk' in the metropolitan area raised two problems which he quickly solved: firstly the prejudice of the London public against 'cooled' milk, and secondly the level of railway transport rates which were very burdensome upon a bulky, low-priced article such as milk. In obtaining specially favourable rates from the railway companies he was doubtless greatly helped by the pressure at the time of the cattle plague crisis referred to above, but once obtained, the concession was permanently retained.

When this was later endangered, however, by the proposal in 1885 for an overall increase in railway rates, the British Dairy Farmers' Association, of which he had become a very active member, appointed him Chairman of a committee set up to watch the interests of the milk traffic, and when some four years later the proposals were tabled in detail and led to the holding of a Board of trade enquiry, he was nominated by the B.D.F.A., and the other dairy associations of the country as their representative to give evidence before the Committee. The success of his presentation of the case, despite the formidable array of Q.C.s briefed by the railways, is evident from the closing comment of the President of the Board of Trade that 'Mr Barham has sustained his objections and the rates must be considerably reduce.' The reduction was not effected without much further effort, but by 1892 the rates were restored to the old level, at which they were more favourable than those on any other article of commerce.

As a side line to his business Barham successfully developed the supply of dairying utensils and equipment and by 1880 this activity had reached such dimensions that a separate company was formed for it under the title 'The Dairy Supply Co. Ltd'. To this company was transferred also the wholesale milk branch of the parent company, the name of which was changed in the following year to the now long familiar 'Express Dairy Cp. Ltd'. On behalf of the new company Barham scoured the Continent for all that was then best in the way of dairy appliances and in 1885 secured the sole agency for that pioneer of mechanical cream separation, the de-Laval (now Alfa-Laval) Separator. He was also influential in the introduction of the Lawrence Capillary Refrigerator, by the use of which the problem of transporting milk over long distances was solved. By equipping with this and other new and improved appliances the working dairies at the various Agricultural Shows he also did much to accelerate the improvement of farm and dairy equipment…

Another activity on behalf of the trade that may e recalled were his efforts to bring under control the trade in butter substitutes which had reached larger dimensions in the closing years of last century, and by the use of names such as 'butterine' and in other ways was sailing very 'near the wind' in its competition with the butter trade.

Two other Bills on the subject were introduced in Parliament and eventually in 1887 the best features of the three were embodied in the Margarine Act of that year.

Throughout the last quarter of the century he was always in the foreground of the defence of the interests of dairying and milk supply when threatened by proposed legislative or administrative measures…

The inclusion in the Act of the 'presumptive' standards of quality by which the 'purity' of milk is assessed for the purposes of the Act was due almost entirely to his forceful and persistent advocacy of this safeguard of the producer's integrity.

On the farming side his interests and aims were implemented through the British Dairy Farmers' Association, on whose Council he served for many years as Chairman of the Finance and general Purposes Committee, followed by election as a Vice-President in 1901 and President in 1908.

The wide spread of his activities did not impair in any way his labours on behalf of the dairy trade of London, as a founder of the Metropolitan Dairyman's Society in 1873, of the Metropolitan Dairymen's Benevolent Institution in 1874 and of the Dairy trade and Can Protection Society in 1890, to mention only a few...

His great record of public service received its crown of recognition when in 1904 he received the honour of knighthood, with nine years of life still left for its enjoyment.

*George Titus Barham*

Passing from father to son – George Titus was born in 1860 in Dean Street, Fetter Lane, E.C., where, as above recorded, his father had recently started his dairy business. As a boy he suffered from a physical infirmity from which he never completely recovered but which fortunately was sufficiently ameliorated to enable him to live out a long and active life.

He was educated at University College, Gower Street, and at the age of eighteen entered his father's business, becoming a Director three years later, rising in due course to be Managing Director, and ultimately on his father's death in 1913 succeeding him as Chairman of the Express Company. Under his control the business expanded rapidly into the great organisation which at his death (1937) owned upwards of 400 dairy and refreshment branches, five processing centres and a chain of creameries and collecting stations covering the whole country and with more than 10,000 employees.

Like his father, his interests were almost inexhaustible, but on the dairying side were more closely confined to the development of his business. He was passionately fond of animals and so was naturally strongly attracted to dairy farming, where he attained high rank as a breeder and judge of cattle, especially Guernseys. Along with five others he founded (1884) the English Guernsey Cattle Society of which he was twice President. He also attained the same honour in the British Kerry Society and the Dexter Society. He was one of the oldest members of the Council of the British Dairy Farmers' Association, and occupied the Presidential chair in the Society's Diamond Jubilee Year (1934). His farming extended over ten farms in Middlesex, and his 'home farm' at Sudbury Park carried a herd of 100 Dairy Shorthorns and as long ago as 1908 was producing T.T. Milk.

In the London trade organisations he continued his father's interest, especially in the Metropolitan Dairymen's Benevolent Institution, of which he was President at the Coming of Age festival in 1895, and again in 1934 at the Society's Diamond Jubilee.

His absorbing interest outside dairying was in gardening, his rose gardens at Sudbury Park, to which he gave the public access, being widely known and admired.

His outstanding personal quality was his genius for organisation and his lasting contribution to dairying was the great organisation passing on his death into the hands of his kinsman the present Chairman, Capt. Walter A Nell.

On Sir George's death his younger son, Col Arthur Barham, succeeded him as Chairman of the Dairy Supply Company of which he had been Managing Director for several years previously. The Company was later brought into United Dairies in the initial amalgamation of the wholesale companies in 1915, and Col. Barham remained a director until his death two years ago."

### Milk Producer Oct 1983
### (Issue on the occasion of 50th Anniversary of the MMB 1933-1983).

This short article *(see overleaf)* in this anniversary issue of the Milk Producer "How the Board began" describes succinctly the chaos which existed before the Board was set up and the really hard times experienced by milk producers. It is a poignant reminder of the incalculable damage done by those responsible for the destruction of what John Empson described as British Farmers Greatest Commercial Enterprise.

# How the Board began

It does no harm to be reminded of the circumstances that existed before
the Board started buying milk from all the farms in England and Wales and
returning the farmers a 'pool' price. Letters like the one below were all too common.

IN the years after the First World War a slump in milk prices hit producers very badly. Those dairy farmers who were situated well away from areas of large consumption like London and Birmingham were even worse hit.

In 1922 a Permanent Joint Milk Committee was set up by the Ministry of Agriculture — this attempted to establish a liquid price and a better manufacturing price. In 1922 the liquid milk price was 16d a gallon and manufacturing price 9½d. By 1932 this had slumped to 13d and 5d respectively.

By the late 1920s some producers had strengthened their bargaining position by joining together. The Bristol Milk Pool and the Yatton Farmers' Co-operative were two small organisations which contained the seeds of a milk marketing scheme.

The Agricultural Marketing Act of 1931 made the formation of a producer-run milk marketing organisation legally possible at last. A Royal Commission and a public enquiry followed before Parliament passed the Milk Marketing Act in July 1933. On 3 August the Board of 18 producers met for the first time, with Somerset producer Thomas Baxter as chairman.

On 2 September 1933 a poll of milk producers in England and Wales resulted in 96 per cent voting for the scheme.

Now that the Board had its mandate to operate it had one month to obtain offices, recruit staff, administer and register contracts and obtain finance in time for the November 'due date', when producers would expect to be paid.

Sidney Foster was appointed general manager, and thanks largely to his efforts, bridging loans were obtained from the banks. Staff were recruited and offices found at Thames House, Westminster. The first monthly cheques were sent out to producers covering the weeks from 6-31 October 1933.

From now on all milk producers in England and Wales were guaranteed a market for their milk. Pouring milk away because the price offered was uneconomic had become a thing of the past.

TABLE 1

*Statistics of the Board's history, 1933–1994: producers and cow numbers*

|  | Number of registered producers as at March | Dairy cow numbers ('000 head) as at June | Average herd size as at June | Yield per cow (litres) April to March |
|---|---|---|---|---|
| 1934 | 141 000 | 2 206 | 17 | na |
| 1940 | 138 490 | 2 349 | 15 | 2 459 |
| 1945 | 158 011 | na | na | 2 355 |
| 1950 | 161 937 | 2 491 | 16 | 2 830 |
| 1955 | 142 792 | 2 415 | 17 | 3 065 |
| 1960 | 123 137 | 2 595 | 21 | 3 320 |
| 1965 | 100 449 | 2 650 | 26 | 3 520 |
| 1970 | 80 265 | 2 714 | 33 | 3 755 |
| 1975 | 60 279 | 2 701 | 46 | 4 070 |
| 1980 | 43 358 | 2 672 | 58 | 4 715 |
| 1985 | 37 815 | 2 580 | 67 | 4 765 |
| 1990 | 31 510 | 2 324 | 71 | 5 070 |
| 1994 | 28 033 | 2 218 | 74 | 5 265 |

TABLE 2

*Sales, utilization and consumption of milk*

| Year ending March | Sales of milk off farms (millions of litres) | Liquid (millions of litres) | Manufacture (millions of litres) | % Manufacture | Liquid milk consumption per head (pints per week) |
|---|---|---|---|---|---|
| 1934 | 3 890 | 2 922 | 968 | 24.9 | 2.99 |
| 1940 | 5 040 | 3 459 | 1 581 | 31.4 | 3.01 |
| 1945 | 5 398 | 4 801 | 596 | 11.0 | 4.00 |
| 1950 | 6 917 | 6 092 | 825 | 11.9 | 4.96 |
| 1955 | 7 516 | 6 104 | 1 413 | 18.8 | 4.87 |
| 1960 | 8 175 | 6 253 | 1 923 | 23.5 | 4.85 |
| 1965 | 9 044 | 6 647 | 2 398 | 26.5 | 4.94 |
| 1970 | 10 022 | 6 664 | 3 358 | 33.5 | 4.80 |
| 1975 | 11 115 | 6 878 | 4 237 | 38.1 | 4.73 |
| 1980 | 12 775 | 6 432 | 6 343 | 49.7 | 4.41 |
| 1985 | 12 604 | 6 084 | 6 505 | 51.6 | 4.12 |
| 1990 | 11 632 | 5 938 | 5 684 | 48.9 | 3.97 |
| 1994 | 11 211 | 5 875 | 5 338 | 47.6 | 3.88 |

Source: The Residuary Milk Marketing Board.

| Calendar Year | 1998 | 1999 | 2000 b | 2001 |
|---|---|---|---|---|
| Average dairy cow numbers ('000) | 2 461 | 2 445 | 2 354 | 2 251 |
| Average yield per cow (litres) | 5 774 | 5 964 | '5 977 | 6 348 |
| | million litres | | | |
| **Gross Production** | **14 210** | **14 581** | **14 071** | **14 285** |
| Fed to stock and waste on farm | 280 | 285 | 277 | 283 |
| **Available for human consumption** | **13 930** | **14 296** | **13 794** | **14 002** |
| **Consumed Liquid** | | | | |
| Wholesale sales | 6 599 | 6 720 | 6 629 | 6 628 |
| Direct sales | 140 | 132 | 139 | 133 |
| On farms | 46 | 46 | 45 | 44 |
| **Total** c | **6 785** | **6 899** | **6 813** | **6 806** |
| **Milk Used for Manufacture** | | | | |
| Wholesale sales | 6 734 | 6 909 | 6 470 | 6 638 |
| Direct sales | 86 | 79 | 80 | 77 |
| **Total** c | **6 821** | **6 988** | **6 550** | **6 715** |
| used for manufacture of: | | | | |
| - Butter | 281 | 290 | 270 | 259 |
| - Cheese | 3 257 | 3 297 | 3 032 | 3 568 |
| - Cream | 263 | 271 | 266 | 259 |
| - Condensed milk d | 643 | 603 | 522 | 536 |
| - Whole milk powder | 809 | 853 | 932 | 781 |
| - Skimmed milk powder | 1 101 | 1 123 | 889 | 663 |
| - Other | 467 | 549 | 640 | 649 |

Reference: 'Dairy Facts and Figures' 2002
Edition, The Dairy Council, London.

## Roland Williams in article
### "The political economy of the common market in milk and dairy products in the European Union, describes the Market Structure in the UK"

### The United Kingdom
#### (a) Market Structure

The UK's unique market and dairy industry structure results from a history that was quite unlike anything elsewhere in the Community at the time of Accession in 1973. In the early 1970s more than 60 percent of the UK's relatively large milk output from farms was used for the liquid market, with the product sold to consumers "as from the cow" (i.e., with natural fat content and "unstandardised"), very little sale of skim as drinking milk, and no sales of reduced-fat milks. A high proportion of the UK's substantial requirements of dairy products was imported, especially butter (90 percent) and cheese. These imports were particularly important in the very narrow "world market", especially to New Zealand, with whom the UK had a special understanding regarding milk policy, but also to Ireland, Denmark, Australia and other countries. The record of understanding on milk policy between the UK and New Zealand and Denmark as traditional suppliers, and complaints by them against "dumping" on the butter market, the UK government introduced limited control on imports in 1962 but not such as to raise prices substantially. Trade in liquid milk was impossible by virtue of sanitary regulations requiring all premises used for pasteurising and packaging liquid milk for sale to UK consumers to be licensed by a local authority who was responsible for hygiene standards. Double pasteurisation was not permitted, and drinking milk had to be pasteurised and packaged (mostly in bottles) in the same premises. Foreign supplies were therefore excluded from the UK liquid market either as bilk milk or packaged milk. While dairy product prices to consumers were low, prices of liquid milk tended to be higher even though they were politically controlled and subsidised for long periods after World War II and before entry into the then EEC in 1973.

Consumption of liquid milk and butter were both high by international standards and the markets were therefore large. Apart from controlling the retail liquid price the government also subsidised the market through a welfare scheme for expectant mothers and children under five years (one pint a day at half price) and a school milk scheme for all school children.

The structure of the dairy industry itself in the UK differed from that of continental Europe in so far as it was dominated by large private and public companies with scarcely any farmer-co-operative enterprises except in Northern Ireland (although consumer co-operatives were strong). The large dairy companies and consumer co-operatives were nearly all involved in a doorstep delivery service for bottled milk, usually by running their own rounds services, and in some parts of the country also selling bottled milk to "bottled milk buyers", who ran small private delivery businesses but were not large enough to run capital-intensive pasteurising and bottling plants efficiently. Sales of milk through shops was very small.

Dairy product manufacture in the UK was a relatively small industry. Large companies made a complete range of dairy products in addition to distributing liquid milk. Priority of manufacture would usually be given to cream when there was a range of products, then cheese, mainly cheddar but also a range of hard, pressed "territorial" varieties. Butter manufacture was very much a residual usage of milk. With the bulk of milk all the year round going to the liquid market, dairy product manufacture tended to be seasonal and fluctuating. As a result it has a history of being relatively high cost, which, together with very low market prices, meant that milk prices to producers for the main products of butter and cheese were very low.

Following the great depression of 1929 and the early 1930s, the UK government enacted the Agriculture Marketing Acts of 1931 and 1932. These Acts enabled farmers' representatives to establish statutory approved Milk Marketing Schemes to constitute Milk Boards that would have the exclusive right to market *all* milk produced in an area, provided the Scheme received the approval at a poll of a two-thirds majority of all producers in the area producing at least two-thirds of the milk. Five Boards were established in the UK under this legislation, one for England and Wales (in 1933), three in Scotland) one main Board plus two independent small ones for Aberdeen and district and the North of Scotland in 1934) and one for Northern Ireland (in 1955). Detailed histories of the activities of the Boards from different perspectives can be found in Baker (1973), Strauss (1972) and Empson (1996).

The most essential features of the Milk Marketing Boards in the Uk were that they controlled *all* milk sold off farms in their areas; and the prices at which they sold milk to the dairy trade varied according to the *usage* of the milk by the company. The highest price was charged for liquid milk, with a range of prices for manufactured products varying from cream at the top to butter and skim powder at the bottom. The receipts from all sales were "pooled" with the average pooled price being paid to farmers. Thus the Milk Marketing Boards were non-govermental, farmer-owned and farmer-controlled, but derived their power and authority from Parliamentary statute. Their constitutions and rules of governance in the form of the Milk Marketing Schemes had to be approved by Parliament. Throughout most of their history the UK government maintained close control over key prices in the system, including liquid wholesale selling prices of the Boards and the retail price by the dairy trade from the beginning of World War II. Farmers were given a *guaranteed price* which was limited after 1954 to a "standard quantity" loosely related to liquid sales. Trade margins were controlled by elaborate costings systems, and the guaranteed price was related to income objectives for agriculture as a whole through negotiations between the government and the National Farmers 'Union. The system was elaborate but completely dependent on the Boards as the policy delivery agents.

## PROPERTY TRUSTEES

As Professor Kay has noted in his early history of the Society, the original Trustees were Mr. Graham Enock, Mr. A. McBride, Mr. W.A. Nell, Mr. F. Procter, Mr. E. White and Mr. Richard Seligman. These Trustees were appointed to enable the fledgling society to hold property - hence the subsequent references to "Property Trustees". At a later date, the late Sir Richard Trehane was appointed a Trustee. At the time of negotiating the lease of the premises at Huntingdon, Sir Richard was the only survivor of this group. Though very ill, Sir Richard completed a Deed of Variation which enabled Mr. Peter Fleming, Mr. Wilf Hipkins and Dr. David Armstrong to be appointed *"Property"* Trustees.

# BRIEF BIBLIOGRAPHY

The dairy industry in N. Ireland 1963 Printed and Published by Nicholson and Bass for the Min of Agriculture for N. Ireland, the MMB for N. Ireland, the Northern Ireland Milk Alliance and the Federation of Producer Co-op creameries.

The UK Milk Marketing Boards:    A concise history Published SDT 2000.

The Royal Association of British Dairy Farmers 1876 - 1976
(Published to commemorate the century of the RABDF).

A Border Co-op, The Town of Monaghan Centenary 1901 - 2001,
John O'Donnell Town of Monaghan Co-op Ltd 2001.

Fruits of a Century.        An illustrated Centenary History 1894 - 1994,
                            Irish Co-operative Organisation Society (ICOS)
                            Editor Maurice Henry 1994.

Journals of the Society of Dairy Technology 1947 –

The political economy of the Common Market in milk and dairy products in the European Union.
Roland E. Williams FAO Economic and Social Development paper 142, Rome 1997.

Milk Marketing Boards of the UK and Milk Marque.    Annual Reports and Accounts.

Dairy Crest (UK) Ltd                                Annual Reports.

Milk Marketing Board                                The Milk Producer

Milk Development Council                            Annual Report 2001-2002

Plunkett Foundation                                 The first 75 years.

First published in the UK by Sweet Cherry Publishing Limited, 2022
Unit 36, Vulcan House, Vulcan Road,
Leicester, LE5 3EF, United Kingdom

Sweet Cherry Europe (Europe address)
Nauschgasse 4/3/2 POB 1017
Vienna, WI 1220, Austria

2 4 6 8 10 9 7 5 3

ISBN: 978-1-78226-901-4

Football Rising Stars: Ansu Fati

Text by Harry Meredith
Illustrations by Sophie Jones

Lexile® code numerical measure L = Lexile® 910L

www.sweetcherrypublishing.com

Printed in India

# ANSU FATI

THE UNOFFICIAL STORY

Written by
HARRY MEREDITH

Sweet Cherry

# CONTENTS

# STARRING AT CAMP NOU

On the first day of the 2019/2020 La Liga season, Barcelona were in a fighting mood. The team had finished second in the previous season, behind their fierce rivals Real Madrid. In order to stop this from happening again, Barcelona's new manager, Ronald

Koeman, sent out an incredibly strong lineup for the starting day match against Villarreal. Their defence included the trusted and talented Gerard Piqué and Jordi Alba, while the hard-working Sergio Busquets led their untouchable midfield.

But it was towards centre field that the stars of the show were placed. The club's star forward, Lionel Messi, was leading a line of four attackers. Behind him were Antoine Griezmann, Philippe Coutinho and ... a 17-year-old.

Alongside the three veteran attackers was a player who looked like they belonged in a classroom rather than on a professional football pitch. But this was no ordinary 17-year-old.

This was Ansu Fati – a forward who deserved to be playing with the world's best.

The match was held at Barcelona's famous Camp Nou stadium. However, the stands were empty due to COVID-19 restrictions. The players emerged from the tunnel, and although there weren't any fans in the

stadium, they knew that millions of fans were watching on their TVs and cheering the team on from afar.

After the coin toss, the players ran to their allocated sides of the pitch and began to warm up. Messi walked over to Ansu.

'Are you ready for this?' asked Messi.

'100% ready!' said Ansu, pumped up.

'That's the spirit.' Messi gave Ansu a fist bump and sprinted on the spot. 'Come on everyone,' he called out to the team. 'We are Barcelona!'

The referee blew the whistle to start the game. The 2019/2020 season

was underway. Barcelona started
the match with the enthusiasm
and determination their captain
demanded. Neat passes and dashing
runs tested the Villarreal defence in
the opening stages.

In the 15th minute, Ansu showed
just why he was on the pitch. Jordi
Alba zoomed down the left flank
and crossed a low ball into the
penalty box. Ansu ran on to the
cross. Using the side of his right
foot, he cushioned the ball under
control and effortlessly directed it
towards the net. The ball zipped past

the goalkeeper so quickly he could hardly react and nestled into the back of the net. *Goal!* Ansu ran behind the goal and celebrated with his teammates. But they quickly fell back into position. There was still a lot of football to be played.

Four minutes later, Coutinho cut open the Villarreal team with a driving run. Ansu sprinted to the left-hand side of the penalty area and Coutinho played in a through ball. Ansu ran on to the pass and slotted the ball into the net once more. The game hadn't even reached twenty

minutes before Ansu had bagged
two goals. On a pitch with the unreal
talent of Lionel Messi, so far it was
the young 17-year-old who was
stealing the show.

In the 35th minute, Ansu found
himself in the Villarreal penalty box
once again. But this time he was too
far to the side to shoot. Keeping the
ball and looking for a passing option,
Ansu ran towards the byline. But
as he did, he was brought down by
Villarreal's right-back. The referee
brought his whistle to his mouth and
pointed his finger to the spot. Penalty!

With Ansu being on a brace, only one goal away from a hat-trick, some teams would have given Ansu the opportunity to score a third. But Ansu was playing on a team with one of the world's best: a player with an unmatched desire for goals and getting the win for his team.

Messi picked up the ball and placed it on the spot. He fired it to the left-hand side and it snuck underneath the outstretched arms of Sergio Asenjo. Messi clenched his fists and cheered as his teammates surrounded him. He had already scored hundreds of goals for this club, but that didn't stop every single one from bringing him the same joy.

Messi's contribution to the match wasn't finished just yet. At the end of the first half, Messi chipped the ball into the penalty box. Busquets ran after the ball, and in an attempt to clear it, Villarreal centre back Pau Torres accidentally prodded the ball into his own net. Barcelona were leading 4-0. The start to their La Liga campaign could hardly have gone any better.

Messi continued to trouble the Villarreal defence in the second half, but no more goals were scored. Ansu was substituted and the entire team clapped as he left the pitch. If Barcelona fans had been in attendance, Ansu's performance would have earned a standing ovation. Even the great Messi applauded as Ansu left the pitch. If one of the greatest footballers of all time was recognising Ansu's talent, it was only a matter of time until the world began to understand his power and potential too.

In a team overflowing with football stars, it was a 17-year-old who was stealing everyone's attention. If he could do this at such a young age, what unknown heights could Ansu Fati reach in his footballing career?

# 2

# LIFE IN GUINEA- BISSAU

Ansu Fati was born on the 31st of October 2002. He grew up in a caring and hard-working family in Guinea-Bissau – one of the poorest countries in West Africa. They lived in tough

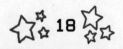

conditions, but Ansu's family always strived to do their best and dreamt of a brighter future. Ansu's parents, Bori and Lurdes, worked hard to look after their family. However, with five mouths to feed, this was no easy task.

One summer, Bori was working on a market stall trying to sell clothes to the thousands of passers-by who made their way through the market. While their father was working, Ansu's older brother, Braima, looked after him. There was only one thing the 10- and 6-year-old siblings did to pass the time:

Play football.

One day, the pair of them dashed away from Bori's stall, dribbling and passing an old drink can through the busy market instead of a ball. All around them were sellers calling out to customers, trying to sell their items. There were stalls brimming with fruits, electrical equipment, clothes and much more. One stall in particular made Ansu stop in his tracks.

'Braima, look!' said Ansu, pointing to the stall ahead of them.

'Woah!' Braima exclaimed.

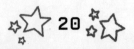

Hanging from the stall was a Barcelona shirt – and not just any shirt. It was Messi's number 10. The classic blue and dark red colours looked like they were glowing in the afternoon sun.

Ansu and Braima gleefully walked towards the stall without taking their eyes off the shirt. They were so awestruck that they bumped into the stand. The seller turned around and glared at the brothers as though they were thieves.

'You two boys,' he said, 'empty your pockets.'

Ansu and Braima showed their empty pockets to the vendor.

'Just as I thought. You can touch the merch when you have money to pay for it. Now go away!'

Ansu and Braima hurried away from the stall. But Ansu glanced back over his shoulder at the shirt. *One day I'll wear a kit like that*, he told himself.

Soon the chaos of the marketplace was quickly replaced with the chaos of a football pitch. On a dusty patch of land, a large crowd of children were playing a match. On one side of the pitch were some goalposts and on

the other was a makeshift
goal, put together using
a tree stump and a mound
of rubbish. The goalposts were
uneven, but none of the kids cared.

One of Braima's friends, Carlos, was
playing. He had brought his football
for the children to use. He spotted
Braima and ran off the pitch.

'We're losing!' said Carlos. 'We
need you on our team.'

'I'm in,' grinned Braima. 'Can my
little brother play?'

'Of course,' said Carlos 'We need all
the help we can get!'

Braima and Ansu shared a smile before charging into the game. To any onlooker this was just a patch of sandy earth, but to the children chasing that football it was their Camp Nou.

Ansu's team won the ball and Braima received a pass. He stopped the ball dead before flicking it past an oncoming player on the opposite team. Ansu was in space, so Braima found him with a pinpoint pass.

Ansu, much smaller than the older boys around him, wasn't fazed one bit. Using his small body, Ansu squeezed between two defenders and dashed up the pitch with the ball.

There was just the goalkeeper left. He tried to steal the ball off Ansu, but he couldn't. Ansu knocked the ball to the right, past the goalkeeper, but the angle wasn't there for him to shoot. Standing on the goal line was Carlos. Ansu passed to him and Carlos fired the ball into the makeshift goal.

'GOOOOOOAL!' he shouted, running up the pitch, pretending to celebrate in front of thousands of spectators.

'Let's go again!' called Ansu, knowing they needed more goals like that to pull ahead of the other team.

The brothers played with their friends until the sun went down, and even then they didn't want to leave. But soon Bori came to pick them up. He knew there was only one way of getting his sons to stop playing football and come home: by ordering them there in person.

When the brothers arrived home, their mother was waiting for them at the kitchen table.

'Sit down please, boys,' she said.

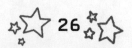

Exhausted from their day, the brothers slumped onto the worn chairs. Lurdes shot a glance at Bori before speaking.

'We've got something to tell you …' she said.

'I'm going to be leaving home for a while,' said Bori. 'I'm moving to Spain so I can find better work.'

'Can't we come too?' asked Ansu.

'Maybe one day, little man,' said Bori. 'But I need to go on my own first.'

Ansu fell silent, his head hung low.

'But maybe we can go on an adventure there together in the future,'

said Bori, stroking Ansu's head.

After a while, Bori left the room to freshen up before dinner. Noticing that his brother was sad, Braima patted Ansu's shoulder.

'Barcelona's in Spain,' said Braima.

'Yeah!' said Ansu. 'Do you think Dad will see a football player?'

'Maybe,' said Braima. 'And if he gets work there, we could see one too.'

Ansu's mood brightened. Maybe one day he'd be watching football heroes in person instead of through a TV screen. Maybe he'd get to meet a real-life professional footballer.

Lurdes turned to them from the stove. 'Clean yourselves up,' she said. 'Or you won't be having any dinner.'

Braima looked to Ansu, his stomach rumbling, and jumped off his chair. Ansu quickly followed and raced his brother – their mother watching them go with a smile.

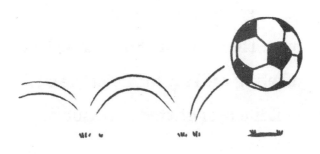

# 3
# WELCOME TO SEVILLE

Life in Seville, Spain, was tough for Ansu's father. Bori struggled to find work and often had to beg for food because he couldn't afford any of his own. Then one day, Bori had a chance meeting with the mayor of the city.

The mayor offered Bori a job as his

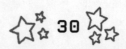

personal driver. This finally gave Bori
a chance to earn money and create a
life for himself. This also set the wheels
in motion for the whole Fati family to
start a new life in Spain. The mayor was
so impressed by Bori's hard work that
he helped the rest of the family move to
Spain. Guinea-Bissau was their home,
but it was time for a new chapter.

Aged 7 and 11, Ansu and Braima
signed up to play for a local team
in the town of Herrera. At Escuela
de Fútbol de Herrera, a free-to-play
football team, the two new boys stood
out from the hundreds of children at

the club. Their ability and flair, plus
an unmatched passion for the game,
drew the coach's attention.

Ansu was so good that the coach
had to check with Bori and Lurdes
that he was telling the truth about
his age. Ansu's parents told them
about Ansu's childhood. When Ansu
was little, the family didn't have a
football, so Ansu would roll up a pair
of socks and practise by kicking them
around the room. Even as a toddler,
Ansu had lived and breathed football,
and that was clear in his ability now
at just 7 years old.

After a year with Herrera, word started to spread about Ansu and his brother. This led to a professional football team approaching them. Sevilla FC, a club competing in La Liga, asked the brothers if they would like to join their academy. Hardly able to contain their excitement, the pair said 'yes'.

Even playing in a professional academy setup, Ansu was the star of the show. Ansu was so good, that many of the other players in the academy learnt more from playing against him than they did from the

coaches. One of Ansu's main traits on the pitch was his selflessness. He often ran all the way back into his half to defend when the opposition attacked. He also had a talent for setting up his teammates to score. Scoring wasn't everything for Ansu. It was a team sport, after all, and he enjoyed helping his teammates to succeed as well as succeeding himself.

In one tournament, Ansu's skill dropped the jaws of everyone watching. A speedy cross was fired across the box, and Ansu leapt into the air, connecting with a thunderous

bicycle-kick. The ball blazed into goal so quickly that it seemed it would burn through the net. This was a skill that one of Ansu's heroes at the time, Real Madrid's Cristiano Ronaldo, had also perfected. However, it was Real Madrid's greatest rivals, Barcelona, who soon came knocking at Sevilla's door.

Barcelona wanted to sign Ansu Fati.

# 4

# LA MASIA

By the time Ansu was 9 years old, he was no longer Sevilla's secret. Clubs from across the country and beyond were aware of a special talent emerging in the region. Scouts from Spain's big two – Barcelona and Real Madrid – frequently came to watch

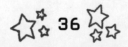

Ansu, and both teams wanted to bring him to their academies.

A representative from Barcelona spoke with Ansu's father several times to persuade him that Barcelona was the place for Ansu. The club were interested in bringing Braima to the academy, too, and offered on-site housing for the brothers.

Real Madrid were so desperate to sign Ansu that they offered to buy the whole family a house in Madrid for

them to move to. But the Fatis' minds and hearts were already set ...

Ansu and Braima were going to Barcelona.

However, Ansu couldn't start straight away because Sevilla refused to let him go. Using the Royal Spanish Football Federation's laws, Sevilla stopped Ansu from joining Barcelona's academy for a year. The same law didn't apply to his brother, so Braima left right away to join Barcelona's academy.

At first this seemed like it was going to slow Ansu's development, but it actually had the opposite effect. Ansu had been training so much that he had hardly had time to be a kid and enjoy life with his family. This one-year pause gave him extra time with his caring, loving family. For a whole year, he didn't have to worry about the pressure of a football academy and the hard work required to succeed within it.

When Ansu turned 10, the ban was over. It was time for Ansu Fati to join one of the most famous academies in world football:

Barcelona's La Masia.

Upon his younger brother's arrival, Braima gave Ansu a tour. He took him around the multiple training pitches, the indoor complex filled with all of the latest gadgets and technology, and the trophy room displaying Barcelona's proud achievements.

'That will be us one day, little brother,' Braima beamed, pointing to a picture of players holding the La Liga trophy in front of thousands of celebrating fans.

'I hope so,' said Ansu, with a smile.

When young players are taken away from the comforts of their

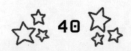

family home, they can often find it difficult to fit into academy life. But Ansu and Braima never had that problem. They had experienced what life was like in the poorest conditions – when they didn't even own a football. The pair greeted every day of training excitedly, without ever complaining.

As he had done in previous teams, Ansu quickly progressed through the ranks at the academy. His combination of speed and technical ability was unmatched. This allowed him to compete against players

much older than him. He stood out in a pool of some of the world's most promising young football players.

At the age of 12, Ansu took part in the MTU Cup – an indoor youth football competition that's held by some of the best coaches in European football. Ansu played exceptionally and was named the player of the tournament. Performances like this were certainly not one-offs.

During his time at the academy, Ansu developed into a clinical goalscorer. He caused opposition teams problems, and no matter

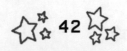

how hard they tried, they couldn't stop him.

Several years of this unstoppable form led to Ansu being offered an improved contract. Barcelona knew that they had something special with Ansu, and they wanted to prevent other clubs from trying to poach him.

One afternoon, Barcelona had a chauffeur drive Ansu, his father and legal team to the academy to discuss a new deal.

'Please take a seat,' said the chairman of the club, as the group entered his office.

Ansu and his team joined the table.

'Ansu, as I'm sure you're aware by now, you are a valued member of our Barcelona family.'

'Thank you,' said Ansu.

'That is why we want to keep you here for many years to come.'

'And I'm happy here,' said Ansu. 'There's nothing I want more than to play for Barcelona as a senior.'

When Ansu heard the €100 million buyout figure to be added to his revised contract, his mouth dropped open. He knew he was quite good at football, but he had never thought

about himself being worth that much money. He was just a 16-year-old who loved to play.

Following discussions with his team, Ansu signed the new contract. After signing, he expected that it was still going to be a long time until he made his way into the Barcelona first team. He thought he was going to continue playing with the academy and progressing through the ranks. Little did Ansu know that the path for him to join the first team was just around the corner.

# 16 YEARS, 9 MONTHS AND 25 DAYS

After signing the contract, Ansu got back to what he loved. He approached every match with the same energy and enthusiasm that he always had. To him, the

teenager running across the grass
was no different to the young boy
sprinting over the sandy earth in
Guinea-Bissau. Worth millions of
euros or not, Ansu only cared about
improving his game and playing
football.

At 16 years old, there were still
many steps a player was expected to
take before progressing to the first
team. Players usually joined academy
sides, and at Barcelona there was
even a B team in the third division
of Spanish football that new players
could hope to play for.

However, things were not going to plan for Barcelona's senior side. Many of their star players were picking up injuries and spaces needed filling in the first-team squad. With Luis Suárez and even Messi on the sidelines, the team had to think differently. The Barcelona manager at the time, Ernesto Valverde, decided to keep an eye on the academy players.

One day, Valverde came down to the training ground to watch a match and see if any of the younger generation looked ready to make the

step up. After watching Ansu for only a few of minutes, Valverde made up his mind.

Sometimes inexperienced players can crumble under the pressure of first-team football. But Valverde was so impressed by Ansu's abilities that, without hesitation, he invited the teenager to train with the first team.

Ansu's heart was racing the next morning before first-team training began. Nervous, he called

his brother before heading to the training ground.

'You've no reason to be nervous,' said Braima. 'Just treat it like any other day.'

Braima's journey at Barcelona was not going as smoothly as Ansu's. Braima had been loaned out to another team in the third division of Spanish football to assist in his development.

'I was nervous for my first training session at CD Calahorra' he said. 'New things are often scary. But look down at your feet for me ...'

Ansu did so.

'Just use those to kick the football and you'll be good,' Braima joked.

'Very funny,' said Ansu, with a smile.

'It's the same game, little brother, only with new faces.'

Having calmed down, Ansu made his way to the training ground to join the first team.

Ansu's confidence grew as the session went on. There may have been famous faces all around, but, just as Braima said, the game was no different. Ansu concentrated on doing his best and finished the day knowing that he had worked as hard as he could.

Despite all of the injuries, the club had a lot of depth in its squad. Ansu expected that he'd be waiting a long time before he could play for the side in a competitive match. Valverde, however, had other ideas.

The 25th August 2019 was a day that the Fati family would remember forever. Back in Guinea-Bissau, one of Ansu's uncles, Bucar, was hard at work. His market was teeming with customers buying everything they needed for the week ahead. But as Bucar waited by his stall, he felt his phone ringing in his pocket.

'Bori!' said Bucar. 'Is everything all right?'

Ansu's father couldn't contain his excitement. He wanted the entire world to know what was happening: that his son's dream was coming true.

'It's Ansu,' said Bori. 'He's been picked to play for Barcelona!'

'Goodness! For the B team?' asked Bucar.

'No,' said Bori. 'Against Real Betis in La Liga. For the senior team!'

Bucar's face lit up. He took a deep breath, then ...

'MY NEPHEW IS GOING TO PLAY

FOR BARCELONA!' he shouted across the market. Everyone around him stared. Most of the passers-by were confused, but the other market workers started to celebrate with him. They knew about Ansu and were happy that someone just like them was going to play for Barcelona.

Later in the day, Bucar and all of those who had known Ansu gathered around a TV to watch the match against Real Betis. The camera panned across the bench. Sitting there, wearing a Barcelona training jacket, was Ansu.

★ ★ ★

Ansu looked up from the bench and across the stadium in amazement. Thousands of fans were packed into the Camp Nou, cheering for their team in a competitive La Liga match. The roar of the fans was so loud that Ansu almost felt like covering his ears. He noticed the cameraman following the players on the pitch. Not all of his family and friends, whether they were in Spain or Guinea-Bissau, could attend the match. But Ansu knew, no matter

where they were, they were watching and cheering him on.

The match did not start as planned. Real Betis took the lead in the 15th minute. Nabil Fekir ran on to a through ball and fired it past the outstretched arms of Barcelona's goalkeeper, Marc-André ter Stegen. But Barcelona were determined not to lose on their own patch.

In the 41st minute, Griezmann latched on to the end of a chipped cross to bring the match level. The team went into the dressing room at half-time and received stern words

from the manager. The team could do better, and all of the players knew it. They needed to up their game in the second half.

The half-time talk had an immediate impact on the team and Griezmann scored his second goal in the 50th minute. Ansu rose from the bench and cheered as his team took the lead. They were fired up and ready to dominate the second half.

Real Betis were not ready for what was about to happen next. Barcelona went on to score three more goals, thanks to efforts from Carles Pérez,

Jordi Alba and Arturo Vidal. Loren Morón managed a consolation goal for Real Betis in the 79th minute, making the score 5-2. However, with the match pretty much in the bag for the home side, it was time for a momentous occasion.

Ansu Fati came onto the pitch to replace the young goalscorer Pérez, and to show the Barca fans  exactly what he could do. In doing so, Ansu became Barcelona's second youngest La Liga debutant. He was only 18

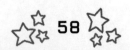

days older than the record holder, Vicenç Martinez, who made his debut for the club in 1941.

As Ansu sprinted onto the pitch, the home crowd erupted into an excited cheer. A 16-year-old was playing for the first team at one of the world's greatest clubs. Fans knew that to have achieved this, Ansu was a special talent.

Ansu didn't disappoint the home fans during his short stint on the pitch. Although he wasn't able to score, Ansu made frequent runs that displayed his lethal combination of

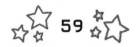

speed, power and intelligence, causing the fans to gasp in amazement.

The referee blew his whistle bringing the match to an end, with thousands of fans standing up and celebrating their team's performance. Ansu stood still for a moment and took it all in. He'd just made his debut for one of the greatest clubs in the world.

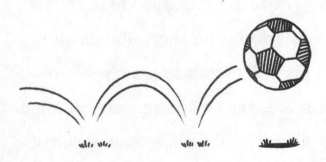

# 6

# RECORD-
# BREAKER

Ansu's impressive performance
meant that he was included in the
matchday squad for the next league
match. Barcelona travelled away to
CA Osasuna – a team that they would
usually be expected to beat without
any difficulty. But having several

injured players on their books meant that the win wasn't guaranteed.

The underdogs got off to the perfect start, with Roberto Torres latching on to a cross in the 7th minute of the match. He pulled back his left foot and smashed the ball into the net. It was 1-0 to Osasuna.

The home side held on to their lead and went into half-time with smiles on their faces. The home fans applauded their team as they left the pitch, while the Barcelona players' heads hung low.

Ansu rose from the bench, but before he could head down the tunnel Valverde caught up to him.

'Ansu – take two minutes, then we'll need you to start warming up,' said the coach.

Ansu beamed. 'No problem,' he said. This wasn't just a handful of minutes played at the end of a match like last time – Valverde was trusting him with an entire second half of football. Valverde had called on Ansu to change the team's fortunes.

Ansu warmed up on the pitch, darting in between brightly coloured

cones and preparing to perform at his best. Half-time was over in a flash. Before Ansu knew it, he was part of the front line with a mission to get Barcelona back in the game.

In the 51st minute, Pérez burst down the wing and fired in a cross. Standing in between two tall defenders, Ansu leapt off the ground and for a moment towered above the pair of them. Ansu made contact with the ball and headed it towards the net. The goalkeeper could do nothing as the ball made its way into the bottom corner. *Goal!*

The cheers of the Osasuna fans were replaced by shouts from ecstatic Barcelona players. Ansu celebrated with his teammates in the corner of the pitch. He had just scored his first ever competitive goal for Barcelona. He had also become the youngest ever goalscorer in La Liga!

Barcelona went on to draw the match 2-2. With the youngest player on the pitch, Ansu Fati, playing a crucial part in the match's outcome. If any Barcelona fans were unaware of his talent a few weeks before, the same could not be said after the matches against Real Betis and Osasuna.

Ansu Fati was a star in the making.

Ansu continued his impressive start to life as a Barcelona first-team player. He had earned his place in the senior squad and no one was going to take it from him.

It was then that Ansu began to break Barcelona's long-held records. He became the youngest starter to play at the Camp Nou during a 5-2 defeat of Valencia. For this match alone, Ansu needs an entire page of the football history books. He also

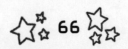

became the youngest player to score and assist in La Liga, and the youngest ever goalscorer at the Camp Nou. The icing on the cake was that all of this came within eight minutes of the match start! Ansu scored in the 2nd minute and provided an assist for Frenkie de Jong in the 7th minute.

Ansu's record-breaking performances earned him a spot in the starting lineup for a Champions League fixture. As expected, Ansu broke yet another record. He became Barcelona's youngest ever Champions League debutant.

On a windy autumn night, Ansu and his teammates emerged from the tunnel in an away match against Borussia Dortmund. As Ansu walked onto the pitch with his teammates, he could hear the famous Champions League anthem booming across the stadium. It was almost drowned out by the cheers of the 80,000+ fans watching. Ansu played for fifty-five minutes. He was a constant threat to Dortmund, and he played like a Champions League veteran.

Ansu looked over to the bench and saw his number – 31 – flashing on the substitution board. Also flashing in green was number 10 for the player replacing him. Ansu made his way to the bench, but before he left the pitch the player replacing him had something to say.

'Well played, Ansu,' said Messi, hugging him. 'You should be proud.'

The match finished as a 0-0 draw, but it was a stalemate that Ansu was going to remember forever. He'd made his Champions League debut.

In the final group stage match of the Champions League, Ansu was brought on in the final five minutes

of an away match against Inter Milan. The score was 1-1, and the thousands of Italian fans at the Stadio San Siro were rooting for their team to defeat Barcelona. Ansu's side had already advanced from the group stages, but Inter Milan needed to win if they were to have any hope of reaching the knockout stages.

Unfortunately for them, Ansu had other plans.

Only a minute after Ansu arrived on the pitch, Suárez cushioned the ball to him outside of a crowded penalty box. Ansu swung his right foot and fired

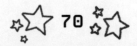

the ball towards goal. It thwacked
against the left-hand post and nestled
into the back of the net. *Goal!* The
home crowd were silenced, their
hopes of progressing from the group
stages dashed.

Ansu celebrated with his
teammates after scoring his first ever
Champions League goal. In contrast
to the thousands of despairing home

fans in the stadium, this
was a moment of pure
joy for the 17-year-
old wonder. And
to no one's surprise,

another record was broken. Ansu Fati
was now Barcelona's youngest ever
Champions League goalscorer.

# 7
# A CLUB IN TRANSITION

Since the early 2000s, Barcelona have been one of the greatest football teams in the world. Year after year, the team became so successful that nothing short of perfection was acceptable. With the likes of Xavi, Andrés Iniesta, Carles Puyol and a

certain Argentinian youngster called Messi coming through the ranks, Barcelona were a team to be feared.

Barcelona were like a bull charging through every competition they played in. During this era, the club's trophy cabinet overflowed with trophies from La Liga, the annual Spanish competition Copa Del Ray and the Champions League. Their roster of players picked up so many winner medals that they could have filled a vault as well as a display case.

Although things continued to go well for Barcelona, the club began to

see changes. One by one, the heroes
of the past hung up their boots and
retired, or chose to end their careers
in a less competitive league.

Camp Nou, once an impenetrable
fortress for visiting teams, started to
leak points. The club tried to replace
the greats of the past with big
money moves and record-breaking
transfer fees. But the players who
were brought in to replace these
greats struggled to deliver. Instead
of first-place finishers, Barcelona
were becoming runners up. This left
a bitter taste in the mouths of fans

who had become so used to victory.

Midway through the 2019/2020 season, the club's bosses had seen enough and understood that it was time for a change. Valverde, the coach who had given Ansu his first opportunity in professional football, was fired from the management role. He was replaced by Quique Setién, in the hope that a new style of leadership would help to improve the club's fading fortunes.

But despite the club's investment, Barcelona still wasn't performing well or improving. In a topsy-turvy season, Ansu was one of the few bright sparks. The club put forward a valiant effort in La Liga, but finished five points behind their rivals Real Madrid, allowing the Galácticos to collect their thirty-fourth title.

Ansu progressed throughout the season, earning twenty-four appearances and scoring seven goals for the first team. These were numbers unheard of for such a young player at such a decorated club. But at the same time, this was seen as evidence of the club's emergency status. Barcelona was known for fielding one of the very best lineups in world football – they usually had players at the peak of their careers. Now they were having to rely on the young talent emerging from their academy. What was first seen as an opportunity for these young players quickly became a heavy burden on their shoulders.

While failing to win the league or the domestic cup during the season was disappointing, it was far from the club's lowest point of the year. Barcelona had progressed to the quarter finals of the Champions

League. They had been drawn against the German champions Bayern Munich – a team filled with quality who were dominating their domestic league and on the hunt for European success.

Matches between the two clubs had always been fierce, exciting encounters. They were two heavyweights of world football, and both had the ability to throw a knockout punch. But this quarter final was different. Now, with Barcelona pitting young players against much more experienced ones, the two sides were no longer on level terms.

Barcelona fell to a humiliating 8-2 defeat – an embarrassing score that showed the world just how far Barcelona had fallen. It was the first time since 1946 that Barcelona had conceded eight goals in a competitive fixture. Only three days later, Setién was removed from his post as manager. Barcelona didn't just need a change in tactics, the club needed a complete restructure.

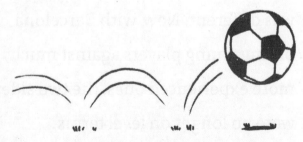

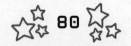

# 8

# LA FURIA ROJA

Despite Barcelona's year to forget, Ansu's talent within the team did not go unnoticed. Luis Enrique, the national team manager of Spain, had seen plenty. He believed Ansu deserved a call-up and decided to name him in the Spanish squad for

two Nations League fixtures.

Ansu wasn't the only member of the Barcelona team to be called up for Spain. His teammate Busquets had also been picked. Upon hearing the news after a training session, Ansu rushed to find his teammate. Busquets was in the car park, searching his pockets for his car keys when he saw Ansu charging towards him.

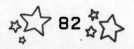

Busquets had been a national team player for years. He had played in one of the country's strongest footballing eras, where they won the Euros and the World Cup. He had played with the likes of Fernando Torres, David Villa and Iker Casillas. However, all three had retired from international football and Spain's golden generation was a thing of the past.

It was time for the Spanish team to build a new identity, and Ansu was going to be key in unlocking the team's next steps. If there was anyone to go to for advice about playing for the national team, it was Sergio Busquets.

'Hey, congratulations!' said Busquets. 'I just heard the news. You earned it.'

'Thanks a million,' said Ansu, catching his breath. 'I've got so many questions about playing international football, though. Do you have time for a chat?'

'Sure thing,' said Busquets, dropping his keys back into his pocket.

Busquets helped Ansu settle into the national team squad. Having someone from his club join him was a big help in easing the jump from club to international football.

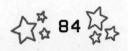

Ansu impressed Enrique during training, but he wasn't named in the starting lineup for the match against Germany. Instead, Ansu joined the rest of the squad members on the bench. No goals were scored in the first half, so Enrique turned to Ansu and substituted him into the game. In doing so, Ansu became the second youngest player to debut for the Spanish national team.

While it was a momentous  occasion for Ansu, the match didn't go entirely to plan. Germany took

the lead thanks to a goal from Timo
Werner, but José Gayà equalised in
the final moments, meaning both
sides earned a point for their draw.
While Ansu hadn't scored on his
debut, he was only getting started
with the national team. Within a
matter of days, Spain were competing
once again.

Enrique was so impressed by
Ansu's determination and skill
that he decided to name him in
the starting lineup for a
match against Ukraine.
Ansu was standing

in the tunnel with his teammates getting ready to head out to the pitch. He ignored the distant noise of the crowd and took a moment for himself.

Ansu looked down at his shirt and the Spanish badge. He was *starting* in a match for the Spanish national team. He was representing the country that had accepted him and his family as one of their own. Ansu felt proud as a member of *La Furia Roja* – The Red Fury. But the time for contemplation was over. Ansu walked out onto the pitch ready to

perform. As suited their fiery red kit, Spain were ready to set this match alight.

In the 2nd minute of the match, Ansu darted down the wing and made his way into the opposition box. Ansu was met by a terrified defender. Unsure of how to stop him, the defender stuck out a leg in hope of catching the ball. Instead, he hit Ansu's legs and sent the forward tumbling to the ground. Penalty!

Spain's captain, Sergio Ramos, tucked away the penalty and ran over to Ansu straight after scoring.

'That's because of you!' Ramos shouted, hugging his teammate.

Ansu was then involved in wave after wave of attack. His constant trickery sent defenders tumbling to the ground as they tried to keep pace. At one point, Ansu attempted an overhead kick, but it was blocked on the line by a defender.

Ramos scored a second goal for the side, but in the 32nd minute, the most magical moment of the match happened. Ansu fearlessly ran at

the opposition and fired from outside of the box. The ball smacked against the post and into the net.

With this goal, Ansu now held the record for the youngest player to ever score for Spain, at the age of 17 years and 308 days. Ansu broke a long-standing record that had been set in 1925.

Spain went on to win the match 4-0 and Ansu played a starring role. No one could doubt Ansu's inclusion in the national team. He may well have been young, but he had every right to be playing on that pitch. This was just

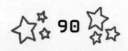

the start for Ansu and Spain. It was obvious that with Ansu in the side, the possibilities were endless.

# 9
# NEW MANAGEMENT

When Ansu returned from a well-
deserved summer break, things had
changed at Barcelona once again.
Leading the team for the 2020/2021
La Liga season, Koeman was going to
be his third manager at the club. To
many observers, this was a surprising

choice. Koeman had managed the Netherlands national team, but it was quite a jump to manage a club as big and as decorated as Barcelona. Did he have what was needed?

Beyond his managerial experience, Koeman had a history with the club. He had played for Barcelona for six years, and during his time there he won four La Liga titles and scored the winning goal in the European Cup final of 1992. The Barcelona board hoped that Koeman could enjoy the same success he had as a player, only now as the manager of Barcelona.

After participating in his first pre-season training session under the new boss, Ansu walked to the sidelines to grab some water. It was a warm day – so warm that instead of drinking his water he poured half of it over his head! While Ansu dripped from head to toe, a blurry figure approached in the distance.

'Ansu,' called Koeman.

Ansu wiped the water from his eyes.

'Hey, coach,' said Ansu.

'Walk with me, son.'

The pair began to walk and talk around the pitch.

'I can see you're already integral to this team,' said Koeman. 'I'm looking forward to working with you. Perhaps there are a few things I can teach you.'

'I'm all ears,' said Ansu.

The pair reached the end of the pitch. Messi had stayed out on his own and was practising his free kicks. He effortlessly fired one over a fake wall. The ball curled over it and shot into the top left-hand corner. But Messi didn't celebrate. He simply picked up the next ball and carried on shooting.

'I'm sure you've heard the rumours that you're Messi's heir,' said Koeman.

'I have,' said Ansu. 'But look at him. No one else can be Messi.'

'That's very true,' said Koeman. 'Messi is Messi.'

Ansu laughed. People weren't usually this honest with him.

'But you're Ansu Fati,' said Koeman. 'You don't need to compare yourself to anyone. You have your talents and other players have theirs. It just turns out you have an awful lot of them.'

'Thank you,' said Ansu, immediately feeling a weight lift from his shoulders.

'You're a star player in your own right,' said Koeman.

The pair shook hands before going their separate ways.

'Let's get the Barcelona fans smiling again,' Koeman called as he walked away.

# 10
# THE TACKLE

Ansu started the 2020/2021 La Liga season with a smile on his face. In the first match of the season against Villarreal, Ansu scored two goals and provided an assist. He aimed to start the season as he meant to carry on: by winning.

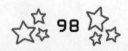

Barcelona started the season well with two wins and a draw. But then their form dropped with two losses against Getafe and Real Madrid. They improved with a draw against Deportivo Alavés, but in their next match they took on Real Betis. A matchday that started with hope ended in despair. It was a terrible day for Barcelona, and an even worse one for Ansu Fati.

Due to COVID-19, there were no cheering fans at the Camp Nou. But Ansu and the team had all of the determination they needed

and were ready for the match. The referee blew the whistle and the game kicked off. In the 22nd minute, Ousmane Dembélé made his way into the opposition's box and struck a fierce strike at goal with his left foot. The ball swerved past the twisting defenders, but they could do nothing as it hit the back of the net. *Goal!*

Ten minutes later, Ansu charged into the Real Betis penalty box. The team were playing with great energy, and Ansu wanted to keep it up. He sprinted forwards towards the goal, jostled with two defenders and broke through them. A cross was fired in from the left-hand side and Ansu reached out to collect it. The two defenders slid to the ground to stop him.

And stop him they did.

The tackle knocked Ansu off balance and he fell to the ground. In that moment he felt a sharp pain, much stronger than anything he had ever felt before. He rolled on the ground clutching his knee. Ansu knew that this was serious. The medical team ran onto the pitch. They stretched out his leg and Ansu

cried out in agony. The two medical assistants looked at one another with concern.

When Ansu had woken up that morning, he had prepared for an intense game. He was ready to give his all for his team to win. But what Ansu wasn't ready for was an injury like this.

The injury would stop Ansu from making another professional football appearance for 323 days.

# 11

# THE
# TREATMENT
# TABLE

Medical staff surrounded Ansu as he lay on a hospital bed.

'Hi Ansu. We now have the results of the scan,' said a doctor. Ansu could sense tension in her body, as though

the news she had was not good. His stomach was in knots.

'And? What's happened to me?' asked Ansu nervously.

He looked down at his knee. It was covered in bandages and an ice pack was tightly strapped to it to help with the swelling. The pain was still fresh, and Ansu was expecting the worst.

'I'm afraid you've torn your internal meniscus,' said the doctor, adjusting her large glasses. 'It helps your knee absorb force when impacting the ground. It also assists with your knee's flexibility.'

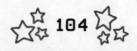

Ansu frowned, fearing the worst. 'What does this mean for my season? How long am I going to be out?'

'I can't say for certain,' said the doctor. 'But if everything goes to plan, following surgery you could be back in up to four months.'

Ansu sank into the pillow behind him. He couldn't bear the thought of missing one week of football, let alone sixteen of them.

Ansu agreed to the surgery and it was booked in. Barcelona provided Ansu with the best doctors and medical teams in world football.

They wanted their star player back as soon as possible, but not at the expense of Ansu's long-term health. They wanted to make sure that Ansu had all of the time he needed to recover properly.

Ansu's surgery appeared to be a success. Afterwards he worked with the medical team to help his knee to heal. However, constant swelling was a problem that stopped Ansu from making a full recovery. It was soon decided that Ansu needed to have another knee surgery. But the second one was not enough either.

In the end, Ansu had four separate surgeries following many setbacks to his recovery. Doctors expected Ansu to recover in four months, but it would take Ansu eleven months to get back to match fitness.

During an incredibly difficult year, Ansu leant on his family and friends for comfort. He had gone from a career high to one of the toughest and most challenging years of his life. But with a determination and unstoppable desire to get back to his best, Ansu didn't allow the injury to define him. He listened to his doctor's advice, took recovery seriously and eventually he was fit again.

It was finally time for Ansu Fati to return to Barcelona. But as Ansu was returning, another player at Barcelona was leaving. It was an exit that shocked the entire footballing world.

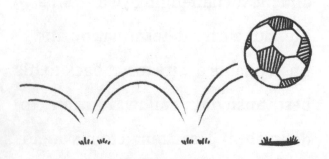

# THE HEIR TO NUMBER 10

On the 8th of August 2021, it seemed as though every single camera and microphone in the world were in one press conference. Messi emerged from a side door, dressed in a luxurious blue suit with a matching face mask. When he arrived at the

podium, Messi removed his mask to reveal a face brimming with sadness.

The room filled with the sounds of camera flashes and jostling spectators. Then the room fell silent. Before Messi could let out a single word, he was overcome with emotion and started to cry. The camera flashes were replaced with supportive applause from all of those in the room.

'I've spent my entire career here,' said Messi, pulling a tissue from his pocket to wipe away his tears. 'Now after twenty-one years, I'm

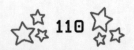

leaving for Paris with my wife and my three children. But no matter what, Barcelona will always be our home.'

Messi was so emotional during his farewell speech because he didn't want to leave. In fact, Barcelona didn't want him to leave either. However, the club had struggled with money during the COVID-19 pandemic. They could no longer afford to pay Messi's high salary. It left no option other than his exit from Barcelona.

Ansu was sitting towards the back of the room with the rest of his Barcelona teammates. During the speech, the majority of them were overcome with sadness too. It really was the end of an era. And what did that mean for the current Barcelona team? What did a Barcelona without Messi, one of the greatest footballers of all time, look like?

A few weeks later, Ansu was called into a meeting room at the club's training complex. Ansu was met there by Koeman and important board members.

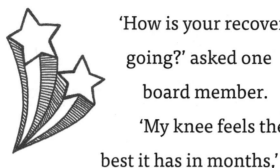

'How is your recovery going?' asked one board member.

'My knee feels the best it has in months,' said Ansu. 'The medical team think I need a couple more weeks and then I'll be ready to play.' As the words came out of Ansu's mouth, he couldn't help but smile. Being back on a professional football pitch was what he'd been dreaming about for months on end.

'That's fantastic news,' said another board member. 'You're

probably wondering why we've asked you here today.'

Koeman walked around the table and presented him with a gift.

'Congratulations,' he said, with a grin. 'I'm sure it will suit you.'

Ansu opened the box and inside was a folded Barcelona home shirt. He held it in the air and turned it around to look at the back.

It read:

## ANSU FATI
# 10

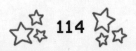

The number 10 shirt had been worn by so many icons at the club. It was Messi's number, and before him it had been worn by Ronaldinho and Rivaldo. To be given this number at a club like Barcelona was a huge deal. It showed that they truly believed Ansu was going to be as great as those players who had come before him.

Ansu was lost for words. He was going to wear the number with pride. It might have been the end of Messi's reign at Barcelona, but it was the beginning of Ansu Fati's.

'Well, I guess it's time to announce this to the world,' said a smiling board member.

Ansu left the room and immediately called his family. Bori, Lurdes and his siblings were so proud of him. Just a few days later, Ansu made his way to the Barcelona club shop with his family. He ordered 100 Barcelona home shirts with 'Ansu Fati 10' on the back. His family planned to send the shirts to friends and family in Spain and Guinea-Bissau.

Ansu smiled to himself. *Just a few years ago, I was staring at the number*

10 shirt in a dusty market in Guinea-Bissau, he thought. And now my name is on it. What would that young boy think if he saw this now?

# 13
# ROAD TO QATAR 2022

After just two wins in their first
six 2021/2022 La Liga matches, the
new Barcelona lineup wasn't to
everyone's taste. The team needed a
hero to help set their season alight.
In their seventh game of the season,
Barcelona were playing at home

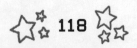

against Levante UD, with the Camp Nou's lively crowd there to cheer on their team. There was a player in the squad – named for the first time in eleven months – who had the fans in the stadium excited for what was to come. It was the return of Ansu Fati.

Ansu had been named on the bench for this match. Over the previous weeks, he had trained hard to regain his match fitness. While he was not named in the starting eleven, Ansu was delighted to be on the team sheet at all. As the announcer read

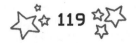

out the names of the players in the squad, the fans cheered each one. But the cheers were louder than any other when she read Ansu's name. Ansu stood up from his seat on the bench and waved to the fans as they applauded him.

Barcelona got off to a flying start thanks to Memphis Depay. The talented Dutch forward tucked away a penalty in the 5th minute, to the joy of the home fans. The day felt like it was Barcelona's for the taking.

In the 14th minute, Luuk de

Jong doubled Barcelona's lead. He coolly fired past Levante's diving goalkeeper, Aitor Fernández. Barcelona kept up the heavy pressure but were unable to score again before half-time.

With the side doing so well, there was no need to change players. Ansu continued to sit on the bench for the second half. He wanted more than anything to be out there on the pitch, but he was also delighted that his team was winning.

The minutes in the second half quickly ticked away, and Ansu's chances of getting to play were fading. But with the team tiring towards the end of the match, Koeman looked to his bench.

'Araujo, Demir, Fati,' said the manager. 'Warm up. I'm bringing you on.'

Ansu had never stood up from a seat so quickly. He jogged up and down the sidelines, and as he did the expectant fans in the stadium clapped for him.

In the 81st minute, Ansu was part of a triple substitution. Ansu felt his entire body cover in goosebumps

as his boots touched that first patch of grass. The cheers for his return seemed even louder than those for the goals that had been scored earlier in the match. Ansu's arrival brought a boost of energy to the team in the final minutes, with Ansu in search of a precious goal on his return to action.

The clock struck ninety minutes, and there were to be four minutes of added time due to the substitutions, stoppages and injuries. Barcelona broke from Levante on the counterattack. Ansu had the ball and charged at the Levante defence.

Ansu's teammates made wide runs, drawing out the defenders and opening a path for Ansu. He reached the edge of the box and was met by the last line of defence. Ansu brought the ball inside. There was nowhere else for him to go, so he pulled his right foot back and released a shot at goal. His positioning made it appear that he was going to shoot to the right, but Ansu had dummied the keeper, Fernández. The ball flew past Fernández's outstretched arms and smacked against the back of the net. *Goooooooooal!*

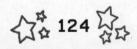

Ansu Fati was back on the pitch and back on the scoresheet!

Ansu sprinted to the bench and hugged a member of the medical team. He thanked them for all of their hard work and help in his recovery. He then made his way into the stands and hugged his family. This had been a moment he had been dreaming of for months. He had finally beaten his injury. Ansu Fati was doing what he loved again.

With Messi having left Barcelona, the team is in search of a new hero. With the same number on his back, and having been developed at the club's famed La Masia academy, it appears that Barcelona will not have to look far for that replacement. Ansu Fati might just be the superstar Barcelona needs.

Ansu may have missed the Euro 2020 tournament because of injury, but if he can stay fit, there will almost certainly be a place for him in the Spanish squad for Qatar 2022. Not only could Ansu be the answer to

Barcelona's prayers, but he could be just what the Spanish national side needs too.

Still only 18 years old, the star has achieved so much at such a young age. To every match he brings the same energy and enthusiasm as the boy running through the market stalls of Guinea-Bissau. Ansu has a long career ahead of him, and no matter how good the holder of the number 10 shirt at Barcelona was

before him, Ansu will be a player who terrifies opposing defences for years to come. There was Rivaldo, Ronaldinho and Messi.

But now, it's time for Ansu Fati.